Phytoplasmas: Plant Pathogenic Bacteria - III

Assunta Bertaccini • Kenro Oshima
Michael Kube • Govind Pratap Rao
Editors

Phytoplasmas: Plant Pathogenic Bacteria - III

Genomics, Host Pathogen Interactions and Diagnosis

Editors

Assunta Bertaccini
Department of Agricultural and Food Sciences
Alma Mater Studiorum – University of Bologna
Bologna, Italy

Kenro Oshima
Department of Clinical Plant Science, Faculty of Bioscience and Applied Chemistry
Hosei University
Tokyo, Japan

Michael Kube
Department of Integrative Infection Biology Crops-Livestock
University of Hohenheim
Stuttgart, Germany

Govind Pratap Rao
Division of Plant Pathology
Indian Agricultural Research Institute
New Delhi, Delhi, India

ISBN 978-981-13-9634-2 ISBN 978-981-13-9632-8 (eBook)
https://doi.org/10.1007/978-981-13-9632-8

This Springer imprint is published by the registered company Springer Nature Singapore Pte Ltd.
The registered company address is: 152 Beach Road, #21-01/04 Gateway East, Singapore 189721, Singapore

Preface

Phytoplasma-associated diseases are severely limiting the quality and productivity of many ornamental, horticultural, and other economically important crops worldwide. Their dangerous impact is economically relevant in sugarcane, sesame, grapevine, fruit trees, and coconut palm, severely reducing the economic revenues for farmers and stakeholders. There is no effective cure for these diseases; therefore, the basic knowledge about their biological and biochemical influence on plant and insect vectors are the basic instruments for designing management options aimed to minimize their spread by insect vectors, propagation materials, and host-plant resistance. The scientific literature concerning genome, effectors, and pathogenicity factors is allowing the understanding of the phytoplasma-host metabolic interaction. It provides comprehensive information on biological, serological, and molecular characterization of the phytoplasmas including the recent approaches for diagnostics. Transcriptomics studies paved the way for analyzing the gene expression pattern in phytoplasmas upon infection and revealed the upregulation of genes associated to hormonal response, transcription factors, and signaling genes. Although phytoplasmas remain the most poorly characterized pathogens, recent studies have identified virulence factors that induce typical disease symptoms and have characterized the unique reductive evolution of the genome. The presentation of the progresses in cultivation in axenic media together with the perspectives for future research that will allow reducing the incidence and the losses due to these pathogens in the agriculture worldwide are also presented. Hence, the major recent research findings are compiled in this book.

We greatly acknowledge all the contributed authors for their efforts in synthesizing the most updated reviews on the subjects. We also like to thank the support and input of the publisher, Springer Nature (Pvt Ltd), Singapore, for its effort to publish this book. We strongly hope that it will be useful to everyone interested in phytoplasma research, plant pathology, microbiology, plant biology, and agriculture and serve as an exhaustive and up-to-date reference on various aspects of phytoplasma-associated diseases in order to help the further progress of this branch of the plant pathology scientific knowledge.

Bologna, Italy — Assunta Bertaccini
Stuttgart, Germany — Michael Kube
Tokyo, Japan — Kenro Oshima
New Delhi, India — Govind Pratap Rao

Contents

About the Editors

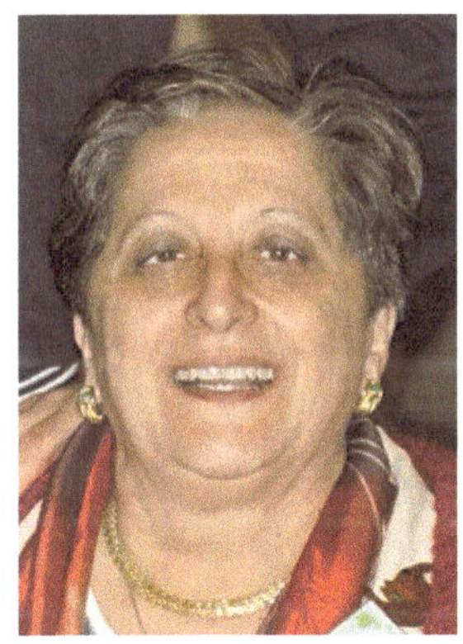

Assunta Bertaccini is a Professor of Plant Pathology at the University of Bologna, Italy. In more than 40 years of research, she has focused on studying plant diseases associated with phytoplasmas and bacteria, particularly their biology and epidemiology. She was an invited speaker at national and international meetings and seminars. Among her major awards is the Emmy Klienenberger-Nobel 2014 for distinguished research in mycoplasmology. She has mentored dozen of students in phytobacteriology and is the author of about 800 publications. She is Editor-in-Chief of *Phytopathogenic Mollicutes*, Senior Editor of *Phytopathologia Mediterranea*, and Founder and Leader of the International Phytoplasmologist Working Group (IPWG) (http://www.ipwgnet.org).

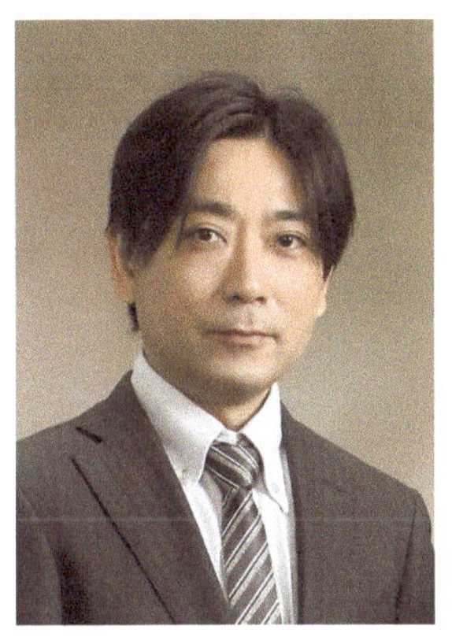

Kenro Oshima is a Professor of Plant Pathology at the Hosei University, Japan. He has 20 years of research experience in the field of plant pathology, especially the genomes and pathogenicity of phytoplasmas. He is an Associate Editor of the Journal of General Plant Pathology. He has published more than 60 peer-reviewed articles in international journals and has delivered numerous oral and poster presentations at national and international meetings. He received the Kitamoto Award from the Japanese Society of Mycoplasmology in 2018.

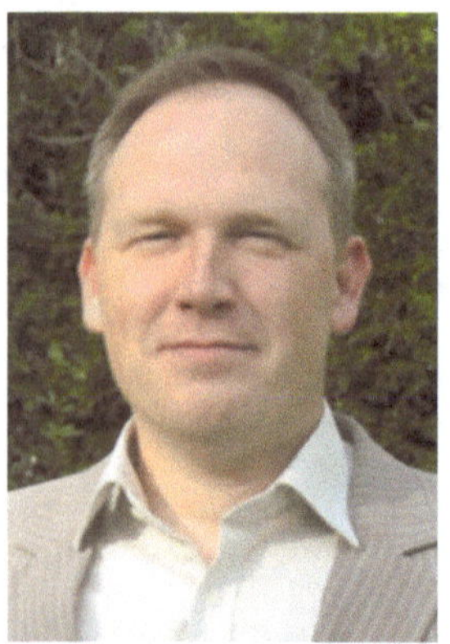

Michael Kube is a Professor and Head of the Department of Integrative Infection Biology Crops Livestock at the University of Hohenheim, Germany. He has been working on phytoplasmas and other bacterial pathogens for more than a decade, and his research covers topics in the disciplines of molecular biology and microbiology with a focus on omics technologies. His findings are documented in more than 100 research publications. He participated as invited speaker at several national and international conferences, meetings, and seminars.

Govind Pratap Rao is a Principal Scientist of Plant Pathology at the Indian Agricultural Research Institute, New Delhi, India. He has 30 years of research experience in the field of plant pathology, especially virus and phytoplasma diseases of several cultivated crops. He has published over 150 research publications and authored or edited 20 books. He is the Editor-in-Chief of *Phytopathogenic Mollicutes* and has received numerous national and international awards. He has given invited talks and training, pursued research, and participated in panel discussions and workshops and conferences around the globe.

Chapter 1
Genome Sequencing

Michael Kube, Bojan Duduk, and Kenro Oshima

Abstract Genome sequences are of major importance to phytoplasma research, as they provide the blueprint for understanding evolution, metabolism and virulence factors of phytoplasmas. Genome projects on these obligate parasites start from metagenomic templates taken from colonised plant- or insect-vector material, meaning that they have to deal with high amounts of untargeted DNA. This problem separates phytoplasmas from the majority of other bacterial genome projects, and methodological approaches deal with it by using strong colonised tissues and enriching phytoplasma DNA. The impact of this situation was severe for the first genome projects using Sanger sequencing, while the most recent phytoplasma genome projects have tried to overcome the problem through huge amounts of reads derived from next-generation sequencing approaches, thus enabling the generation of draft sequences or even complete phytoplasma genomes. Genomic sequence determination is hampered by their repeat-rich content, resulting in conflicts during the sequence assemblies in addition. An overview is provided of the strategies applied to phytoplasma genome sequencing and data processing, as well as currently available data on these particular bacteria.

Keywords NGS · Complete genomes · Draft sequences · Genome instability

M. Kube (✉)
Integrative Infection Biology Crops-Livestock, University of Hohenheim, Stuttgart, Germany

B. Duduk
Institute of Pesticides and Environmental Protection, Belgrade, Serbia

K. Oshima
Department of Clinical Plant Science, Faculty of Bioscience and Applied Chemistry, Hosei University, Tokyo, Japan

A. Bertaccini et al. (eds.), *Phytoplasmas: Plant Pathogenic Bacteria - III*,
https://doi.org/10.1007/978-981-13-9632-8_1

1.1 General Features

Genome sequences are highly influential in the understanding of the biology of organisms. The importance of genomes is still increasing, in particular for the bacteria which cannot be cultivated in axenic media or for which the cultivation is not established or available for a species. This situation has hampered for decades research into the cell wall-less bacteria of the taxon '*Candidatus* Phytoplasma'. Knowledge obtained from genomes has provided the blueprint for life necessary for the analyses of the functional genetic repertoire, highlighting the host dependency of phytoplasmas and their interactions with insect vectors and plants. Subsequent ground-breaking experiments have been made possible by these genome analyses, thereby expanding the knowledge, in particular, on the actions in host manipulation by phytoplasma-encoded virulence factors; however, genome data have also been used in many other phytoplasma research projects. Early studies provided estimations on the size of those phytoplasma genomes in about 530–1,350 kb with a GC content of 21–33 mol% (Neimark and Kirkpatrick 1993; Marcone et al. 1999; IRPCM 2004). The small size of phytoplasma genomes highlights a reduction process that is reported also for many other bacteria in the *Mollicutes* class. However, *Acholeplasma* spp. and '*Ca.* Phytoplasma' species are distinguished from the other *Mollicutes* by the usage of UGA as codon for tryptophan instead of as a termination signal (Razin et al. 1998) and by the distinct phylogenetic position of the *Acholeplasmataceae* within the *Mollicutes* (Hicks et al. 2014). Seven complete phytoplasma genomes have been determined to date.

1.2 Metagenomic DNA Templates

Reconstruction of a complete bacterial genome starts ideally with a DNA template obtained from an axenic culture. This kind of culture was not available for any of the previously determined phytoplasma genomes, so infected and highly colonised material is needed initially, due to the unfavourable ratio of phytoplasmas carrying small genomes to the comparable huge amounts of plant host, endophytic bacteria or insect vector DNA. As a consequence, phloem sap (Bai et al. 2006), phloem (Kube et al. 2008) and insect tissue (Orlovskis et al. 2017) with relatively high amounts of phytoplasma cells are used. Several genome projects did not apply the original host as source for DNA extraction. For example, '*Ca.* P. australiense' was transmitted to *Catharanthus roseus* G Don via grafting (Tran-Nguyen et al. 2008), and '*Ca.* P. mali' was transmitted by dodder (*Cuscuta* sp.) to *C. roseus* and *Nicotiana* spp. (Kube et al. 2008). While infected plant material enabled the complete genome reconstruction of several phytoplasmas, the DNA template for the '*Ca.* P. asteris' M3 genome project was the only one obtained from an insect vector using infected *Dalbulus maidis*, which transmits the maize bushy stunt disease in Brazil (Orlovskis et al. 2017).

Further enrichment of phytoplasma DNA is a crucial step in the template generation. Several phytoplasma genome projects obtained their starting DNA from phytoplasma-enriched fractions by centrifugation in (bisbenzimide-)cesium chloride buoyant density gradients or from pulsed field gel electrophoresis (PFGE) experiments enabling its separation from the host's DNA (Kirkpatrick et al. 1987; Oshima et al. 2001; Bai et al. 2006; Kube et al. 2008; Tran-Nguyen et al. 2008; Andersen et al. 2013). A drastic decrease in the percentage of the host DNA was reached, but the projects have still to deal with an untargeted background of plant DNA characterizing them initially as metagenomic. The impact of a successful enrichment process resulting in content of more than 50% phytoplasma DNA can be estimated; however, other approaches gaining phytoplasma-enriched DNA templates, e.g. preferential amplification and enrichment of genomic phytoplasma DNA (Mitrovic et al. 2014), are not available and should be established.

Despite the problem of sequencing large amounts of unwanted DNA background in phytoplasma genome sequencing (plant or vector derived but also endophytes), it is notable that the analysis of this untargeted microbial sequences by catch was analysed successfully in some projects, for instance, resulting in the reconstruction of the complete genome sequences of '*Candidatus* Sulcia muelleri' and '*Candidatus* Nasuia deltocephalinicola', which are endosymbionts of the phytoplasma vector *Macrosteles quadripunctulatus* (Bennett et al. 2016).

1.3 Sequencing and Assembly

The small genome size of phytoplasmas makes them, in principle, the perfect targets for genome determination by the whole genome shotgun (WGS) approach (Bodenteich et al. 1994). WGS follows a simple and striking strategy of redundant random sequencing of small genomic fragments covering all parts of a genome (Fraser et al. 1995), while the extraction provides the template containing multiple copies of the targeted DNA. Within the first steps, the DNA is randomly fragmented and size-selected, and DNA ends are modified by ligating known DNA sequences to them, such as adapters or cloning vectors (Frangeul et al. 1999). Primers complementary to these new fragment ends can be used for sequencing the unknown target DNA from both ends in separated reactions (end sequencing of templates). Assuming that the fragments randomly and redundantly represent each part of the genome (physical coverage), random sequencing of the fragments will provide overlapping sequence information, thereby enabling the reconstruction of a contiguous genome sequence by a multiple read alignment (assembly). In an ideal case, increasing the sequencing of each position of a genome multiple times (sequencing coverage) will result in higher sequence quality and completeness of the sequence, thus ending in a complete genome sequence. Read-pair information obtained from and assigned to each sequencing template enables the verification of the assembly by their assignment, and consequently, this information is used during the assembly of WGS data.

Unfortunately, the sequence information obtained will not represent the genome equally, due to several technical reasons resulting from sequencing templates and reads showing bias. In the case of sequencing templates derived from clone libraries, additional bias will be introduced during ligation and transformation, and by clone instability within the bacterial host. Furthermore, additional bias in coverage will be introduced, due to differences in the sequence quality of reads resulting from extremes in GC content (AT or GC rich stretches), or in DNA template structure (folding) resulting in weak sequence quality or loss of reads. These problems have a significant impact on Sanger sequencing approaches (*synonim* of dye terminator sequencing) using plasmids as templates, but they are also visible within next-generation sequence (NGS) data using amplified templates. The problem of unequal representation of a genome by sequence reads is managed by high sequencing coverage (> eight fold). However, even high coverage cannot guarantee the complete genome determination, due to read quality and conflicts within the assembly process (Al-Okaily 2016) for building the alignment of sequence reads for the construction of continuous genomic sequences (contigs). Conflicts in the assembly process result from miss-assemblies incorporating wrong reads, or competing options in a possible alignment. These scenarios prevent successful assembly and are characterised by contigs with weak or wrong terminal sequences. Poor read quality and/or repetitive sequence elements are the main sources in this regard. Furthermore, long (perfect) repeats present in phytoplasmas resulting from multi-copy genes or operons (*e.g.* rRNA operons) of longer size than the read sequence result in conflicts during assembly. Several strategies are applied to solve such conflicts, but the main one is based on manual editing in the database using read-pair information (reads spanning or anchored by one end sequence in the conflict region) and removal of all ambiguous incorporated reads.

Other options comprise sequencing on templates physically covering the conflict region (plasmids, fosmids and PCR products), the separated assembly of conflict regions or matrix-assisted assembly using a homolog-conserved sequence synteny of a closely related strain. The so-called finishing experiments for improving the sequence quality and gap closure represent the most time-consuming and cost-intensive ways of attaining a complete genome sequence. Several phytoplasma projects have included large insert libraries in this respect, thus ensuring an efficient gap-closure process. These comprise plasmid libraries (insert a size above >2.5 kb), lambda libraries (>10 kb) and low copy vector libraries (e.g. fosmids, average insert size 37–40 kb). However, a strong negative correlation of cloning success by increasing fragment size has been observed for phytoplasma DNA compared to other bacterial genome, resulting in the low physical coverage by large inserts of phytoplasma genomes.

WGS sequencing resulting in complete genome sequences was performed on clonal templates by dye-terminator cycle sequencing and, in most cases, subsequent read determination by applying capillary electrophoresis technology. Reads undergoing sequence quality assignment by Phred (Ewing et al. 1998) were assembled by

Phrap (www.phrap.org). Target-assigned assemblies were performed in many projects by excluding non-phytoplasma-originated reads or reads assigned to phytoplasmas by BLAST approaches, achieving improved results on the way to gaining complete genomes or improving assemblies in general (Mitrovic et al. 2014). The manual curation process was essential for a clone-based complete genome reconstruction applying different types of database editors. This editing process comprises mainly read inspections and the addition of further sequence data for gap closure. The gap4 platform (Bonfield et al. 1995) and Consed (Gordon et al. 1998) were used for complete genome determination, except for the '*Ca.* P. asteris' strain M3 (Orlovskis et al. 2017), which limited this process to the application of the short-read assembler Velvet (Zerbino 2010). In contrast to several phytoplasma sequencing projects driven by NGS short-read approaches failing to reconstruct complete genomes *de novo*, the short read assembly of strain M3 successfully resulted in a complete sequence, applying the available complete genomes of closely related '*Ca.* P. asteris' strains as assistants during the genome reconstruction process (Orlovskis et al. 2017). The power of long sequence read-based NGS assemblies has been recently demonstrated for jujube witches' broom (Wang et al. 2018). The team applied single-molecule real-time sequencing (SMRT) provided by PacBio in combination with Illumina's short read technologies for the complete genome determination.

1.4 Analysis of Genomic Content

Genomic sequence annotation of phytoplasmas apply state-of-the-art software packages for bacterial genomes, including, for the prediction of structural RNAs, rRNA and tRNA genes, RNAmmer (Lagesen et al. 2007) and tRNAscan-SE (Lowe and Eddy 1997). These software packages applied for predicting protein-coding genes differ between individual projects. The established prediction by Glimmer (Delcher et al. 1999a) of protein-coding genes was applied for '*Ca.* P. mali' strain AT, '*Ca.* P. asteris' strain AY-WB and '*Ca.* P. australiense' strains PA and SLY, while Prodigal (Hyatt et al. 2010) was applied for '*Ca.* P. asteris' strain M3.

The functional assignment of the deduced proteins comprised the application of various software packages, depending on the required depth of the analysis. General analysis for manual annotation and curation includes BLAST analysis (Altschul et al. 1997) against the non-redundant database of NCBI (blast.ncbi.nlm.nih.gov). This initial step in analysis was supplemented by screening the protein content against functional databases such as the signature database Pfam (Finn et al. 2016) and the integrative database InterPro (Hunter et al. 2009), thereby improving the assignments. The comprehensive database of *Mollicutes*, MolliGen (Barre et al. 2004), was also introduced in the analysis of '*Ca.* P. australiense' strain SLY (Andersen et al. 2013). Cellular localisation of phytoplasma proteins with respect to the secretome was predicted applying SignalP, Secretome (Bendtsen et al. 2004) and/or Phobius (Kall et al. 2004). The curated phytoplasma protein sets were used for the functional reconstruction of pathways often including the KEGG pathway database (Kanehisa and Goto 2000) or MetaCyc (Caspi et al. 2018) as central references alongside the published analyses of phytoplasma genomes. Through their functional assignment, clusters of orthologous groups (COGs) can be generated by additionally comparing the deduced proteins (Tatusov et al. 2000). However, the numbers of orthologous proteins were calculated by reciprocal BLAST or by advanced approaches such as OrthoMCL (Li et al. 2003; Wang et al. 2018), while the first approach was also applied for estimating paralogous gene content (Kube et al. 2012). Genomic re-arrangements and repeats were examined in dot plot analysis by applying MUMmer (Delcher et al. 1999b), and other tools implemented in software such as Artemis (Carver et al. 2008) were used for the calculation of cumulative GC skews.

1.5 Complete Phytoplasma Genomes

Due to technical difficulties, only seven complete phytoplasma genome sequences have been published to date (Table 1.1), and they cover just four '*Candidatus* Phytoplasma' taxa, comprising three '*Ca.* P. asteris' strains (OY-M, AY-WB and M3), two '*Ca.* P. australiense' strains (PAa, NZSb11), the '*Ca.* P. mali' strain AT and the '*Ca.* P. ziziphi' strain jwb-nky. Moreover, five of the fully sequenced strains

Table 1.1 Complete phytoplasma chromosomes

Phytoplasma strain	Size (bb)	GC%	Protein	rRNA operons	tRNA	GenBank Accession number (literature)
'*Candidatus* Phytoplasma asteris' (16SrI)						
OY-M	860,631	28	754	2	32	AP006628 (Oshima et al. 2004)
AY-WB	707,569	27	671	2	31	CP000061 (Bai et al. 2006)
M3	576,118	28	485	2	32	CP015149 (Orlovskis et al. 2017)
'*Candidatus* Phytoplasma mali' (16SrX)						
AT	601,943	21	497	2	32	CU469464 (Kube et al. 2008)
'*Candidatus* Phytoplasma australiense' (16SrXII)						
PA	879,324	27	839	2	35	AM422018 (Tran-Nguyen et al. 2008)
NZSb11	959,779	27	1126	2	35	CP002548 (Andersen et al. 2013)
'*Candidatus* Phytoplasma ziziphi' (16SrV)						
nky	750,803	23	694	2	31	CP025121 (Wang et al. 2018)

belong to the same phylogenetic cluster, leaving several phylogenetic phytoplasma cluster uncovered (Fig. 1.1). Additional incomplete genome information on 19 strains is provided, albeit as draft sequences. The first complete phytoplasma genome, the '*Ca.* P. asteris' strain OY-M from *Chrysanthemum* (onion yellow mild), was determined in 2004 (Oshima et al. 2004), followed by the '*Ca.* P. asteris' AY-WB from infected lettuce (Bai et al. 2006), the '*Ca.* P. australiense' strain PAa from cotton bush (*Gomphocarpus physocarpus* L.) (Tran-Nguyen et al. 2008) and the '*Ca.* P. mali' strain AT from apple (Kube et al. 2008). Later, the complete genomes of the strawberry lethal yellows strain SLY of '*Ca.* P. australiense' (Andersen et al. 2013), the maize bushy stunt pathogen '*Ca.* P. asteris' strain M3 (Orlovskis et al. 2017) and, recently, the '*Ca.* P. ziziphi' strain jwb-nky were added (Wang et al. 2018). Besides the '*Candidatus* Phytoplasma' species assignment, phytoplasmas are grouped by the 16S rRNA gene RFLP profile, which in majority corresponds to each other (Lee et al. 1998). The complete genomes are from phytoplasmas enclosed in the subgroups 16SrI-A, 16SrI-B (aster yellows), 16SrXII-B (Australian grapevine yellows), 16SrX-A (apple proliferation) and 16SrV-B (jujube witches' broom).

Available information on complete genome sequences is still poor with respect to the species assignment of the strains, so there is a strong need for additional complete sequences due not only to known basic differences in genome architecture and organisation, but also with respect to the coding repertoire underlying the well-known rapid phytoplasma evolution (Bai et al. 2006). The complete chromosome sequences range between 576 and 960 kb in size (Kube et al. 2012), and size variation observed within a species is impressive with respect to the different chromosome sizes of '*Ca.* P. asteris' strains, ranging from 853 kb for strain OY-M to 498 kb for strain M3 (Bai et al. 2006; Orlovskis et al. 2017). However, differences within the groups are not limited to size, since phytoplasmas also differ in chromosome architecture. A striking example in this regard is '*Ca.* P. mali' strain AT, which carries a linear chromosome with terminal inverted repeats of 43 kb (Kube et al. 2008). The chromosome ends are protected by hairpin structures. Based on the genome analysis of '*Ca.* P. mali' strain AT, it has been shown that the strains AP15, 1/93 and 12/93 of the apple proliferation phytoplasma, and the agent of pear decline, '*Ca.* P. pyri', also carry linear chromosomes, thereby supporting the hypothesis that linear chromosomes might be a particular feature of members of this phylogenetic branch. Genome content also differs in a number of extrachromosomal elements (eDNA) from two and four plasmids in '*Ca.* P. asteris' strains OY-M (Oshima et al. 2004) and AY-WB (Bai et al. 2006), respectively. The eDNA presence was not determined in '*Ca.* P. mali' strain AT (Kube et al. 2008) or '*Ca.* P. australiense' strain SLY (Andersen et al. 2013), though plasmids have been observed in other '*Ca.* P. australiense' strains (Liefting et al. 2006). The missing detection of small plasmids needs cautious interpretation with respect to the enrichment procedures applied to purify phytoplasma DNA and the unknown stability of these plasmids in the strains over time. In addition, the presence of a circulative eDNA originated from the complex transposons (potential mobile units) carrying possible virulence factors was proven for '*Ca.* P. asteris' (Toruno et al. 2010).

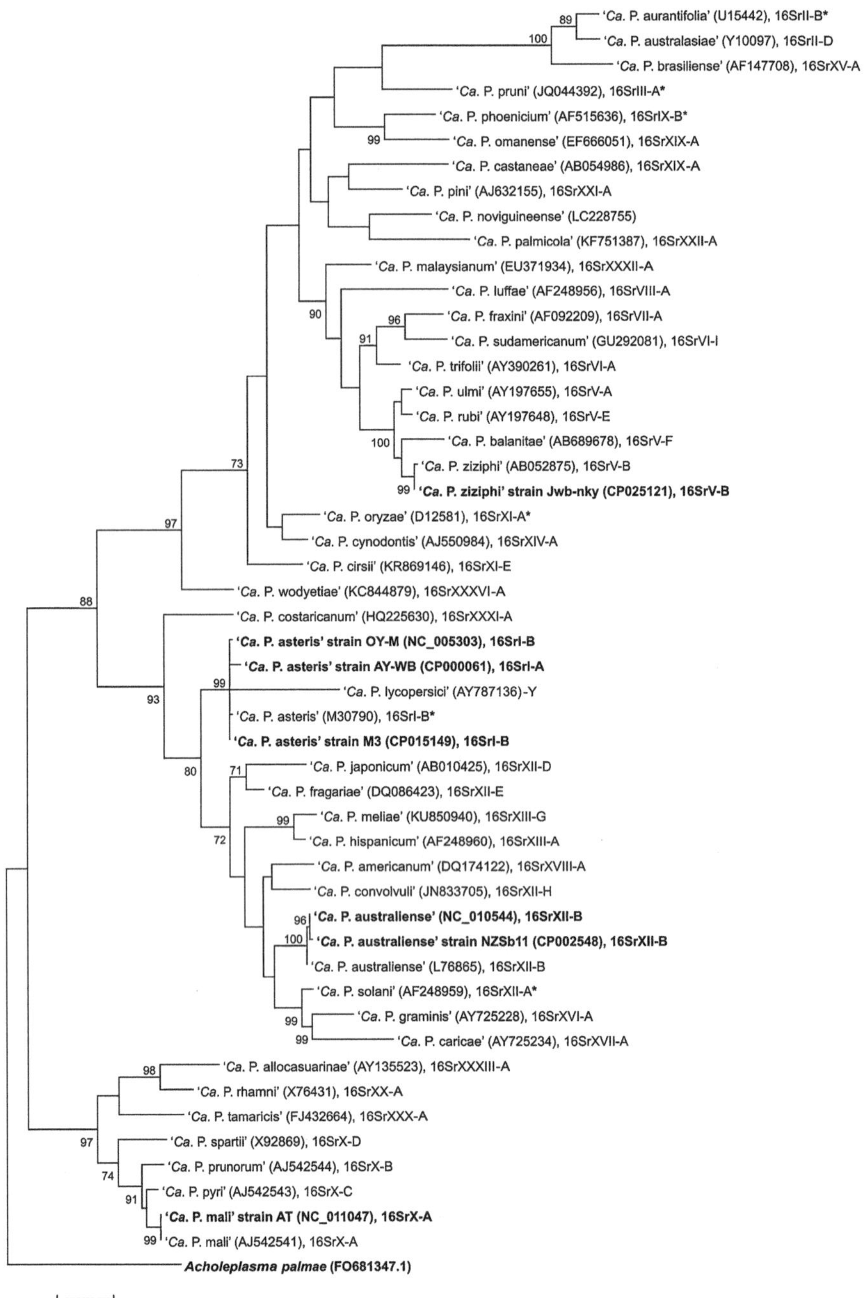

Fig. 1.1 Phylogenetic tree constructed by parsimony analyses of 16S rDNA sequences of 50 phytoplasma strains, belonging to all '*Candidatus* Phytoplasma' species described employing *Acholeplasma palmae* as outgroup. The completely sequenced strains are in bold. CLUSTAL W

Genomic re-arrangements and integration events limit the ability to use conserved sequence synteny as a map for *de novo* genome sequencing among the species, but it could be applied successfully for the reconstruction of '*Ca.* P. asteris' strains (Bai et al. 2006; Orlovskis et al. 2017). The genomic instability of phytoplasmas can be made visible by applying cumulative GC-skew analysis (Grigoriev 1998), thereby highlighting its extreme in '*Ca.* P. asteris' AY-WB and its lower degree in '*Ca.* P. mali' (Kube et al. 2012). The situation is also critical for the chromosomal encoded repertoire. The genome condensation process in the evolution of phytoplasmas has resulted in fragmented or lack of pathways for amino acid synthesis and the absence of pathways such as the tricarboxylic acid cycle, oxidative phosphorylation, pentose phosphate pathway or F_0F_1-type ATPase supported by complete genome sequences (Oshima et al. 2004; Kube et al. 2012). This increase in host dependency is also highlighted by the incomplete glycolysis encoded by '*Ca.* P. mali' lacking the energy-yielding part (Kube et al. 2008). The sequencing of complete genomes and subsequent bioinformatical analysis has enabled these results.

1.6 Potential Mobile Units in Phytoplasma Genomes

Phytoplasma genomes contain clusters of repeated gene sequences, named potential mobile units (PMUs) (Bai et al. 2006) or sequence-variable mosaics (SVMs) (Jomantiene and Davis 2006; Jomantiene et al. 2007; Wei et al. 2008). PMUs and SVMs have similar compositions and contain similar genes; henceforth, these gene clusters are referred here to as PMUs. In the AY-WB genome, PMUs are about 20 kb in size and consist of genes with similarities to *sigF*, *hflB*, *dnaG*, *dnaB*, *tmk*, *ssb*, *himA* and the IS3 family insertion sequence tra5 that are organised in a conserved order (Bai et al. 2006). These genes are also found in multiple copies, singly or in clusters, in other phytoplasma genomes (Oshima et al. 2004; Lee et al. 2005; Jomantiene and Davis 2006; Jomantiene et al. 2007; Arashida et al. 2008). The repeated presence of PMUs, their gene contents, including genes for recombination (*tra5*, *ssb*, *himA*) and replication (*dnaG*, *dnaB*), and their conserved gene orders suggest that the PMUs are replicative composite transposons (Bai et al. 2006;

Fig. 1.1 (continued) from the Molecular Evolutionary Genetics Analysis program MEGA7 was used for multiple alignment (Kumar et al. 2016). The maximum parsimony tree was obtained using the subtree–pruning-regrafting algorithm, implemented in the MEGA7, with search level 5 in which the initial trees were obtained with the random addition of sequences (ten replicates). The most parsimonious tree is shown. Numbers on the branches are bootstrap values obtained for 1,000 replicates (only values above 70% are shown). GenBank accession numbers are given in parentheses, and 16Sr group classification (when available) is shown on the right side of the tree. Asterisk indicates 16Sr groups for which draft sequences are publicly available. The tree is drawn to scale with branch lengths calculated using the average pathway method and represents the number of changes over the whole sequence. The scale bar represents 20 nucleotide substitutions

Arashida et al. 2008). It has been reported that the PMU exists as linear chromosomal and circular extrachromosomal elements in AY-WB phytoplasma (Toruno et al. 2010), implying that the PMU has the ability to transpose within the genome. The presence of multiple PMUs or apparently degenerated PMU-like sequences, on one hand, and the dramatic loss of basic metabolic pathways in phytoplasma genomes, on the other hand, suggest that PMUs are likely to be somehow important for phytoplasma fitness (Oshima et al. 2004; Bai et al. 2006; Wei et al. 2008).

1.7 Complete Genome Versus Draft Sequences

Gapless complete genomes are separated from available genomic drafts through the organisation of genetic information in (un)ordered DNA fragments and/or genome sequences of comparable weak quality. The term draft suggests the need for further improvement. Genomic data from both categories have a significant impact on expanding the knowledge on marker genes and providing valuable insights by enabling *in silico* functional reconstruction. Depending on hypotheses and questions, complete or draft sequences are generated. The major benefit of a complete genome sequence lies within its possibilities in analysis. While a genomic draft is limited to the determined information, a complete genome can also be interpreted with respect to absent genetic content in comparison to other genomes, fact that can become crucial, keeping in mind that small genes can be lost during analyses because of the low chance of identifying them even as fragments in a draft. Small-sized genes comprise genes such as tRNAs, house-keeping genes coding protein (ribosomal proteins) and also effector proteins of particular importance for virulence, so statements on their absence rely on gapless genome sequences in general. Apart from these limitations, phytoplasma draft sequences are of high influence, since they have enhanced the knowledge by providing information on so far unknown genetic content in phytoplasma strains (Chung et al. 2013). In addition, obtained genomic information is not limited to functional reconstruction; in fact, it provides a profound treasure trove, for example, in the design of new primer sets used in population genetics (Mitrovic et al. 2015). Today, numerous draft sequences differing in levels of information are available (Table 1.2). The degree of fragmentation/contiguity of a genomic draft can be estimated by the N50 value, which defines assembly quality according to the size of the contig which, together with the other larger contigs of the assembly, comprises 50% of the assumed genome (Miller et al. 2010). In accordance, a high N50 value (*e.g.* obtained for the wheat blue dwarf phytoplasma) hints at high assembly quality, while low values indicate a weak assembly with respect to contiguity; however, the quality of the assembly by itself with respect to miss-assemblies or individual positions cannot be estimated from the N50 value. Genomic drafts vary from close to complete (Chen et al. 2014), to the availability of unassembled sequence reads of several strains (Orlovskis et al. 2017) with respect to the different objectives of the studies.

Table 1.2 Phytoplasma draft sequences

16Sr subgroup	Strain/acronym (disease)	Contigs	N50 value	Size (kb)	GC%	CDS	ncbi.nlm.nih.gov (literature)
***'Candidatus* Phytoplasma asteris'**							
16SrI-B	WBD (wheat blue dwarf)	6	345,813	611	27	471	NZ_AVAO00000000 (Chen et al. 2014)
	OY-V (onion yellows)	170	10,878	740	27	901	NZ_BBIY00000000 (Kakizawa et al. 2014)
	CYP (chrysanthemum yellows)	252	6,555	660	28	546	NZ_JSWH00000000 (Pacifico et al. 2015)
	LD1 (rice orange leaf)	8	163,230	599	28	532	NZ_MIEP00000000 (Zhu et al. 2017)
	TW1 (*Brassica napus* phytoplasma)	5	373,899	744	28	457	QGKT00000000 (Town et al. 2018)
16SrI-A	NJAY (New Jersey aster yellows)	75	21,929	652	27	371	MAPF00000000 (Sparks et al. 2018)
***'Candidatus* Phytoplasma aurantifolia'**							
16SrII-A	AMWZ (peanut witches' broom)	14	71,463	567	24	425	AMWZ00000000 (Chung et al. 2013)
	NCHU2014 (purple coneflower witches' broom)	28	46,697	545	23	460	NZ_LKAC00000000 (Chang et al. 2015)
16SrII-B	WBDL (witches' broom disease of lime)	98	15,999	475	24	390	NZ_MWKN00000000 (submitted by Foissac and Carle in 2017)
***'Candidatus* Phytoplasma pruni'**							
16SrIII-H	JR1 (poinsettia branch-inducing)	185	8,188	631	27	515	AKIK00000000 (Saccardo et al. 2012)
16SrIII-F	MW1 (milkweed yellows disease)	158	7,972	584	27	454	AKIL00000000 (Saccardo et al. 2012)
16SrIII-B	MA1 (Italian clover phyllody)	197	12,309	597	27,1	495	AKIM00000000 (Saccardo et al. 2012)
16SrIII-F	VAC (vaccinium witches' broom)	272	6,572	648	27,3	500	AKIN00000000 (Saccardo et al. 2012)
16SrIII-A	CX (X-disease)	46	38,825	599	27	554	NZ_LHCF00000000 (Lee et al. 2015)
16SrIII-J	Vc33	475	45,393	687	28	696	LLKK00000000 (Zamorano and Fiore 2016)
***'Candidatus* Phytoplasma phoenicium'**							
16SrIX-B	SA213 (almond witches' broom)	78	5,782	346	25	327	JPSQ01000000 (Quaglino et al. 2015)

(continued)

Table 1.2 (continued)

16Sr subgroup	Strain/acronym (disease)	Contigs	N50 value	Size (kb)	GC%	CDS	ncbi.nlm.nih.gov (literature)
'Candidatus **Phytoplasma oryzae'**							
16SrXI	Mbita1 (Napier grass stunt)	28	24,446	533	19	425	NZ_LTBM00000000 (Fischer et al. 2016)
'Candidatus **Phytoplasma solani'**							
16SrXII-A	284/09 ("stolbur" of tobacco)	128	–	558	28	448	FO393427 (Mitrovic et al. 2014)
	231/09 ("stolbur" of parsley)	298	–	516	29	346	FO393428 (Mitrovic et al. 2014)
	SA-1 ("bois noir")	19	76,256	709	28	821	MPBG00000000 (Seruga-Music et al. 2018)

1.8 Outlook

Besides other advances in phytoplasma research, the ongoing decrease in sequencing costs (Muir et al. 2016) enables to generate phytoplasma genomes for a wider research community. This development in the cost base also changes the impact of the unwanted host DNA, to the point that it will become affordable and commonplace to skip >95% of all data and enable the generation of high(er)-quality draft sequences from infected tissue, without previous enrichment steps, thereby focusing on the remaining percentage of targeted phytoplasma data. Technical development will not be limited to genomic drafts, due to the increasing success of long sequence reads, providing technologies such as single-molecule read sequencing (PacBio). This approach is of major significance to the full genome sequencing of phytoplasmas – as shown for other bacteria (Teng et al. 2017; Wang et al. 2018). Long-read spanning repeat regions will result in improved assemblies and more complete genomes, which are needed not only for economically important strains and so-far unexplored phytoplasma groups but also for the examination of evolutionary processes on the genome level – as already performed for other bacteria (Kim et al. 2018). While the low sequencing costs needed for NGS techniques providing long reads for full genome sequencing are not yet achievable, data processing is no longer a bottleneck for phytoplasmas. Freely available software includes tools for the automated assembly of NGS reads not requiring large computing power (Zerbino 2010), and free services for annotation enable genome analysis (Overbeek et al. 2014), thereby also providing data for labs not specialised in genome research in general. As a consequence, it is predictable that not only is phytoplasma research blooming (Strauss 2009) but also phytoplasma genomics.

References

Al-Okaily AA (2016) HGA: de novo genome assembly method for bacterial genomes using high coverage short sequencing reads. *BMC Genomics* **17**, 193.

Altschul SF, Madden TL, Schaffer AA, Zhang J, Zhang Z, Miller W, Lipman DJ (1997) Gapped BLAST and PSI-BLAST: a new generation of protein database search programs. *Nucleic Acids Research* **25**, 3389–3402.

Andersen MT, Liefting LW, Havukkala I, Beever RE (2013) Comparison of the complete genome sequence of two closely related isolates of '*Candidatus* Phytoplasma australiense' reveals genome plasticity. *BMC Genomics* **14**, 529.

Arashida R, Kakizawa S, Hoshi A, Ishii Y, Jung H-Y, Kagiwada S, Yamaji Y, Oshima K, Namba S (2008) Heterogeneic dynamics of the structures of multiple gene clusters in two pathogenetically different lines originating from the same phytoplasma. *DNA and Cell Biology* **27**, 209–217.

Bai X, Zhang J, Ewing A, Miller SA, Jancso Radek A, Shevchenko DV, Tsukerman K, Walunas T, Lapidus A, Campbell JW, Hogenhout SA (2006) Living with genome instability: the adaptation of phytoplasmas to diverse environments of their insect and plant hosts. *Journal of Bacteriology* **188**, 3682–3696.

Barre A, De Daruvar A, Blanchard A (2004) MolliGen, a database dedicated to the comparative genomics of Mollicutes. *Nucleic Acids Research* **32**, D307–D310.

Bendtsen JD, Nielsen H, Von Heijne G, Brunak S (2004) Improved prediction of signal peptides: signalP 3.0. *Journal of Molecular Biology* **340**, 783–95.

Bennett GM, Abba S, Kube M, Marzachì C (2016) Complete genome sequences of the obligate symbionts '*Candidatus* Sulcia muelleri' and '*Ca.* Nasuia deltocephalinicola' from the pestiferous leafhopper *Macrosteles quadripunctulatus* (Hemiptera: Cicadellidae). *Genome Announcements* **4**, e01604–15.

Bodenteich A, Chissoe S, Wang YF, Roe BA (1994) Shotgun cloning as the strategy of choice to generate templates for high-throughput dideoxynucleotide sequencing. In: Automated DNA Sequencing and Analysis Techniques. Eds Adams M, Fields C, Venter JC, Academic Press, San Diego, California United States of America.

Bonfield JK, Smith K, Staden R (1995) A new DNA sequence assembly program. *Nucleic Acids Research* **23**, 4992–4999.

Carver T, Berriman M, Tivey A, Patel C, Bohme U, Barrell BG, Parkhill J, Rajandream MA (2008) Artemis and ACT: viewing, annotating and comparing sequences stored in a relational database. *Bioinformatics* **24**, 2672–2676.

Caspi R, Billington R, Fulcher CA, Keseler IM, Kothari, Krummenacker M, Latendresse M, Midford PE, Ong Q, Ong WK, Paley S, Subhraveti P, Karp PD (2018) The MetaCyc database of metabolic pathways and enzymes. *Nucleic Acids Research* **46**, D633–D639.

Chang SH, Cho ST, Chen CL, Yang JY, Kuo CH (2015) Draft genome sequence of a 16SrII-A Subgroup phytoplasma associated with purple coneflower (*Echinacea purpurea*) witches' broom disease in Taiwan. *Genome Announcements* **3**, e01398–15.

Chen W, Li Y, Wang Q, Wang N, Wu Y (2014) Comparative genome analysis of wheat blue dwarf phytoplasma, an obligate pathogen that causes wheat blue dwarf disease in China. *Plos One* **9**, e96436.

Chung WC, Chen LL, Lo WS, Lin CP, Kuo CH (2013) Comparative analysis of the peanut witches' broom phytoplasma genome reveals horizontal transfer of potential mobile units and effectors. *Plos One* **8**, e62770.

Delcher AL, Harmon D, Kasif S, White O, Salzberg SL (1999a) Improved microbial gene identification with GLIMMER. *Nucleic Acids Research* **27**, 4636–4641.

Delcher AL, Kasif S, Fleischmann RD, Peterson J, White O, Salzberg SL (1999b) Alignment of whole genomes. *Nucleic Acids Research* **27**, 2369–2376.

Ewing B, Hillier L, Wendl MC, Green P (1998) Base-calling of automated sequencer traces using phred. I. Accuracy assessment. *Genome Research* **8**, 175–185.

Finn RD, Coggill P, Eberhardt RY, Eddy SR, Mistry J, Mitchell AL, Potter SC, Punta M, Qureshi M, Sangrador-Vegas A, Salazar GA, Tate J, Bateman A (2016) The Pfam protein families database: towards a more sustainable future. *Nucleic Acids Research* **44**, D279–D285.

Fischer A, Santana-Cruz I, Wambua L, Olds C, Midega C, Dickinson M, Kawicha P, Khan Z, Masiga D, Jores J, Schneider B (2016) Draft genome sequence of '*Candidatus* Phytoplasma oryzae' strain Mbita1, the causative agent of napier grass stunt disease in Kenya. *Genome Announcements* **4,** e00297–16.

Frangeul L, Nelson KE, Buchrieser C, Danchin A, Glaser P, Kunst F (1999) Cloning and assembly strategies in microbial genome projects. *Microbiology* **145**, 2625–2634.

Fraser CM, Gocayne JD, White O, Adams MD, Clayton RA, Fleischmann RD, Bult CJ, Kerlavage AR, Sutton G, Kelley JM, Fritchman RD, Weidman JF, Small KV, Sandusky M, Fuhrmann J, Nguyen D, Utterback TR, Saudek DM, Phillips CA, Merrick JM, Tomb JF, Dougherty BA, Bott KF, Hu PC, Lucier TS, Peterson SN, Smith HO, Hutchison 3th CA, Venter JC (1995) The minimal gene complement of *Mycoplasma genitalium. Science* **270**, 397–403.

Gordon D, Abajian C, Green P (1998) Consed: a graphical tool for sequence finishing. *Genome Research* **8**, 195–202.

Grigoriev A (1998) Analyzing genomes with cumulative skew diagrams. *Nucleic Acids Research* **26**, 2286–2290.

Hicks CA, Barker EN, Brady C, Stokes CR, Helps CR, Tasker S (2014) Non-ribosomal phylogenetic exploration of Mollicute species: new insights into haemoplasma taxonomy. *Infectious Genetic Evolution* **23**, 99–105.

Hunter S, Apweiler R, Attwood TK, Bairoch A, Bateman A, Binns D, Bork P, Das U, Daugherty L, Duquenne L, Finn RD, Gough J, Haft D, Hulo N, Kahn D, Kelly E, Laugraud A, Letunic I, Lonsdale D, Lopez R, Madera M, Maslen J, McAnulla C, McDowall J, Mistry J, Mitchell A, Mulder N, Natale D, Orengo C, Quinn AF, Selengut JD, Sigrist CJ, Thimma M, Thomas PD, Valentin F, Wilson D, Wu CH, Yeats C (2009) InterPro: the integrative protein signature database. *Nucleic Acids Research* **37**, D211–D215.

Hyatt D, Chen GL, Locascio PF, Land ML, Larimer FW, Hauser LJ (2010) Prodigal: prokaryotic gene recognition and translation initiation site identification. *BMC Bioinformatics* **11**, 119.

IRPCM (2004) '*Candidatus* Phytoplasma', a taxon for the wall-less, non-helical prokaryotes that colonize plant phloem and insects. *International Journal of Systematic and Evolutionary Microbiology* **54**, 1243–1255.

Jomantiene R, Davis RE (2006) Clusters of diverse genes existing as multiple, sequence-variable mosaics in a phytoplasma genome. *FEMS Microbiology Letters* **255**, 59–65.

Jomantiene R, Zhao Y, Davis RE (2007) Sequence-variable mosaics: composites of recurrent transposition characterizing the genomes of phylogenetically diverse phytoplasmas. *DNA and Cell Biology* **26**, 557–564.

Kakizawa S, Makino A, Ishii Y, Tamaki H, Kamagata Y (2014) Draft genome sequence of '*Candidatus* Phytoplasma asteris' strain OY-V, an unculturable plant-pathogenic bacterium. *Genome Announcements* **2**, e00944–14

Kall L, Krogh A, Sonnhammer EL (2004) A combined transmembrane topology and signal peptide prediction method. *Journal of Molecular Biology* **338**, 1027–1036.

Kanehisa M, Goto S (2000) KEGG: Kyoto encyclopedia of genes and genomes. *Nucleic Acids Research* **28**, 27–30.

Kim J, Lindsey RL, Garcia-Toledo L, Loparev VN, Rowe LA, Batra D, Juieng P, Stoneburg D, Martin H, Knipe K, Smith P, Strockbine N (2018) High-quality whole-genome sequences for 59 historical *Shigella* strains generated with PacBio sequencing. *Genome Announcements* **6**, e00282–18.

Kirkpatrick BC, Stenger DC, Morris J, Purcell AH (1987) Cloning and detection of DNA from a nonculturable plant pathogenic mycoplasma-like organism. *Science* **238**, 197–200.

Kube M, Schneider B, Kuhl H, Dandekar T, Heitmann K, Migdoll AM, Reinhardt R, Seemüller E (2008) The linear chromosome of the plant-pathogenic mycoplasma '*Candidatus* Phytoplasma mali'. *BMC Genomics* **9**, 306.

Kube M, Mitrovic J, Duduk B, Rabus R, Seemüller E (2012) Current view on phytoplasma genomes and encoded metabolism. *Scientific World Journal* **2012**, 185942.

Kumar S, Stecher G, Tamura K (2016) MEGA7: Molecular evolutionary genetics analysis version 7.0 for bigger datasets. *Molecular Biology and Evolution* **33**, 1870–1874.

Lagesen K, Hallin P, Rodland EA, Staerfeldt HH, Rognes T, Ussery DW (2007) RNAmmer: consistent and rapid annotation of ribosomal RNA genes. *Nucleic Acids Research* **35**, 3100–3108.

Lee I-M, Gundersen-Rindal DE, Davis RE, Bartoszyk IM (1998) Revised classification scheme of phytoplasmas based on RFLP analyses of 16S rRNA and ribosomal protein gene sequences. *International Journal of Systematic and Evolutionary Microbiology* **48**, 1153–1169.

Lee I-M, Zhao Y, Bottner KD (2005) Novel insertion sequence-like elements in phytoplasma strains of the aster yellows group are putative new members of the IS3 family. *FEMS Microbiology Letters* **242**, 353–360.

Lee I-M, Shao J, Bottner-Parker KD, Gundersen-Rindal DE, Zhao Y, Davis RE (2015) Draft genome sequence of '*Candidatus* Phytoplasma pruni' strain CX, a plant-pathogenic bacterium. *Genome Announcements* **3**, e01117–15.

Li L, Stoeckert CJ Jr., Roos DS (2003) OrthoMCL: identification of ortholog groups for eukaryotic genomes. *Genome Research* **13**, 2178–2189.

Liefting LW, Andersen MT, Lough TJ, Beever RE (2006) Comparative analysis of the plasmids from two isolates of '*Candidatus* Phytoplasma australiense'. *Plasmid* **56**, 138–144.

Lowe TM, Eddy SR (1997) tRNAscan-SE: a program for improved detection of transfer RNA genes in genomic sequence. *Nucleic Acids Research* **25**, 955–964.

Marcone C, Neimark H, Ragozzino A, Lauer U, Seemüller E (1999) Chromosome sizes of phytoplasmas composing major phylogenetic groups and subgroups. *Phytopathology* **89**, 805–810.

Miller JR, Koren S, Sutton G (2010) Assembly algorithms for next-generation sequencing data. *Genomics* **95**, 315–327.

Mitrovic J, Siewert C, Duduk B, Hecht J, Molling K, Broecker F, Beyerlein P, Buttner C, Bertaccini A, Kube M (2014) Generation and analysis of draft sequences of "stolbur" phytoplasma from multiple displacement amplification templates. *Journal of Molecular Microbiological Biotechnology* **24**, 1–11.

Mitrovic J, Smiljkovic M, Seemüller E, Reinhardt R, Hüttel B, Büttner C, Bertaccini A, Kube M., Duduk B (2015) Differentiation of '*Candidatus* Phytoplasma cynodontis' based on 16S rRNA and *groEL* genes and identification of a new subgroup, 16SrXIV-C. *Plant Disease* **99**, 1578–1583.

Muir P, Li S, Lou S, Wang D, Spakowicz DJ, Salichos L, Zhang J, Weinstock GM, Isaacs F, Rozowsky J, Gerstein M (2016) The real cost of sequencing: scaling computation to keep pace with data generation. *Genome Biology* **17**, 53.

Neimark H, Kirkpatrick BC (1993) Isolation and characterization of full-length chromosomes from non-culturable plant-pathogenic mycoplasma-like organisms. *Molecular Microbiology* **7**, 21–28.

Orlovskis Z, Canale MC, Haryono M, Lopes JRS, Kuo CH, Hogenhout SA (2017) A few sequence polymorphisms among isolates of maize bushy stunt phytoplasma associate with organ proliferation symptoms of infected maize plants. *Annals of Botany* **119**, 869–884.

Oshima K, Shiomi T, Kuboyama T, Sawayanagi T, Nishigawa H, Kakizawa S, Miyata S, Ugaki M, Namba S (2001) Isolation and characterization of derivative lines of the onion yellows phytoplasma that do not cause stunting or phloem hyperplasia. *Phytopathology* **91**, 1024–1029.

Oshima K, Kakizawa S, Nishigawa H, Jung H-Y, Wei W, Suzuki S, Arashida R, Nakata D, Miyata S, Ugaki M, Namba S (2004) Reductive evolution suggested from the complete genome sequence of a plant-pathogenic phytoplasma. *Nature Genetics* **36**, 27–29.

Overbeek R, Olson R, Pusch GD, Olsen GJ, Davis JJ, Disz T, Edwards RA, Gerdes S, Parrello B, Shukla M, Vonstein V, Wattam AR, Xia F, Stevens R (2014) The SEED and the rapid annotation of microbial genomes using subsystems technology (RAST). *Nucleic Acids Reseach* **42**, D206–D214.

Pacifico D, Galetto L, Rashidi M, Abbà S, Palmano S, Firrao G, Bosco D, Marzachì C (2015) Decreasing global transcript levels over time suggest that phytoplasma cells enter stationary phase during plant and insect colonization. *Applied and Environmental Microbiology* **81**, 2591–2602.

Quaglino F, Kube M, Jawhari M, Abou-Jawdah Y, Siewert C, Choueiri E, Sobh H, Casati P, Tedeschi R, Molino Lova M, Alma A, Bianco PA (2015) '*Candidatus* Phytoplasma phoenicium' associated with almond witches' broom disease: from draft genome to genetic diversity among strain populations. *BMC Microbiology* **15**, 148.

Razin S, Yogev D, Naot Y (1998) Molecular biology and pathogenicity of mycoplasmas. *Microbiology and Molecular Biology Reviews* **62**, 1094–1156.

Saccardo F, Martini M, Palmano S, Ermacora P, Scortichini M, Loi N, Firrao G (2012) Genome drafts of four phytoplasma strains of the ribosomal group 16SrIII. *Microbiology* **158**, 2805–2814.

Seruga-Music M, Samarzija I, Hogenhout SA, Haryono M, Cho ST, Kuo CH (2018) The genome of '*Candidatus* Phytoplasma solani' strain SA-1 is highly dynamic and prone to adopting foreign sequences. *Systematic and Applied Microbiology* **42**, 117–127.

Sparks ME, Bottner-Parker KD, Gundersen-Rindal DE, Lee I-M (2018) Draft genome sequence of the New Jersey aster yellows strain of '*Candidatus* Phytoplasma asteris'. *Plos One* **13**, e0192379.

Strauss E (2009) Phytoplasma research begins to bloom. *Science* **325**, 388–390.

Tatusov RL, Galperin MY, Natale DA, Koonin EV (2000) The COG database: a tool for genome-scale analysis of protein functions and evolution. *Nucleic Acids Research* **28**, 33–36.

Teng JLL, Yeung ML, Chan E, Jia L, Lin GH, Huang Y, Tse H, Wong SSY, Sham PC, Lau SKP, Woo PCY (2017) PacBio but not Illumina technology can achieve fast, accurate and complete closure of the high GC, complex *Burkholderia pseudomallei* two-chromosome genome. *Frontieres in Microbiology* **8**, 1448.

Toruno TY, Seruga-Music MS, Simi S, Nicolaisen M, Hogenhout SA (2010) Phytoplasma PMU1 exists as linear chromosomal and circular extrachromosomal elements and has enhanced expression in insect vectors compared with plant hosts. *Molecular Microbiology* **77**, 1406–1415.

Town JR, Wist T, Perez-Lopez E, Olivier CY, Dumonceaux TJ (2018) Genome sequence of a plant-pathogenic bacterium '*Candidatus* Phytoplasma asteris' strain TW1. *Microbiology Resources Announcements* **7**, e01109–18.

Tran-Nguyen LT, Kube M, Schneider B, Reinhardt R, Gibb KS (2008) Comparative genome analysis of '*Candidatus* Phytoplasma australiense' (subgroup tuf-Australia I; rp-A) and '*Ca.* Phytoplasma asteris' strains OY-M and AY-WB. *Journal of Bacteriology* **190**, 3979–3991.

Wang J, Song L, Jiao Q, Yang S, Gao R, Lu X, Zhou G (2018) Comparative genome analysis of jujube witches' broom phytoplasma, an obligate pathogen that causes jujube witches' broom disease. *BMC Genomics* **19**, 689.

Wei W, Davis RE, Jomantiene R, Zhao Y (2008) Ancient, recurrent phage attacks and recombination shaped dynamic sequence-variable mosaics at the root of phytoplasma genome evolution. *Proceedings of the National Academy of Sciences United States of America* **105**, 11827–11832.

Zamorano A, Fiore N (2016) Draft genome sequence of 16SrIII-J phytoplasma, a plant pathogenic bacterium with a broad spectrum of hosts. *Genome Announcements* **4**, e00602–16.

Zerbino DR (2010) Using the Velvet *de novo* assembler for short-read sequencing technologies. *Current Protocols in Bioinformatics* **11**, Unit 11 5.

Zhu Y, He Y, Zheng Z, Chen J, Wang Z, Zhou G (2017) Draft genome sequence of rice orange leaf phytoplasma from Guangdong, China. *Genome Announcements* **5**, e00430–17.

Chapter 2
Phytoplasma Effectors and Pathogenicity Factors

Assunta Bertaccini, Kenro Oshima, Kensaku Maejima, and Shigetou Namba

Abstract For the study and the management of phytoplasma-associated diseases, the most relevant knowledge needed is the one related to their pathogenicity. After the availability of full and draft genome sequences of some of the phytoplasmas, a mining search allowed identifying a number of possible virulence factors. Their possible pathogenic action was verified mainly by their expression in transgenic plants such as *Arabidopsis* spp. and *Nicotiana* spp. Several possible pathogenicity factors such as TENGU and SAP11 and/or effector molecules were shown to be related to metabolic and or phenotypic modifications indistinguishable from those present in the phytoplasma-infected plants such as phyllody and witches' broom. The possible pathogenicity factors or disease effectors studied enclosing extrachromosomal DNAs, phloem structural modifications, and very recently miRNAs are also described.

Keywords TENGU · SAP11 · Phyllody · Witches' broom · Disease symptomatology · Transgenic plants

2.1 Plant Host Responses to Phytoplasma Presence

Phytoplasma infection often induces unique morphological changes particularly in floral organs such as phyllody, virescence, or proliferation (Fig. 2.1) (Arashida et al. 2008; Bertaccini and Duduk 2009), and thus, it is biologically very intriguing and

---

A. Bertaccini (✉)
Department of Agricultural and Food Sciences, *Alma Mater Studiorum* – University of Bologna, Bologna, Italy

K. Oshima
Department of Clinical Plant Science, Faculty of Bioscience and Applied Chemistry, Hosei University, Tokyo, Japan

K. Maejima · S. Namba
Department of Agricultural and Environmental Biology, Graduate School of Agricultural and Life Sciences, The University of Tokyo, Tokyo, Japan

A. Bertaccini et al. (eds.), *Phytoplasmas: Plant Pathogenic Bacteria - III*,
https://doi.org/10.1007/978-981-13-9632-8_2

Fig. 2.1 Symptoms in phytoplasma-infected flowers from left of seed cabbage (virescence), tomato (floral structure malformation and giant ovarium), and purple coneflower (phyllody)

also very important from the phytopathological point of view to understand the mechanism of interaction of these bacteria with their host plants.

The phytoplasma genome lacks homologues of the type III secretion system, which are essential for the virulence of most phytopathogenic bacteria (Abramovitch et al. 2006); moreover they have no cell walls and reside within the host cells; therefore, their secreted proteins function directly in the host plant or insect cells. With advances in the understanding of the molecular genetic mechanisms underlying floral organ identity, the first studies have focused on the relationships between the phytoplasma infection and the expression of floral homeotic genes. It has been reported that "stolbur" phytoplasma infection affected the expression of some floral development genes in tomato plants that showed flower malformations (Pracros et al. 2006). Abnormal growth in phytoplasma-infected tomato corresponds to a distinct phase in meristem fate derailment. A WUS ortholog, LeWUS, had decreased activity in the meristems of lateral shoot apices from phytoplasma-infected plants inducing the apical dominance release. Depending upon the developmental stage when it became affected by the phytoplasma, the meristem produced distinct abnormal structures, and disruption of the apical dominance by the phytoplasma infection resulted in repetitive initiation and outgrowth of axial shoots and witches' broom growth (Wei et al. 2013). The expression levels of homeotic genes were also analyzed using each flower organ of phytoplasma-infected *Petunia* plants showing phyllody and virescence symptoms, and it was demonstrated that these symptoms were clearly related with a decreased expression levels of several homeotic genes (Himeno et al. 2011). Additionally, several global transcriptional profiling have provided interesting aspects of host responses to phytoplasma infection (Albertazzi et al. 2009; Hren et al. 2009; Margaria and Palmano 2011).

2.2 Extrachromosomal DNA

The phytoplasma genome consists of one chromosome and several extrachromosomal DNAs, mainly plasmids. Several of them were cloned in the late 1980s (Davis et al. 1988; Nakashima et al. 1991) and used as targets of DNA hybridization to compare phytoplasma strains (Nakashima and Hayashi 1995) in the earlier phytoplasma pathogenicity studies. By such studies, the loss of the insect transmissibility of a phytoplasma strain in phyllody-infected clover shoots after 2 years in micropropagation was linked to the loss of an extrachromosomal DNA fragment (Denes and Sinha 1991, 1992). The complete nucleotide sequence analyses of some plasmids revealed that they may be products of DNA recombination (Nishigawa et al. 2002), and that each plasmid encodes a replication initiation protein (Rep) involved in rolling-circle replication similar to those encoded by bacterial plasmids and geminivirus (Nakashima and Hayashi 1997; Kuboyama et al. 1998; Nishigawa et al. 2001, 2003) and circovirus replicases (Oshima et al. 2001a). The functions of other genes in the phytoplasma plasmids remain unknown; however, some of them appear to be related to the adaptation to the insect host.

2.3 Virulence by Duplicated Glycolytic Genes

The high dependence of some phytoplasmas on the glycolytic pathway is supported by the analysis of about 80 kb genomic DNA of OY-W phytoplasma, which is associated with severe symptoms (Oshima et al. 2007). Interestingly, an approximately 30 kb region was tandemly duplicated in the OY-W genome. Two sets of five glycolytic enzymes were encoded in this genomic region, which is a unique gene structure not identified in any other bacterial genomes. The gene organization of the glycolytic genes of AY-WB phytoplasma (Bai et al. 2006) is similar to that of OY-M (mild pathogenic strain) rather than OY-W, suggesting that the duplication of glycolytic genes is an OY-W-specific event not a general feature common to phytoplasmas. In the OY-M genome, all genes of this genomic region seem to be encoded as intact genes because the alignment analysis indicated that the length of each gene was not so different from that of its homologue in other bacteria. However, in the OY-W genome, one of two copy genes in the duplicated regions became a pseudogene, which had a frameshift or a stop-codon mutation within the gene, except for in five duplicated genes (Oshima et al. 2007). This is probably because of their functional redundancy. However, five types of genes, including two glycolytic genes, retained full-length ORFs, suggesting that it is advantageous for the phytoplasma to retain these genes in its lifestyle. In particular, 6-phosphofructokinase is known as a rate-limiting enzyme of glycolysis, implying that the different number of glycolytic genes between OY-W and OY-M may influence their respective glycolytic activity. It was reported that the phytoplasma population of OY-W was larger than that of OY-M in their infected plants (Oshima et al. 2001b). In addition, as described above, the ATP synthesis in phytoplasma might be strongly dependent on the glycolytic pathway (Oshima et al. 2004). Taking this result into account, the higher consumption of the carbon source may affect the growth rate of phytoplasmas and may also directly or indirectly cause more severe symptoms in the OY-W-infected plant (Oshima et al. 2007).

2.4 Phloem Blockage

Phytoplasma infection severely impairs the assimilate translocation and might be responsible for massive changes in phloem physiology including signaling components. In phytoplasma diseases in pome and stone fruits due to the presence of '*Ca.* P. mali,' '*Ca.* P. pyri,' and '*Ca.* P. prunorum' it was shown that the infection brings Ca^{2+} into sieve tubes, leading to their occlusion by callose deposition or protein plugging (Musetti et al. 2013). Effectors may induce the opening of Ca^{2+} channels leading to sieve-tube occlusion with effects on phytoplasma spread, photoassimilate distribution, and the whole phloem physiology. As sieve elements need a permanent input of energy to ensure their viability, sugar metabolism and the associate energy production of the companion cells have a dramatic impact on the physiological fitness of phloem function. It is presumptive that signaling substances are produced prior to sieve-element occlusion, and the analyses of diverse phytohormones in response to the challenge with the '*Ca.* P. mali,' apple proliferation phytoplasma, show a strong increase of salicylic acid concentration accompanied by a decrease of jasmonic acid. It is a matter of debate whether mechanisms involved in phloem impairment could differ between pathosystems and vary with the plant susceptibility to infection. There is no direct molecular or biochemical proof for the presence of virulence proteins or effectors in '*Ca.* P. mali.' However, based on phenotypic symptoms, several strains were classified as being strongly, moderately, or mildly virulent (Seemüller and Schneider 2007; Seemüller et al. 2011), and molecular data from genes encoding AAA+ proteins which are present in an unusually high number in phytoplasmas, namely, the hflB- and AAA+ ATPase genes, supported these observations. Analysis of nucleic acid sequences or derived amino acid sequences of some of the genes revealed virulence-related substitutions. Cluster analysis clearly separated the virulent from non-virulent strains. The AAA+ superfamily is a large, functionally diverse group of proteins and comprehends various types of ATPases (*e.g.* ClpC, ClpV, p97) and proteases (*e.g.* HflB, Clp group, Lon) possessing the ATPase module (Snider et al. 2008; Langklotz et al. 2012). Both are integral membrane proteins with the long, catalytic relevant C-terminal domain facing the cytosol. Topology prediction programs, however, indicated that the C-tail of four AAA+ ATPases and two of three HflBs of '*Ca.* P. mali' face the outside, hence the sieve-tube cytoplasm (Seemüller et al. 2013). The AAA+ proteins are of crucial importance for bacterial virulence. The ClpC ATPase from *Staphylococcus aureus*, for example, regulates transcription, and is a major virulence factor (Luong et al. 2011). Similarly, the AAA+ proteins are essential for the function of the secretion systems. Several bacterial species such as *Salmonella enterica*, *Agrobacterium tumefaciens*, and *Pseudomonas syringae* use type III secretion system (T3SS) to inject virulence factors into the eukaryotic host (van Melderen and Aertsen 2009; Alix and Blanc-Potard 2008). This unusual finding on C-tail orientation is a new aspect in understanding phytoplasma pathogenicity at the sieve-tube level. It is conceivable that the powerful AAA+ proteins attack structures or components of sieve elements, in particular if the pathogens are attached to the membrane. Adherence to host membranes is well established for most mycoplasmas pathogenic to humans and animals and it is considered to be a prerequisite for colonization and infection (Razin et al. 1998). However, such an attachment, which is based on protein/protein interaction, is not clearly established for phytoplasmas. It was supposed that after infection by different phytoplasma strains and the release of effectors/proteases from plasma membrane proteins (*e.g.* AAA+ proteins), the first reaction is the Ca^{2+} increase in the sieve elements, resulting in filament formation of phloem proteins and, later on, callose deposition at sieve plates and sieve pores. A subsequent reaction is the change of phytohormone dispersion and of mass flow direction (Zimmermann et al. 2015).

2.5 Phytoplasma Protein Secretion System

Many Gram-negative bacteria pathogenic of plants and animals possess type III secretion systems (T3SS) that can inject virulence effector proteins into host cells (Cornelis and van Gijsegem 2000). The T3SS are important for the pathogenicity in many bacteria including *Pseudomonas*, *Xanthomonas*, *Ralstonia*, *Erwinia*, and *Pantoea*, together with the flagella they are evolutionarily related and share a remarkably similar basal structure. However, both T3SS and flagella are restricted to Gram-negative bacteria; thus, phytoplasmas, which belong phylogenetically to Gram-positive bacteria, possess no T3SS. Bacterial-type IV secretion systems (T4SS) are also important secretion mechanisms in plant and animal pathogens. T4SS comprise pilus-like structures to transfer DNA or proteins to bacterial or eukaryotic cells (Grohmann et al. 2003; Tanaka and Sasakawa 2005). Several component proteins of the conjugative transfer system in Gram-positive and Gram-negative bacteria have sequence similarities to the component proteins of T4SS; therefore, the conjugative transfer systems and the T4SS are ancestrally related (Grohmann et al. 2003) and widely distributed among bacteria. However, four of the completely sequenced phytoplasma genomes contain no genes with similarities to those encoding the component proteins of pili or T4SS (Oshima et al. 2004; Bai et al. 2006; Kube et al. 2008; Tran-Nguyen et al. 2008), and electron microscopic analyses have not detected the existence of pilus-like structures in phytoplasmas. In contrast, pilus-like structures and four distinct *traE* genes that bear similarities to genes typically present in type IV secretory pathways have been reported in *Spiroplasma kunkelii* (Ammar el et al. 2004) and *S. citri* (Joshi et al. 2005). Since phytoplasmas are cell wall-less and reside inside the host cells, their membrane proteins and secreted proteins function in the cytoplasm of the host plant or insect cell and are predicted to have some important roles in host-parasite interactions and/or virulence. Thus, the identification of secreted proteins encoded in the phytoplasma genome is important for understanding the biology of phytoplasmas. Phytoplasmas are known to have the Sec system for the secretion of proteins into the host cell cytoplasm.

2.6 Identification of Phytoplasma Virulence Factors

Since membrane and secreted proteins are thought to be potential virulence factors, the complete phytoplasma genomes have been mined for the presence of these proteins. In OY phytoplasma genome, more than 30 putative secreted proteins were identified using the SignalP and SOSUI programs (Hirokawa et al. 1998; Bendtsen et al. 2004), and each of them was analyzed to function as a virulence factor. As a result, TENGU gene, one of the secreted proteins of OY phytoplasma, was confirmed to induce clear symptoms of phytoplasma infection, including witches' broom (development of numerous shoot branches) and dwarfism (Hoshi et al. 2009). When TENGU gene was expressed in *Nicotiana benthamiana* plants by a viral vector, clear symptoms were observed. In particular, the number of shoots and leaves was dramatically increased, while plant height was reduced. Similar symptoms were observed in TENGU-expressing transgenic *Arabidopsis thaliana* plants. These phenotypes resembled typical phytoplasma disease symptoms, indicating that TENGU is the virulence factor responsible for witches' broom and dwarfism in plants. TENGU encodes a very small protein (4.5 kDa), and the mature protein, after the cleavage of its N-terminal signal peptide, is 38 amino acids in length. Localization of virulence factors is also an important problem, because the phytoplasmas reside only in the phloem tissues, but plant development starts from apical meristem. By immuno-histochemical analysis, TENGU protein was confirmed to localize not only in phloem tissues but also in other tissues, *e.g.* parenchyma, the branching region of axillary buds, and meristem. Therefore, TENGU is transported from phloem tissues into other tissues and induces symptoms (Hoshi et al. 2009). Plant cells are connected to each other through channels called plasmodesmata, and there is the size exclusion limit (SEL) where certain size of molecules can pass through plasmodesmata or not. The SEL of plasmodesmata between a sieve element and a companion cell ranges from 10 to 40 kDa (Kempers and van Bel 1997), which is much higher than the SEL of plasmodesmata between non-phloem cells such as mesophyll cells (*ca.* 1 kDa) (Goodwin 1983). However, GFP proteins (27 kDa) was shown to be transported through plasmodesmata from the phloem into developing tissues, *e.g.* young rosette leaves, petals, and root tips (Imlau et al. 1999). Since TENGU (4.5 kDa) is quite smaller than GFP, the transportation of TENGU from phloem tissues into other tissues is reasonable. Thus, although phytoplasmas cannot invade apical buds, they may secrete TENGU to perturb plant metabolism in apical buds by remote manipulation.

2.7 Mechanism of Virulence Induction by TENGU

How does TENGU induce symptoms in plants? To answer this question, microarray analyses were performed to compare the expression levels of plant genes between *tengu*-transgenic and *gus*-transgenic (control) *Arabidopsis* plants. As a result, the expression levels of many auxin-related genes were significantly downregulated in the *tengu*-transgenic plants, suggesting that TENGU would suppress auxin signaling or biosynthesis pathways (Hoshi et al. 2009) (Fig. 2.2). Auxin is known to be involved in apical dominance, which is where an apical bud inhibits development of an axillary bud growth. Auxin is biosynthesized in the apical bud and transported to the root and inhibits the growth of axillary buds (Mori et al. 2005). It has been

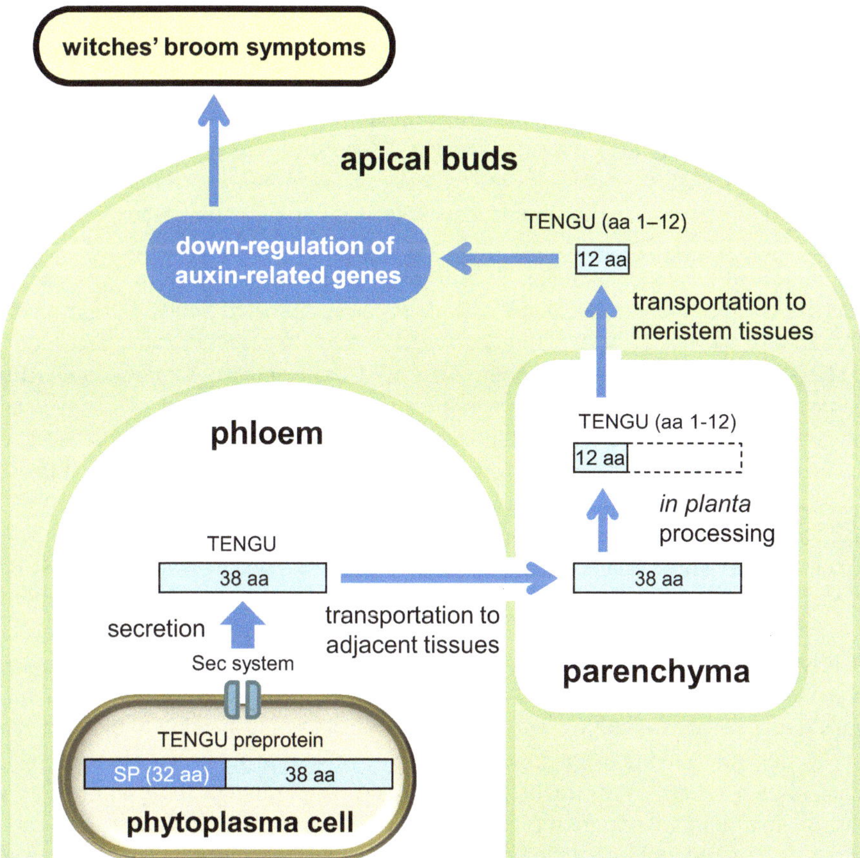

Fig. 2.2 Hypothetical mechanism of witches' broom symptoms induced by TENGU, a small peptide secreted from the phytoplasma. The TENGU pre-protein is secreted into plant phloem, and it is transported from phloem into parenchyma and apical buds. During the transportation, TENGU is enzymatically processed into the active form consisting of N-terminus 12 aa. The active form of TENGU inhibits auxin-related pathways, resulting in induction of witches' broom symptoms

reported that the *A. thaliana* mutant of IAA/AXR2, which is an early auxin-responsive gene, exhibited severe dwarfism (Timpte et al. 1994). Therefore, TENGU suppresses plant auxin responses, resulting in promoting the growth of axillary buds and the growth inhibition of apical buds in *tengu*-transgenic or OY-infected plants.

Some phytopathogenic fungi and bacteria are known to produce phytohormones that cause disease symptoms (Crespi et al. 1992; Nishida and Sugiyama 1993; de Manes et al. 2004), for example, the bacterium *Rhodococcus fascians*, which affects carnation plants (de Manes et al. 2004), and the fungus *Taphrina wiesneri*, which affects cherry trees (Nishida and Sugiyama 1993). These pathogens have biosynthetic genes for auxin or cytokinin and produce these phytohormones to cause disease symptoms. However, phytoplasmas do not have genes for the biosynthesis of phytohormones (Oshima et al. 2004). Instead, through secretion of TENGU, phytoplasmas are thought to perturb the auxin pathway of the host plant. This represents a mechanism by which these pathogenic bacteria may affect plant development.

TENGU homologues from several phytoplasma strains induce similar symptoms and were found to be processed *in planta* into a small functional peptide, similar to the proteolytic processing of plant endogenous peptide signals (Sugawara et al. 2013). TENGU may also act as an inducer of sterility. Transgenic expression of TENGU induced both male and female sterility in *Arabidopsis thaliana* flowers, similar to those observed in double knockout mutants of auxin response factor 6 (ARF6) and 8 (ARF8), which are known to regulate the flower development in a jasmonic acid-dependent manner. Transcripts of ARF6 and ARF8 were significantly decreased in both tengu-transgenic and phytoplasma-infected plants. Furthermore, JA and auxin levels were actually decreased in tengu-transgenic buds, suggesting that TENGU reduces the endogenous levels of phytohormones by repressing ARF6 and ARF8, resulting in impaired flower maturation (Minato et al. 2014). Why does phytoplasma possess TENGU and induce abnormal organ formation in plants? The answer may be to increase their own evolutionary fitness by modifying plants. Disease symptoms of phytoplasma-infected plants, including witches' broom, phyllody, and virescence, have common characteristics, *i.e.* the production of young and green organs. These characteristic symptoms may be related to the life cycle of phytoplasmas. Because phytoplasmas are transmitted by insect vectors, sap-feeding by insects is one of the most important steps in the phytoplasma life cycle (Christensen et al. 2005). Leafhoppers, which are the main insect vector of phytoplasmas, prefer young and green/yellow tissues for feeding, as well as for laying eggs (Hogenhout et al. 2008). Therefore, phytoplasmas that are able to increase the production of young and green/yellow leaves in plants would increase their own transmission efficiency by making the infected plants appear more attractive to insects, in turn increasing their own survival (Hogenhout et al. 2008). Thus, disease symptoms by phytoplasma infection might be a benefit increasing their fitness and extending their ecological niche.

2.8 SAP11 and SAP11-Like Effectors

Finding virulence factors was achieved also in other phytoplasma strains. In aster yellows witches' broom (AY-WB) phytoplasma, secreted proteins were mined from its genome sequence, and 56 proteins were predicted as secreted. Among them, four genes contain eukaryotic nuclear localization signals (NLSs), and one of them, coding the SAP11 protein, was confirmed to localize in plant cell nuclei (Bai et al. 2009). Recently, it was demonstrated that SAP11-expressing plants showed crinkled leaves and produced more stems, and the presence of this protein also downregulates the jasmonic acid synthesis. Moreover, the fecundity of insect vectors was increased on the SAP11-expressing plants compared to normal plants (Sugio et al. 2011). The '*Ca.* P. solani' strain SA-1 genome also contained putative secreted protein/effector genes, including a homologue of SAP11, found in many other phytoplasmas (Seruga-Music et al. 2018). Thus, phytoplasma-secreted proteins could manipulate hosts and are involved in virulence, similar to other pathogens (Higgins 2001; Boutareaud et al. 2004).

Other effectors of phytoplasma strains related to '*Ca.* P. aurantifolia' were identified in witches' broom disease of lime (WBDL) genome (SAP11, SAP21, Eff64, Eff115, Eff197, Eff211, and EffSAP67), five (SAP11, SAP21, Eff64, Eff99, and Eff197) in crotalaria phyllody (CrP) and two (SAP11, Eff64) in faba bean phyllody (FBP). No homologues to Eff64, Eff197, and Eff211 in phytoplasmas of other phylogenetic groups were found, while the SAP11 and Eff64 homologues of 'Ca. P. aurantifolia' strains shared at least 95.9% identity and were detected in the three phytoplasmas. Five of the putative effectors (SAP11, SAP21, Eff64, Eff115, and Eff99) were transcribed from total RNA extracts of periwinkle plants infected with these phytoplasmas. Transcription profiles of selected putative effectors of CrP, FBP, and WBDL indicated that the SAP11 transcripts were the most abundant in the three phytoplasmas. SAP21 transcript levels were comparable to those of SAP11 for CrP and not measurable for the other phytoplasmas. Eff64 had the lowest transcription level irrespective of sampling date and phytoplasma strain while the Eff115 transcript levels were the highest in WBDL-infected plants (Anabestani et al. 2017).

A SAP11-like effector was also detected in the wheat blue dwarf (WBD) phytoplasma. The SWP1-expressing transgenic *A. thaliana* plants showed typical witches' broom symptom, and the overexpression of SWP1 truncation mutants in *N. benthamiana* indicates that the coiled-coil domain and nuclear localization as responsible for the induction of the witches' broom symptoms. Additionally, the SWP1 interaction with *A. thaliana* transcription factor TCP18 (BRC1), the key negative regulator of branching signals in various plant species, was reported. Moreover, in planta co-expression analysis showed that SWP1 promoted the degradation of BRC1 via a proteasome system, suggesting that the SWP1 induced witches' broom symptom targeting and degradation of the BRC1 factor (Wang et al. 2017, 2018). Two TCP (TEOSINTE-BRANCHED/CYCLOIDEA/PROLIFERATING CELL FACTOR) transcription factors of *Malus* x *domestica* were detected as binding partners of the '*Ca.* P. mali' SAP11-like effector ATP_00189, a protein which shares homology to the AY-WB effector SAP11 (Siewert et al. 2014). It contains an N-terminal sequence-variable mosaic (SVM) protein signal sequence and shares 41% identity to SAP11.

The presence of '*Ca.* P. mali' infection was demonstrated to interfere with jasmonates, salicylic acid, and abscissic acid levels. The presence of ATP_00189 allows to indicate it as effector, assuming its binding to the TCP transcription factor. It was shown that the increased LOX activity in '*Ca.* P. mali'-infected apple tree leaves correlates with the recovery phenomenon (Seemüller et al. 1984; Carraro et al. 2004; Musetti et al. 2013; Patui et al. 2013), and this emphasizes the importance of decreased jasmonate levels for the success of phytoplasma infection. Moreover, since the results of a study with SAP11-transgenic *A. thaliana* plants indicate a role of SAP11 in the downregulation of salicylic acid responses (Lu et al. 2014a, 2014b) and it was demonstrated that salicylic acid accumulation in '*Ca.* P. mali'-infected apple trees is reduced, it is possible to consider a similar function for ATP_00189 (Janik et al. 2017).

The TB/CYC (TEOSINTE-BRANCHED/CYCLOIDEA)-TCPs are destabilized by phytoplasma SAP11 effectors, leading to the proliferation of axillary meristems. Although a high degree of sequence diversity was observed among putative SAP11 effectors identified from evolutionarily distinct clusters of phytoplasmas, these effectors acquired fundamental activity in destabilizing TB/CYC-TCPs. In addition, the miR156/SPLs and miR172/AP2 modules, which represent key regulatory hubs involved in plant phase transition, were modulated by SAP11 AY-WB. A late-flowering phenotype with significant changes in the expression of flowering-related genes was observed in transgenic *Arabidopsis* plants expressing SAP11 AY-WB and correlated with the ability of SAP11 effectors to destabilize CIN (CINCINNATA)-TCPs (Chang et al. 2018).

Although not all putative SAP11 effectors display broad-spectrum activities in modulating morphological and physiological changes in host plants, they serve as core virulence factors responsible for the witches' broom symptom. As SAP11 homologues are found in several phytoplasmas infecting a broad range of different plants, SAP11-like proteins seem to be key players in phytoplasma infection.

2.9 Virulence Factor for Phyllody

PHYL1 and SAP54 were found as homologous-secreted proteins that induce phyllody in floral organs of *A. thaliana* (MacLean et al. 2011; Maejima et al. 2014). PHYL1 actually interacts with and degrades the MADS-domain transcription factors (Maejima et al. 2014) and it was shown to be genetically and functionally conserved among other phytoplasma strains and species. Therefore, PHYL1 and its homologues were designated as members of the phyllody-inducing gene family phyllogen. Transgenic *A. thaliana* plants expressing PHYL1 exhibit phyllody-like morphological changes in their flowers. PHYL1 interacts directly with APETALA1 (AP1) and SEP1-SEP4 [A- and E-class MADS-domain transcription factors (MTFs), respectively] of *A. thaliana* (Maejima et al. 2014, 2015) and induces their degradation in a proteasome-dependent manner (Maejima et al. 2014) (Fig. 2.3). Additionally, PHYL1 was shown to cause phyllody phenotypes in several eudicots species belonging to three different families. Moreover PHYL1 can interact with MTFs of not only angiosperm species including eudicots and monocots but also gymnosperms and a fern, and induce their degradation, suggesting that it induces phyllody in angiosperms and inhibits MTF function in diverse plant species (Kitazawa et al. 2017).

Leafy flowers of peanut witches' broom (PnWB) phytoplasma infected *Catharanthus roseus* were used to study the relationship of the orthologs of the phyllody symptoms (PHYL1) effector with other possible effectors. The flowering negative regulator gene SHORT VEGETATIVE PHASE (SVP) was upregulated in PnWB-infected *C. roseus* plants, but most of the microRNA (miRNA) genes showed a repressed expression. Coincidentally, transgenic *A. thaliana* plants expressing the PHYL1 gene of PnWB upregulate SVP of *Arabidopsis*, but repress the putative regulatory action of miR396; it directly or indirectly interferes with miR396-mediated SVP mRNA decay and synergizes with other effects (*e.g.* MADS-box transcription factor degradation), resulting in abnormal flower formation. It was concluded that this miR396-targeted SVP is required to repress flowering and it is related to the development of the phyllody symptoms (Yang et al. 2015).

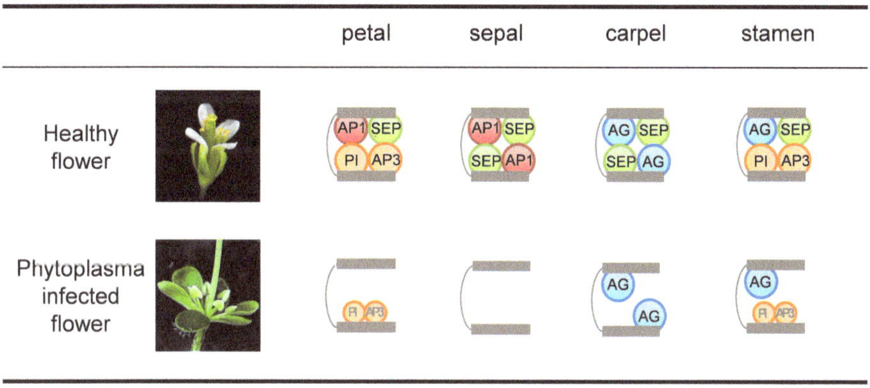

Fig. 2.3 Model of induction of phyllody-like symptoms by PHYL1. Phytoplasma secretes PHYL1 into plant cells. PHYL1 binds to the class A and E proteins AP1 and SEP1–4, respectively, and induces their degradation leading to downregulation of the class B genes AP3 and PI. According to the floral quartet model, floral organ identity is determined by organ-specific combinational quaternary complexes consisting of SEP–AP1, SEP–AG, and AP3–PI heterodimers of ABCE class MADS-domain transcription factors. Degradation of the class A and E proteins results in a significant reduction in the frequency of formation of the SEP3–AP1 and SEP3–AG heterodimers. Likewise, downregulation of the class B genes decreases the frequency of AP3–PI heterodimerization. As a consequence of these molecular effects, the phytoplasma-infected flower exhibits a severe leaf-like phenotype

Another phyllogen, SAP54, was also reported to bind to and induce degradation of A- and E-class MTFs, but it did not bind to APETALA3 (AP3) and PISTILLATA (PI; B-class MTFs) nor to AGAMOUS (AG; C-class MTF) (MacLean et al. 2014). Therefore, phyllogen-mediated degradation of the A- and E-class MTFs is regarded as the molecular mechanism responsible for phyllody symptoms in *A. thaliana*. SAP54 and his homologues, collectively named phyllogen, degrade MADS-box transcription factors (MTFs), including those involved in flower development. For SAP54 AY-WB, this degradation process requires the 26S proteasome shuttle factors RAD23, particularly RAD23C and RAD23D. The SAP54 AY-WB binds the MTF K-domains, which facilitates dimerization of MTFs. SAP54/phyllogen homologues were identified in at least 17 diverse phytoplasmas, and for those tested, interactions with MTFs were shown that may have evolved to selectively target the plant MTFs (that contain a K-domain) and not those of insects (that lack this domain), an important characteristic given that phytoplasmas effectively colonize many organs of their insect vectors (Sugio et al. 2011). SAP54 may act directly as a protease to catalyze the proteolysis of select MTFs, or alternatively, this effector may exploit a host mechanism, such as the ubiquitin/26S proteasome system (UPS) to degrade MTFs. SAP54 interacts with RADIATION SENSITIVE23 (RAD23) family isoforms RAD23C and RAD23D. Thus, the SAP54-mediated degradation of MTFs is dependent predominantly on RAD23C and RAD23D, whereas other RAD23 isoforms may be involved depending on SAP54 abundance (MacLean et al. 2014). The developmental reprogramming relies on specific interactions of the phytoplasma protein SAP54 with a small subset of MADS-domain transcription factors. It was proposed that SAP54 folds into a structure that is similar to that of the K-domain, a protein-protein interaction domain of MADS-domain proteins (Forest et al. 2015). Also, in the 16SrIII-J genome of Chile phytoplasmas (Zamorano and Fiore, 2016), a SAP54 (AYWB_224) homologue protein was identified after mining and detected also in the *Bellis* virescence phytoplasma strain Vc33 from Argentina that belong to the same ribosomal subgroup. Its structural and phylogenetic analyses confirmed that it is highly conserved and that its co-divergence among phytoplasmas is not directly consistent with the evolutionary trajectories derived from 16S rRNA gene analyses leading to hypothesize its association with selective pressure and active genomic drift (Fernandez et al. 2018).

2.10 Perspectives

Micro(mi)RNAs play crucial roles in plant developmental processes and in defense responses to biotic and abiotic stresses. Their study is also now blooming for the phytoplasma-associated pathogenicity/metabolic studies. Conserved and novel miRNAs were used for analyzing *Vitis vinifera* plants infected by "flavescence dorée" (FD) phytoplasma, associated with one of the most severe phytoplasma diseases affecting grapevine. The analysis of small RNAs from healthy, recovered (plants showing spontaneous and stable remission of symptoms), and FD-infected "Barbera" grapevines showed that FD altered the expression profiles of several miRNAs, including those involved in cell development and photosynthesis, jasmonate signaling, and disease resistance response, and revealed key roles of miRNAs in photosynthesis and jasmonate signaling (Chitarra et al. 2018). Paulownia witches' broom (PaWB) key miRNAs associated with the formation of PaWB symptoms were detected by the miRNA and degradome sequencing exploration of healthy and diseased *Paulownia tomentosa*, *P. fortunei*, and *P. tomentosa* × *P. fortunei* seedlings. Differentially expressed miRNAs and degradome sequencing were verified by the expression patterns, and the miR156 was reported as related to witches' broom (Cao et al. 2018).

The draft genome sequence of '*Ca.* P. solani' strain SA-1 corresponding to more than 80% of the estimated genome size was shown to contain a high occurrence of repetitive sequences. The majority of repeats consisted of gene arrangements characteristic of phytoplasma potential mobile units (PMUs). These regions showed variation in gene orders intermixed with genes of unknown functions and lack of similarity to other phytoplasma genes, suggesting that they were prone to rearrangements and acquisition of new sequences via recombination. Phylogenetic analyses provided evidence of horizontal transfer for PMU-like elements from various phytoplasmas, including distantly related ones (Seruga-Music et al. 2018). Maize bushy stunt phytoplasma (MBS)-infected maize plants with diversity of symptoms severity were studied for their possible correlation with differences in their genome sequences. A statistically significant correlation between variations in disease symptoms of infected maize plants and MBS sequence polymorphisms was detected. It was also reported that MBS strains contributed consistently to organ proliferation symptoms, while the maize genotype contributes to the leaf necrosis and reddening and yellowing of infected maize plants. The symptom differences are associated with the presence of polymorphisms in a phase-variable lipoprotein, which is a candidate effector, and an ATP-dependent lipoprotein ABC export protein, whereas no polymorphisms were observed in other candidate effector genes. Lipoproteins and ABC export proteins activate the host defense responses, regulate the pathogen attachment to host cells, and activate effector secretion systems in other pathogens (Orlovskis et al. 2017).

Infection with OY-W phytoplasma affects the expression of the flower homeotic genes of petunia plants in an organ-specific manner. The expression levels of several homeotic genes required for organ development, such as PFG, PhGLO1, and FBP7, were significantly downregulated by the phytoplasma infection in floral organs, except the stamens. Moreover, the expression levels of TER, ALF, and DOT genes, which are known to participate in floral meristem identity, were significantly downregulated in the phytoplasma-infected petunia meristems, implying that the phytoplasma would affect an upstream signaling pathway of floral meristem identity (Himeno et al. 2011). More recently, the activation of the host anthocyanin biosynthesis pathway in response to phytoplasma infection was shown to be responsible for purple top symptoms and also associated with a reduction of leaf cell death. By using plant mutants defective in anthocyanin biosynthesis, it was demonstrated that anthocyanin accumulation is directly responsible for the purple top symptoms, and it is also associated with the reduction of leaf cell death. Phytoplasma infection led moreover to significant activation of the anthocyanin biosynthetic pathway and the dramatic accumulation of sucrose by about 1,000-fold, and this can activate the anthocyanin biosynthetic pathway (Himeno et al. 2014). This last finding opens the possibility to also verify beneficial effects related to the phytoplasma presence in plants that could be an indication of the long-time interaction of these prokaryotes with plant and insect hosts. This interaction could lead to some kind of permanent relationship implying phenotypic modification in same cases useful as a new step in coevolution.

References

Abramovitch RB, Anderson JC, Martin GB (2006) Bacterial elicitation and evasion of plant innate immunity. *Nature Revue on Molecular Cell Biology* **7**, 601–611.

Alix E, Blanc-Potard AB (2008) Peptide-assisted degradation of the *Salmonella* MgtC virulence factor. *Embo Journal* **27**, 546–557.

Anabestani A, Izadpanah K, Abbà S, Galetto L, Ghorbani A, Palmano S, Siampour M, Veratti F, Marzachì C (2017) Identification of putative effector genes and their transcripts in three strains related to '*Candidatus* Phytoplasma aurantifolia'. *Microbiological Research* **199**, 57–66.

Albertazzi G, Milc J, Caffagni A, Francia E, Roncaglia E, Ferrari F, Tagliafico E, Stefani E, Pecchioni N (2009) Gene expression in grapevine cultivars in response to "bois noir" phytoplasma infection. *Plant Science* **176**, 792–804.

Ammar el D, Fulton D, Bai X, Meulia T, Hogenhout SA (2004) An attachment tip and pili-like structures in insect- and plant-pathogenic spiroplasmas of the class Mollicutes. *Archives of Microbiology* **181**, 97–105.

Arashida R, Kakizawa S, Ishii Y, Hoshi A, Jung H-Y, Kagiwada S, Yamaji Y, Oshima K, Namba S (2008) Cloning and characterization of the antigenic membrane protein (Amp) gene and in situ detection of Amp from malformed flowers infected with Japanese hydrangea phyllody phytoplasma. *Phytopathology* **98**, 769–775.

Bai X, Zhang J, Ewing A, Miller SA, Jancso Radek A, Shevchenko DV, Tsukerman K, Walunas T, Lapidus A, Campbell JW, Hogenhout SA (2006) Living with genome instability: the adaptation of phytoplasmas to diverse environments of their insect and plant hosts. *Journal of Bacteriology* **188**, 3682–3696.

Bai X, Correa VR, Toruno TY, Ammar el D, Kamoun S, Hogenhout SA (2009) AY-WB phytoplasma secretes a protein that targets plant cell nuclei. *Molecular Plant Microbe Interactions* **22**, 18–30.

Bendtsen JD, Nielsen H, von Heijne G, Brunak S (2004) Improved prediction of signal peptides: signalP 3.0. *Journal of Molecular Biology* **340**, 783–795.

Bertaccini A, Duduk B (2009) Phytoplasma and phytoplasma diseases: a review of recent research. *Phytopathologia Mediterranea* **48**, 355–378.

Boutareaud A, Danet J-L, Garnier M, Saillard C (2004) Disruption of a gene predicted to encode a solute binding protein of an ABC transporter reduces transmission of *Spiroplasma citri* by the leafhopper *Circulifer haematoceps*. *Applied Environmental Microbiology* **70**, 3960–3967.

Cao X, Zhai X, Zhang Y, Cheng Z, Li X, Fan G (2018) Comparative analysis of microRNA expression in three paulownia species with phytoplasma infection. *Forests* **9**, 302.

Carraro L, Ermacora P, Loi N, Osler R (2004) The recovery phenomenon in apple proliferation-infected apple trees. *Journal of Plant Pathology* **86**, 141–146.

Chang SH, Tan CM, Wu CT, Lin TH, Jiang SY, Liu RC, Tsai MC, Su LW, Yang JY (2018) Alterations of plant architecture and phase transition by the phytoplasma virulence factor SAP11. *Journal of Experimental Botany* **22**, 5389–5401.

Chitarra W, Pagliarani C, Abbà S, Boccacci P, Birello G, Rossi M, Palmano S, Marzachì C, Perrone I, Gambino G (2018) MiRVIT: a novel miRNA database and its application to uncover *Vitis* responses to "flavescence dorée" infection. *Frontiers in Plant Science* **9**, 1034.

Christensen NM, Axelsen KB, Nicolaisen M, Schulz A (2005) Phytoplasmas and their interactions with hosts. *Trends in Plant Science* **10**, 526–535.

Cornelis GR, van Gijsegem F (2000) Assembly and function of type III secretory systems. *Annual Revue of Microbiology* **54**, 735–774.

Crespi M, Messens E, Caplan AB, van Montagu M, Desomer J (1992) Fasciation induction by the phytopathogen *Rhodococcus fascians* depends upon a linear plasmid encoding a cytokinin synthase gene. *EMBO Journal* **11**, 795–804.

Davis MJ, Tsai JH, Cox RL, McDaniel LL, Harrison NA (1988) Cloning of the chromosomal and extrachromosomal DNA of the mycoplasma-like organisms that causes maize bushy stunt disease. *Molecular Plant Microbe Interaction* **4**, 295–302.

Denes AS, Sinha RC (1991) Extrachromosomal DNA elements of plant pathogenic mycoplasma-like organisms. *Canadian Journal of Plant Pathology* **13**, 26–32.

Denes AS, Sinha RC (1992) Alteration of clover phyllody mycoplasma DNA after *in vitro* culturing of phyllody diseased clover. *Canadian Journal of Plant Pathology* **14**, 189–196.

Fernandez FD, Debat HJ, Conci LR (2018) Molecular characterization of effector protein SAP54 in *Bellis* virescence phytoplasma (16SrIII-J). *BioRxiv* **411140**, doi: https://doi.org/10.1101/411140.

Forest ER, Rümpler F, Gramzow L, Theißen G, Melzer R (2015) Did convergent protein evolution enable phytoplasmas to generate "zombie plants"? *Trends in Plant Science* **20**, 798–806.

Goodwin P (1983) Molecular-size limit for movement in the symplast of the elodea leaf. *Planta* **157**, 124–130.

Grohmann E, Muth G, Espinosa M (2003) Conjugative plasmid transfer in gram-positive bacteria. *Microbiology and Molecular Biology Revue* **67**, 277–301.

Higgins CF (2001) ABC transporters: physiology, structure and mechanism - an overview. *Research in Microbiology* **152**, 205–210.

Himeno M, Neriya Y, Minato N, Miura C, Sugawara K, Ishii Y, Yamaji Y, Kakizawa S, Oshima K, Namba S (2011) Unique morphological changes in plant pathogenic phytoplasma-infected petunia flowers are related to transcriptional regulation of floral homeotic genes in an organ-specific manner. *Plant Journal* **67**, 971–979.

Himeno M, Kitazawa Y, Yoshida T, Maejima K, Yamaji Y, Oshima K, Namba S (2014) Purple top symptoms are associated with reduction of leaf cell death in phytoplasma-infected plants. *Science Reporter* **4**, 4111.

Hirokawa T, Boon-Chieng S, Mitaku S (1998) SOSUI: classification and secondary structure prediction system for membrane proteins. *Bioinformatics* **14**, 378–379.

Hogenhout SA, Oshima K, Ammar E-D, Kakizawa S, Kingdom HN, Namba S (2008) Phytoplasmas: bacteria that manipulate plants and insects. *Molecular Plant Pathology* **9**, 403–423.

Hoshi A, Oshima K, Kakizawa S, Ishii Y, Ozeki J, Hashimoto M, Komatsu K, Kagiwada S, Yamaji Y, Namba S (2009) A unique virulence factor for proliferation and dwarfism in plants identified from a phytopathogenic bacterium. *Proceeding of the National Academy of Science United States of America* **106**, 6416–6421.

Hren M, Nikolic P, Rotter A, Blejec A, Terrier N, Ravnikar M, Dermastia M, Gruden K (2009) "Bois noir" phytoplasma induces significant reprogramming of the leaf transcriptome in the field grown grapevine. *BMC Genomics* **10**, 460.

Imlau A, Truernit E, Sauer N (1999) Cell-to-cell and long-distance trafficking of the green fluorescent protein in the phloem and symplastic unloading of the protein into sink tissues. *Plant Cell* **11**, 309–322.

Janik K, Mithöfer A, Raffeiner M, Stellmach H, Hause B, Schlink K (2017) An effector of apple proliferation phytoplasma targets TCP transcription factors — a generalized virulence strategy of phytoplasma? *Molecular Plant Pathology* **18**, 435–442.

Joshi BD, Berg M, Rogers J., Fletcher J, Melcher U (2005) Sequence comparisons of plasmids pBJS-O of *Spiroplasma citri* and pSKU146 of *S. kunkelii*: implications for plasmid evolution. *BMC Genomics* **6**, 175.

Kempers R, van Bel A (1997) Symplasmic connections between sieve element and companion cell in the stem phloem of *Vicia faba* L have a molecular exclusion limit of at least 10 kDa. *Planta* **201**, 195–201.

Kitazawa Y, Iwabuchi N, Himeno M, Sasano M, Koinuma H, Nijo T, Tomomitsu T, Yoshida T, Okano Y, Yoshikawa N, Maejima K, Oshima K, Namba S (2017) Phytoplasma-conserved phyllogen proteins induce phyllody across the *Plantae* by degrading floral MADS domain proteins. *Journal of Experimental Botany* **68**, 2799–2811.

Kube M, Schneider B, Kuhl H, Dandekar T, Heitmann K, Migdoll AM, Reinhardt R, Seemüller E (2008) The linear chromosome of the plant-pathogenic mycoplasma '*Candidatus* Phytoplasma mali'. *BMC Genomics* **9**, 306.

Kuboyama T, Huang C-C, Lu X, Sawayanagi T, Kanazawa T, Kagami T, Matsuda I, Tsuchizaki T, Namba S (1998) A plasmid isolated from phytopathogenic onion yellows phytoplasma and its heterogeneity in the pathogenic phytoplasma mutant. *Molecular Plant Microbe Interactions* **11**, 1031–1037.

Langklotz S, Baumann U, Narberhaus F (2012) Structure and function of the bacterial AAA protease FtsH. *Molecular Cell Research* **1823**, 40–48.

Lu Y-T, Cheng K-T, Jiang S-Y, Yang J-Y (2014a) Post-translational cleavage and self-interaction of the phytoplasma effector SAP11. *Plant Signalling Behaviour* **9**, 1–3.

Lu Y-T, Li M-Y, Cheng K-T, Tan CM, Su L-W, Lin W-Y, Shih HT, Chiou TJ, Yang JY (2014b) Transgenic plants that express the phytoplasma effector SAP11 show altered phosphate starvation and defense responses. *Plant Physiology* **164**, 1456–1469.

Luong TT, Sau K, Roux C, Sau S, Dunman PM, Lee CY (2011) *Staphylococcus aureus* ClpC divergently regulates capsule via sae and codY in strain Newman but activates capsule via codY in strain UAMS-1 and in strain Newman with repaired saeS. *Journal of Bacteriology* **193**, 686–694.

Manes de OCL, Beeckman T, Ritsema T, van Montagu M, Goethals K, Holsters M (2004). Phenotypic alterations in *Arabidopsis thaliana* plants caused by *Rhodococcus fascians* infection. *Journal of Plant Research* **117**, 139–145.

Margaria P, Palmano S (2011) Response of the *Vitis vinifera* L. cv. Nebbiolo proteome to "flavescence dorée" phytoplasma infection. *Proteomics* **11**, 212–224.

MacLean AM, Sugio A, Makarova OV, Findlay KC, Grieve VM, Tóth R, Nicolaisen M, Hogenhout SA (2011) Phytoplasma effector SAP54 induces indeterminate leaf-like flower development in *Arabidopsis* plants. *Plant Physiology* **157**, 831–841.

MacLean AM, Orlovskis Z, Kowitwanich K, Zdziarska AM, Angenent GC, Immink RGH, Hogenhout SA (2014) Phytoplasma effector SAP54 hijacks plant reproduction by degrading MADS-box proteins and promotes insect colonization in a RAD23-dependent manner. *Plos Biology* **12**, e1001835.

Maejima K, Iwai R, Himeno M, Komatsu K, Kitazawa Y, Fujita N, Ishikawa K, Fukuoka M, Minato N, Yamaji Y, Oshima K, Namba S (2014) Recognition of floral homeotic MADS domain transcription factors by a phytoplasmal effector, phyllogen, induces phyllody. *Plant Journal* **78**, 541–554.

Maejima K, Kitazawa Y, Tomomitsu T, Yusa A, Neriya Y, Himeno M, Yamaji Y, Oshima K, Namba S (2015) Degradation of class E MADS-domain transcription factors in *Arabidopsis* by a phytoplasmal effector, phyllogen. *Plant Signal Behaviour* **10**, e1042635.

Minato N, Himeno M, Hoshi A, Maejima K, Komatsu K, Takebayashi Y, Kasahara H, Yusa A, Yamaji Y, Oshima K, Kamiya Y, Namba S (2014) The phytoplasmal virulence factor TENGU causes plant sterility by downregulating of the jasmonic acid and auxin pathways. *Scientific Reports* **4**, 7399–7405.

Mori Y, Nishimura T, Koshiba T (2005) Vigorous synthesis of indole-3-acetic acid in the apical very tip leads to a constant basipetal flow of the hormone in maize coleoptiles. *Plant Science* **168**, 467–473.

Musetti R, Buxa SV, De Marco F, Loschi A, Polizzotto R, Kogel K-H, van Bel AJE (2013) Phytoplasma-triggered Ca^{2+} influx is involved in sieve-tube blockage. *Molecular Plant Microbe Interactions* **26**, 379–386.

Nishida H, Sugiyama J (1993) Phylogenetic relationships among *Taphrina*, *Saitoella*, and other higher fungi. *Molecular Biology Evolution* **10**, 431–436.

Nakashima K, Hayashi T (1995) Extrachromosomal DNAs of rice yellow dwarf and sugarcane white leaf phytoplasmas. *Annals of the Phytopathological Society of Japan* **61**, 456–462.

Nakashima K, Hayashi T (1997) Sequence analysis of extrachromosomal DNA of sugarcane white leaf phytoplasma. *Annals of the Phytopathological Society of Japan* **63**, 21–25.

Nakashima K, Kato S, Iwanami S, Murata N (1991) Cloning and detection of chromosomal and extrachromosomal DNA from mycoplasma-like organisms that cause yellow dwarf disease of rice. *Applied and Environmental Microbiology* **57**, 3570–3575.

Nishigawa H, Miyata SI, Oshima K, Sawayanagi T, Komoto A, Kuboyama T, Matsuda I, Tsuchizaki T, Namba S (2001) *In planta* expression of a protein encoded by the extrachromosomal DNA of a phytoplasma and related to geminivirus replication proteins. *Microbiology* **147**, 507–513.

Nishigawa H, Oshima K, Kakizawa S, Jung HY, Kuboyama T, Miyata S, Ugaki M, Namba S (2002) A plasmid from a non-insect-transmissible line of a phytoplasma lacks two open reading frames that exist in the plasmid from the wild-type line. *Gene* **298**, 195–201.

Nishigawa H, Oshima K, Miyata S, Ugaki M, Namba S (2003) Complete set of extrachromosomal DNAs from three pathogenic lines of onion yellows phytoplasma and use of PCR to differentiate each line. *Journal of General Plant Pathology* **69**, 194–198.

Orlovskis Z, Canale M, Haryono M, Lopes J, Kuo C-H, Hogenhout S (2017) A few sequence polymorphisms among isolates of maize bushy stunt phytoplasma associate with organ proliferation symptoms of infected maize plants. *Annals of Botany* **119**, 869–884.

Oshima K, Kakizawa S, Nishigawa H, Kuboyama T, Miyata S, Ugaki M, Namba S (2001a) A plasmid of phytoplasma encodes a unique replication protein having both plasmid- and virus-like domains: clue to viral ancestry or result of virus/plasmid recombination? *Virology* **285**, 270–277.

Oshima K, Shiomi T, Kuboyama T, Sawayanagi T, Nishigawa H, Kakizawa S, Miyata S, Ugaki M, Namba S (2001b) Isolation and characterization of derivative lines of the onion yellows phytoplasma that do not cause stunting or phloem hyperplasia. *Phytopathology* **91**, 1024–1029.

Oshima K, Kakizawa S, Nishigawa H, Jung H-Y, Wei W, Suzuki S, Arashida R, Nakata D, Miyata S, Ugaki M, Namba S (2004) Reductive evolution suggested from the complete genome sequence of a plant-pathogenic phytoplasma. *Nature Genetics* **36**, 27–29.

Oshima K, Kakizawa S, Arashida R, Ishii Y, Hoshi A, Hayashi Y, Kagiwada S, Namba S (2007) Presence of two glycolytic gene clusters in a severe pathogenic line of '*Candidatus* Phytoplasma asteris'. *Molecular Plant Pathology* **8**, 481–489.

Patui S, Bertolini A, Clincon L, Ermacora P, Braidot E, Vianello A, Zancani M (2013) Involvement of plasma membrane peroxidases and oxylipin pathway in the recovery from phytoplasma disease in apple (*Malus domestica*). *Physiology of Plants* **148**, 200–213.

Pracros P, Renaudin J, Eveillard S, Mouras A, Hernould M (2006) Tomato flower abnormalities induced by "stolbur" phytoplasma infection are associated with changes of expression of floral development genes. *Molecular Plant-Microbe Interactions* **19**, 62–68.

Razin S, Yogev D, Naot Y (1998) Molecular biology and pathogenicity of mycoplasmas. *Microbiology and Molecular Biology Revue* **62**, 1094–1156.

Seemüller E, Schneider B (2007) Differences in virulence and genomic features of strains of '*Candidatus* Phytoplasma mali', the apple proliferation agent. *Phytopathology* **97**, 964–970.

Seemüller E, Kunze L, Schaper U (1984) Colonization behavior of MLO, and symptom expression of proliferation-diseased apple trees and decline-diseased pear trees over a period of several years. *Zeitschrift Pflanzenkrankh Pflanzenchutz* **91**, 525–532.

Seemüller E, Kampmann M, Kiss E, Schneider B (2011) HflB gene-based phytopathogenic classification of '*Candidatus* Phytoplasma mali' strains and evidence that strain composition determines virulence in multiply infected apple trees. *Molecular Plant-Microbe Interactions* **24**, 1258–1266.

Seemüller E, Sule S, Kube M, Jelkmann W, Schneider B (2013) The AAA plus ATPases and HflB/FtsH proteases of '*Candidatus* Phytoplasma mali': phylogenetic diversity, membrane topology, and relationship to strain virulence. *Molecular Plant-Microbe Interactions* **26**, 367–376.

Seruga-Music M, Samarzija I, Hogenhout SA, Haryono M, Cho S-T, Kuo C-H (2018) The genome of '*Candidatus* Phytoplasma solani' strain SA-1 is highly dynamic and prone to adopting foreign sequences. *Systematic and Applied Microbiology* **42**, 117–127.

Siewert C, Luge T, Duduk B, Seemüller E, Büttner C, Sauer S, Kube M (2014) Analysis of expressed genes of the bacterium '*Candidatus* Phytoplasma mali' highlights key features of virulence and metabolism. *Plos One* **9**, e94391.

Snider J, Thibault G, Houry WA (2008) The AAA+ superfamily of functionally diverse proteins. *Gene Biology* **9**, 216.

Sugawara K, Honma Y, Komatsu K, Himeno M, Oshima K, Namba S (2013) The alteration of plant morphology by small peptides released from the proteolytic processing of the bacterial peptide TENGU. *Plant Physiology* **162**, 2005–201410.

Sugio A, MacLean AM, Kingdom HN, Grieve VM, Manimekalai R, Hogenhout SA (2011) Diverse targets of phytoplasma effectors: from plant development to defense against insects. *Annual Revue of Phytopathology* **49**, 175–195.

Tanaka J, Sasakawa C (2005) Type IV secretion system in *Helicobacter pylori*: comparison with bacterial type III secretion apparatus. *Tanpakushitsu Kakusan Koso* **50**, 36–43.

Timpte C, Wilson AK, Estelle M (1994) The axr2-1 mutation of *Arabidopsis thaliana* is a gain-of-function mutation that disrupts an early step in auxin response. *Genetics* **138**, 1239–1249.

Tran-Nguyen LT, Kube M, Schneider B, Reinhardt R, Gibb KS (2008) Comparative genome analysis of '*Candidatus* Phytoplasma australiense' (subgroup tuf-Australia I; rp-A) and '*Ca.* Phytoplasma asteris' strains OY-M and AY-WB. *Journal of Bacteriology* **190**, 3979–3991.

van Melderen L, Aertsen A (2009) Regulation and quality control by Lon-dependent proteolysis. *Research in Microbiology* **160**, 645–651.

Wang N, Li Y, Chen W, Yang HZ, Zhang PH, Wu YF (2017) Identification of wheat blue dwarf phytoplasma effectors targeting plant proliferation and defence responses. *Plant Pathology* **67**, 603–609.

Wang N, Yang H, Yin Z, Liu W, Sun L, Wu Y (2018) Phytoplasma effector SWP1 induces witches' broom symptom by destabilizing the TCP transcription factor BRANCHED1. *Molecular Plant Pathology* **19**, 2623–2634.

Yang C-Y, Huang Y-H, Lin C-P, Lin Y-Y, Hsu H-C, Wang C-N, Daisy Liu L-Y, Shen B-N, Lin S-S (2015) Micro RNA396-targeted short vegetative phase is required to repress flowering and is related to the development of abnormal flower symptoms by the phyllody symptoms effector. *Plant Physiology* **168**, 1702–1716.

Wei W, Davis RE, Nuss DL, Zhao Y (2013) Phytoplasmal infection derails genetically preprogrammed meristem fate and alters plant architecture. *Proceedings of the National Academy of Sciences United States of America* **110**, 19149–19154.

Zamorano A, Fiore N (2016) Draft genome sequence of 16SrIII-J phytoplasma, a plant pathogenic bacterium with a broad spectrum of hosts. *Genome Announcements* **4**, e00602–16.

Zimmermann MR, Schneider B, Mithöfer A, Reichelt M, Seemüller E, Furch ACU (2015) Implications of '*Candidatus* Phytoplasma mali' infection on phloem function of apple trees. *Journal of Endocytobiosis and Cell Research* **26**, 67–75.

Chapter 3
Transcriptomic and Proteomic Studies of Phytoplasma-Infected Plants

Marina Dermastia, Michael Kube, and Martina Šeruga-Music̀

Abstract Recent advances in the development of high-throughput techniques and the corresponding software tools have enabled novel -omics approaches that are aimed at a better understanding of the mechanisms underlying phytoplasma pathogenicity and interactions with their hosts. In this chapter, the literature on transcriptomic and proteomic studies on phytoplasma-infected plants are outlined and summarised. Although data are available only for a few plant species infected with phytoplasmas belonging to different taxonomic groups, some general conclusions on interactions with their plant hosts can be deduced. Some of the most studied effects on phytoplasma-infected plants include (i) down-regulation of a wide array of genes associated with photosynthesis and changes in the corresponding protein levels; (ii) alterations to carbohydrate metabolism at the transcriptome and proteome levels; (iii) differential expression of plant secondary metabolites, as mainly up-regulation of genes involved in flavonoid biosynthesis; and (iv) changes in expression of genes related to auxin, jasmonic acid and salicylic acid signalling pathways involved in plant defence responses. Furthermore, studies on the roles of micro-RNAs in post-transcriptional gene regulation during plant responses to phytoplasmas, and on the functions of long noncoding RNAs during phytoplasma infection, are also reviewed.

Keywords Gene expression · Metabolome · Phytoplasma infection · Proteome · Transcriptome

---

M. Dermastia (✉)
National Institute of Biology, Ljubljana, Slovenia

M. Kube
Integrative Infection Biology Crops-Livestock, University of Hohenheim, Stuttgart, Germany

M. Šeruga-Musić
Department of Biology, Faculty of Science, University of Zagreb, Zagreb, Croatia

A. Bertaccini et al. (eds.), *Phytoplasmas: Plant Pathogenic Bacteria - III*,
https://doi.org/10.1007/978-981-13-9632-8_3

3.1 Introduction

Phytoplasmas are assigned to the class *Mollicutes*. These bacteria colonise and multiply in plant hosts and insect vectors (Bertaccini 2017). As it was not yet possible to obtain pure phytoplasma cultures (Contaldo et al. 2012, 2016, 2019), the knowledge about their biology and pathogenicity is still very rudimentary. However, with the molecular biology approaches, and especially those that use high-throughput techniques, information can now be acquired that will gradually uncover the life of these plant pathogens through the use of appropriate bioinformatics tools. Based on detailed analyses of known whole or draft phytoplasma genome sequences, some of the general interactions between phytoplasmas and their hosts can be deduced. The knowledge of relationships among phytoplasmas and their insect and plant hosts has been recently greatly improved by functional analyses of transcriptomic, proteomic and metabolomic data. Recent evidence of the important roles of small non-coding endogenous micro-RNAs (miRNAs) in post-transcriptional gene regulation during plant responses to biotic stress is showing to be crucial for the understanding of phytoplasma interactions with their host plants (Ehya et al. 2013; Yang et al. 2015a; Fan et al. 2015a, b, 2016; Snyman et al. 2017; Gai et al. 2018). In addition, expression levels of long noncoding RNAs (lncRNAs) have been reported to be associated with phytoplasma infections in plants (Fan et al. 2018).

Phytoplasma-induced symptoms in host plants can include, for example, yellowing, phyllody, stunting, growth proliferation and witches' broom. It has been shown that most of these symptoms are mirrored in specific alterations to gene expression and/or changed protein levels in the infected plants, in comparison with the healthy plant. In addition, these data have revealed that phytoplasmas can affect the plant hormonal, nutritional and stress signalling pathways and the complex interactions between these.

Transcriptomic and proteomic data are available for only a few plant species infected with phytoplasmas. The majority of these data are from economically important plants, such as grapevine (*Vitis vinifera* L.) (Dermastia 2017), paulownia (*Paulownia* spp.) (Fan et al. 2015a, b, 2016) and mulberry (*Morus alba*) (Ji et al. 2009; Gai et al. 2014a, b, 2018), and thus not from model plants, as for data on most of other plant-microbe interaction studies. Although the phytoplasmas that have been studied belong to taxonomically different groups, some general conclusions on interactions with their host plants can be drawn based on the existing information. This chapter reviews the available transcriptomic and proteomic data on plants affected by phytoplasmas.

3.2 Methodologies

Early studies that were aimed at the identification of global transcription profiles of plants infected with phytoplasmas used oligonucleotide microarrays (Albertazzi et al. 2009; Hren et al. 2009; Oshima et al. 2011; Punelli et al. 2016). The development of next-generation sequencing platforms and their increasing affordability have led to their successful application to genome-wide gene-expression profiling of phytoplasma-infected plants. Such studies have involved small RNAs (sRNAs), microRNAs (miRNAs) and long noncoding RNAs (lncRNAs) (Ehya et al. 2013; Abbà et al. 2014; Gai et al. 2014a, 2018; Fan et al. 2014, 2015a, b, c, 2018; Liu et al. 2014; Nejat et al. 2015; Yang et al. 2015a, b; Mardi et al. 2015; Niu et al. 2016). Usually, selected genes from several differentially expressed pathways were additionally analysed using quantitative PCR analyses. Studies of quantitative changes in the proteome of phytoplasma-infected plants started with two-dimensional (Ji et al. 2009; Margaria et al. 2013) or one-dimensional sodium dodecyl sulphate polyacrylamide gel electrophoresis and nanocapillary liquid chromatography-tandem mass spectrometry (Ji et al. 2010). However, in these studies, only a few proteins could be quantified, mainly those that were most abundant. Comprehensive proteomic analyses started with the introduction of label-free quantitative shotgun mass spectrometry workflow (Monavarfeshani et al. 2013). Although this provides greater throughput, it is relatively difficult to control, less precise in terms of quantification and vulnerable for the detection of medium-to-low abundance proteins (Bantscheff et al. 2007). More recently the proteome analyses of the changes in the proteome of healthy and phytoplasma-infected plants have used isobaric tags for relative and absolute quantitation and liquid chromatography coupled with tandem mass spectrometry (Luge et al. 2014; Wang et al. 2017; Fan et al. 2017).

To investigate the potential biological pathways involved in such changes, annotated genes and proteins have been mapped, *e.g.* against Enzyme Commission numbers in the Kyoto Encyclopedia of Genes and Genomes database (Kanehisa et al. 2010). However, transcriptomic and proteomic analyses are still challenging. To solve this problem, different software tools have been developed. For instance, the MapMan (Thimm et al. 2004; Rotter et al. 2009) has been successfully applied to the interpretation and visualisation of transcriptomic datasets obtained from grapevines and lime plants infected by phytoplasmas (Hren et al. 2009; Mardi et al. 2015). In general, transcriptomic and proteomic datasets are often combined with metabolomics approaches (Margaria et al. 2014; Luge et al. 2014; Prezelj et al. 2016a; Fan et al. 2017).

3.3 Phytoplasma Infection and Plant Host Photosynthesis

Transcriptomic and proteomic data on phytoplasma-infected grapevine, mulberry, apple, coconut, paulownia and jujube trees have indicated that several genes in photosynthesis are down-regulated during the phytoplasma infection (Fig. 3.1) (Albertazzi et al. 2009; Hren et al. 2009; Ji et al. 2009; Margaria et al. 2013; Luge et al. 2014; Nejat et al. 2015; Fan et al. 2015b; Wei et al. 2017; Wang et al. 2018). Furthermore, transcriptomic analysis of coconut (*Cocos nucifera*) leaves infected with a yellow decline phytoplasmas (group 16SrXIV) showed down-regulation of a wide array of genes associated with photosynthesis (*i.e.* ATP synthase, plastocyanin chloroplast precursor, chlorophyll a/b binding protein, photosystem II reaction

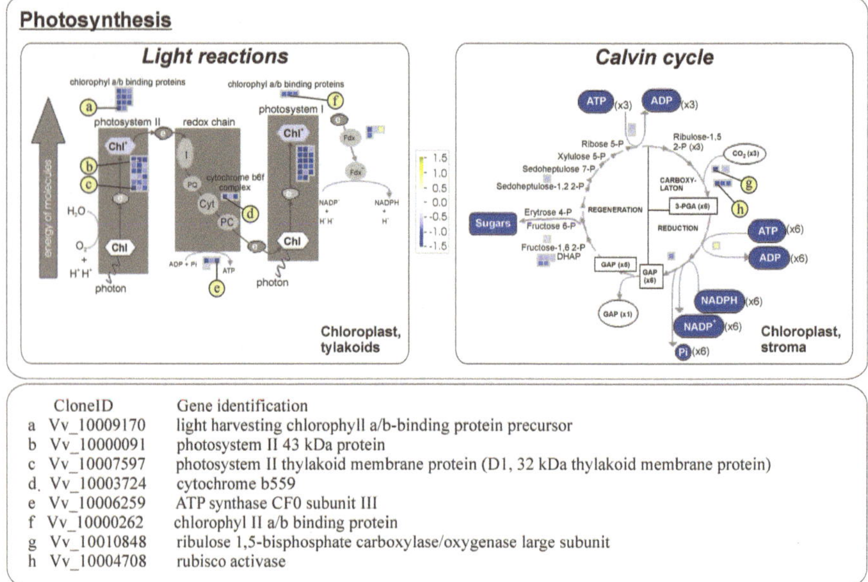

	CloneID	Gene identification
a	Vv_10009170	light harvesting chlorophyll a/b-binding protein precursor
b	Vv_10000091	photosystem II 43 kDa protein
c	Vv_10007597	photosystem II thylakoid membrane protein (D1, 32 kDa thylakoid membrane protein)
d.	Vv_10003724	cytochrome b559
e	Vv_10006259	ATP synthase CF0 subunit III
f	Vv_10000262	chlorophyl II a/b binding protein
g	Vv_10010848	ribulose 1,5-bisphosphate carboxylase/oxygenase large subunit
h	Vv_10004708	rubisco activase

Fig. 3.1 A schematic representations of genes involved in photosynthesis of grapevine infected with '*Ca.* P. solani'. Each gene is represented by a coloured square. Yellow colour represents up-regulation and blue colour down-regulation in infected *vs.* healthy samples. A few most interesting genes are pointed out. Schemes are slightly adapted visualisations of MapMan pathways. Only differentially expressed genes are presented (adapted from Hren et al. 2009)

centre N chloroplast precursor, photosystem II core complex proteins PsbY chloroplast precursor) (Nejat et al. 2015).

Analysis of different proteomes of *Paulownia fortunei* infected with paulownia witches' broom (PaWB) phytoplasmas (subgroup 16SrI-D) identified 27 PaWB-related proteins that were classified in four groups based on their protein-protein interactions; one of these groups is photosynthesis related (Wei et al. 2017). Here, nine proteins involved in photosynthesis were lower than in the healthy plants: light-harvesting complex I chlorophyll a/b binding protein 3; light-harvesting complex II chlorophyll a/b binding protein 5; photosystem I P700 chlorophyll a apoprotein A1; cytochrome b6f Rieske iron-sulphur subunit; photosystem II oxygen-evolving enhancer protein; photosystem II 10 kDa protein; ribulose bisphosphate carboxylase small chain family protein; thioredoxin Mtype 4; and cytosolic pyruvate kinase. At the same time, two proteins showed greater abundance: cytochrome b6 and photosystem II Psb27 (Wang et al. 2017).

In mulberry (*Morus alba* variety *Multicaulis*) infected with the mulberry witches' broom phytoplasmas (subgroup 16SrI-B) associated with the mulberry dwarf disease, the photosynthetic proteins rubisco large subunit, rubisco activase and sedoheptulose-1,7-bisphosphatase showed enhanced degradation in the infected leaves (Ji et al. 2009). Also, in transcriptomic-assisted proteomic analysis of *Nicotiana occidentalis* infected with '*Candidatus* Phytoplasma mali' strain AT (subgroup 16SrX-A), the porphyrin and chlorophyll metabolism showed a strong down-regulation (Luge et al. 2014).

In periwinkle (*Catharantus roseus* G. Don), where the infection with '*Ca.* P. solani' (subgroup 16SrXII-A) was first studied, Jagoueix-Eveillard et al. (2001) reported the repression of the genes of subunit III of photosystem I, ribulose 1,5-biphosphate carboxylase-oxygenase, delta-C-methyltransferase and transketolase. In infected leaves of the grapevine cultivar Chardonnay, 3 and 11 genes that encode chlorophyll a/b binding proteins in photosystem I and photosystem II, respectively, were significantly down-regulated (Hren et al. 2009). Significant repression of the *rbcL* gene that encodes rubisco large subunit (Hren et al. 2009) is consistent with the loss of this protein in the infected grapevines (Bertamini et al. 2002). Moreover, the gene that encodes rubisco activase was significantly down-regulated in infected plants, and the expression of genes involved in cytochrome b6f complex in the electron-transport pathway decreased in infected leaves (Hren et al. 2009), which correlated with the observed down-regulation of the genes that encode the ATP synthase. These data support the biochemical studies of infected grapevines that showed decreased chlorophyll, carotenoid and soluble protein contents and decreased activities of rubisco and nitrate and nitrite reductase (Bertamini and Nedunchezhian 2001a, b; Bertamini et al. 2002; Endeshaw et al. 2012; Rusjan et al. 2012a). Decreases in photosynthesis were also detected at the protein level in grapevine cultivar Barbera infected with "flavescence dorée" phytoplasmas (subgroups 16SrV-C/-D) (Margaria and Palmano 2011; Margaria et al. 2013). Several proteins were down-regulated, including two related to the dark reactions (*i.e.* rubisco, and rubisco activase) and four related to the light-dependent reactions of photosystem II (*i.e.* chloroplastic ATP synthase CF1 α subunit, ATP synthase CF1 β subunit, oxygen-evolving enhancer protein 2, and Mn-stabilising protein).

3.4 Phytoplasma Infection and Plant Host Carbohydrate Metabolism

There is a growing evidence that the feedback inhibition of photosynthesis results in leaf yellowing, or chlorosis, because of the carbohydrate accumulation in the source leaves, as reported for periwinkle, tobacco, papaya, coconut, maize, apple tree, mulberry and grapevines infected with phytoplasmas from several ribosomal groups (Lepka et al. 1999; Guthrie et al. 2001; Maust et al. 2003; Junqueira et al. 2004; Christensen et al. 2005; Giorno et al. 2013; Gai et al. 2014b; Covington Dunn et al. 2016; Prezelj et al. 2016a). It has also been suggested that this increase in carbohydrate concentration is a consequence of the manipulation of the host metabolism by the phytoplasmas, thus turning these infected tissues into a carbohydrate sink that provides the phytoplasmas with hexoses (Hren et al. 2009; Santi et al. 2013a; Luge et al. 2014). This source-to-sink transition is usually characterised by the increased activities of invertases, which irreversibly hydrolyse sucrose to glucose and fructose

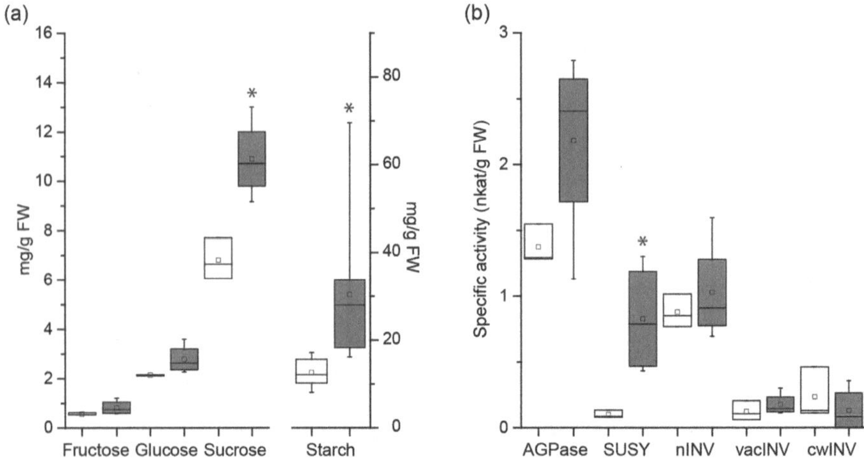

Fig. 3.2 Box plot of the carbohydrate concentrations **(a)** and the specific activities of ADP-glucose pyrophosphorylase (AGPase), sucrose synthase (SuSy) and neutral (nINV), vacuolar (vacINV) and cell wall (cwINV) invertases **(b)** in summer whole-leaf samples of uninfected (white boxes) and "flavescence dorée"-infected (gray boxes) grapevines. Line across the box, median; square, mean; box, 25th and 75th percentiles; whiskers, minimum and maximum values. $^{*}p < 0.05$ (Student's t-tests) (From Prezelj et al. 2016a)

(Roitsch and González 2004). Indeed, there was significant transcript increase of the acid-soluble vacuolar invertase *vacINV2* gene in leaf vein-enriched samples of grapevines infected with '*Ca.* P. solani' (Hren et al. 2009), as also with the "flavescence dorée" phytoplasmas (Fig. 3.2) (Prezelj et al. 2016a).

However, the activity of the enzyme itself was not significantly higher than in the healthy control grapevines (Covington Dunn et al. 2016; Prezelj et al. 2016a). In contrast, in periwinkle and tomato (*Solanum lycopersicum* L.) infected with '*Ca.* P. solani', the enzyme activity of the acid-soluble vacuolar invertase vacINV was increased, although there was no differential expression of the *vacINV* gene between healthy and infected plants (Machenaud et al. 2007). In addition, there was a small increase in the expression of the gene that encodes the acid-insoluble invertase that is bound to the cell wall, *cwINV*, in grapevines infected with '*Ca.* P. solani' (Santi et al. 2013a). However, this was not confirmed in another study on the same interactions (Hren et al. 2009).

Little information is available on the sugar transporters that are involved in phytoplasma pathogenicity. In micropropagated apple shoots infected with '*Ca.* P. mali', there was a significant increase in expression of the gene that encodes the sorbitol transporter SOT5, together with decreased levels of sorbitol (Giorno et al. 2013). In the apple, sorbitol is the main carbohydrate that is translocated in the phloem of the tree, and it is therefore the main carbon source imported by the apple sink tissues (Li et al. 2012). In grapevine leaf vein-enriched samples of cultivar Blaufränkisch infected with the "flavescence dorée" phytoplasmas, the transcript levels were significantly increased for the gene *SWEET17a* (Prezelj et al. 2016a), which encodes a protein from the family of SWEET hexose and sucrose transporters (Eom et al. 2015). On the other hand, in the cultivar Chardonnay leaves infected with '*Ca.* P. solani', the transcript of the gene that encodes sucrose transporter SUC27 was significantly down-regulated in comparison with uninfected leaves. This has been suggested to be associated with the establishment of the source-to-sink transition in the leaf phloem (Santi et al. 2013b). During the same interaction, two other sucrose transporter genes, *SUC11* and *SUC12*, were differentially expressed in comparison with healthy leaves: they were down-regulated in the infected leaves, but their transcript abundance increased during the recovery process from '*Ca.* P. solani' (Santi et al. 2013a, 2013b).

Several evidences have confirmed the association of the sucrose synthase with the carbohydrate accumulation in the mesophyll of the phytoplasma-infected leaves and the physical obstruction of phloem loading and transport due to callose deposition in the sieve tubes (Musetti et al. 2013a). The sucrose synthase reversibly catalyses the sucrose breakdown to UDP-glucose and fructose, and this enzyme is localised in both companion cells and sieve elements of the phloem (Koch 2004). The localisation of the sucrose synthase in the phloem sieve tubes might facilitate its role in the direct supply of the UDP-glucose for a rapid biosynthesis of callose plugs in the sieve pores. Increases in callose synthase gene expression have been confirmed in Mexican lime trees in response to infection with '*Ca.* P. aurantifolia' (Mardi et al. 2015), as also at the proteome level in a label-free quantitative shotgun proteomics study (Monavarfeshani et al. 2013). Among the genes that encode callose synthase in the grapevine genome, the transcript abundances of *CAS2* (Rotter et al. 2009; Santi et al. 2013a, 2013b; Dermastia et al. 2015) and *CAS7* (Santi et al. 2013a) genes were greater in grapevine leaves infected with '*Ca.* P. solani', in comparison with healthy grapevines. Callose deposition is a dynamic process that is coordinated through the activities of callose synthase and the callose hydrolysing enzyme β-1-3-glucanase. Significant increases in transcription of β-1-3-glucanase genes in grapevine leaves infected with '*Ca.* P. solani'(Hren et al. 2009; Landi and Romanazzi 2011) and "flavescence dorée" phytoplasmas (Margaria et al. 2013; Prezelj et al. 2016a) supported the suggested short span of callose molecules shifted towards catabolism (Zabotin et al. 2002). An additional role for sucrose synthase in phytoplasma pathogenicity has also been proposed. This relates to providing fructose for the direct use by the phytoplasmas, as also proposed for another genus in the *Mollicutes* class, the *Spiroplasma* (Prezelj et al. 2016a; Dermastia 2017). This was based on significantly increased *SUSY4* gene transcript in the vein-enriched tissues of grapevine leaves infected with '*Ca.* P. solani' (Santi et al. 2013a, 2013b) and with "flavescence dorée" phytoplasmas (Prezelj et al. 2016a). The transcript of a sucrose synthase gene and its protein product were also increased in Mexican lime trees infected with '*Ca.* P. aurantifolia' (Monavarfeshani et al. 2013; Mardi et al. 2015).

In micropropagated apple shoots, a considerable increase in starch is associated with the increased expression of the *GBSSIa* gene, which is involved in the synthesis of the starch component amylose (Giorno et al. 2013). Indeed, a key feature of the grapevine leaves infected with '*Ca.* P. solani' and "flavescence dorée" phytoplasmas is the abundance of the *AGPL* gene transcript (Hren et al. 2009; Dermastia et al. 2015; Rotter et al. 2018); this encodes the large subunit of ADP-glucose pyrophosphorylase, which is a rate-limiting enzyme in starch biosynthesis (Ballicora et al. 2004). In agreement with the increased *AGPL* gene transcript levels in the leaves of the grapevine cultivar Blaufränkisch, there are also trends towards a higher ADP-glucose pyrophosphorylase activity and significantly higher starch concentrations (Prezelj et al. 2016a). The high starch concentrations in mulberry with yellow dwarf disease have been explained by the lower expression of genes and/or the lower activity of enzymes involved in the degradation of starch (Gai et al. 2014b). In contrast, the gene encoding the α-amylase in grapevine infected with '*Ca.* P. solani' was up-regulated (Hren et al. 2009).

High-throughput transcriptome sequencing of Mexican lime trees in response to the infection with '*Ca.* P. aurantifolia' revealed the up-regulation of the genes involved in the sucrose metabolism, cell wall biogenesis and degradation (Mardi et al. 2015). Alterations in host sugar metabolism have also been confirmed at the proteome level, in a label-free quantitative shotgun proteomics study (Monavarfeshani et al. 2013).

3.5 Phytoplasma Infection and Plant Host Secondary Metabolism

Plants have developed sophisticated processes to enhance their survival, which include the production of various secondary metabolites (Croteau et al. 2000). These compounds serve diverse purposes, such as in the defence responses against herbivores and pathogens and in the protection against ultraviolet light. Several studies of phytoplasma-infected grapevine, Mexican lime, coconut and jujube trees have shown that the secondary metabolism is changed upon infection (Hren et al. 2009; Landi and Romanazzi 2011; Margaria et al. 2014; Mardi et al. 2015; Nejat et al. 2015; Prezelj et al. 2016a; Ye et al. 2017; Wang et al. 2018).

During the interactions between Mexican lime tree and '*Ca.* P. aurantifolia', several transcripts related to the phenylpropanoid biosynthesis pathway were differentially expressed (Mardi et al. 2015). This pathway contributes to the production of lignin, suberin and condensed tannins, along with other secondary metabolites. In jujube plants infected with jujube witches' broom (JWB) phytoplasmas (subgroup 16SrV-B), 22 genes are involved in the phenylpropanoid biosynthesis, and their metabolism was up-regulated (Ye et al. 2017). Similar data have been reported for paulownia infected with the PaWB phytoplasmas (Fan et al. 2015c).

The most studied plant secondary metabolites during the phytoplasma infections have been the flavonoids. These comprise a large class of >10,000 compounds with very diverse roles, which include the plant defence against microbial pathogens in general (Treutter 2005) and phytoplasmas in particular (Albertazzi et al. 2009; Hren et al. 2009; Landi and Romanazzi 2011; Rusjan et al. 2012a, 2012b; Margaria et al. 2014; Rusjan and Mikulic-Petkovsek 2015; Prezelj et al. 2016a; Wang et al. 2018).

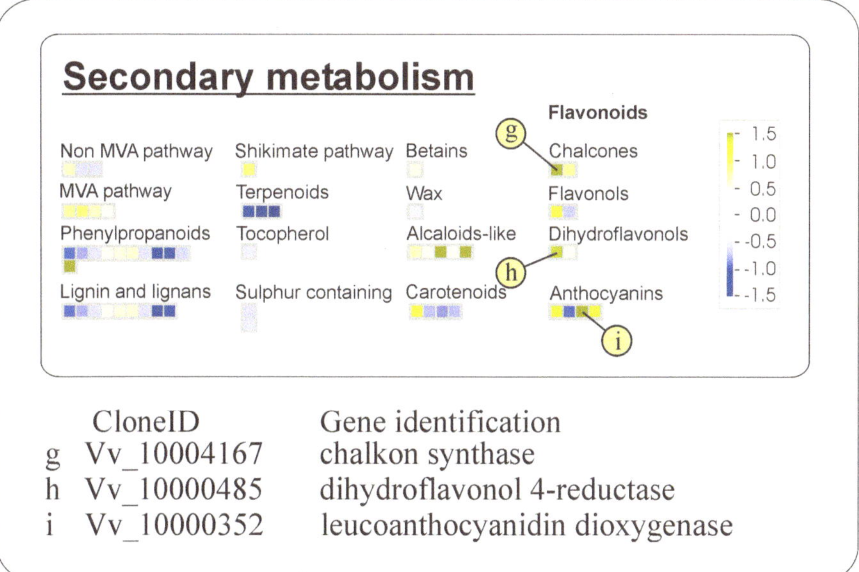

Fig. 3.3 A schematic representations of the genes involved in secondary metabolism of grapevine infected with '*Ca.* P. solani'. Each gene is represented by a coloured square. Yellow colour represents up-regulation and blue colour down-regulation in infected *versus* healthy samples. A few most interesting genes are pointed out. Schemes are slightly adapted visualisations of MapMan pathways. Only differentially expressed genes are presented (adapted from Hren et al. 2009)

After infection of grapevine cultivar Chardonnay with '*Ca.* P. solani', the transcript levels of several genes that encode the enzymes involved in the biosynthesis of phenolic compounds were increased, as well as the activities of their enzyme products. In particular, the genes that encode the phenylalanine ammonia lyase, chalcone synthase, flavanone 3-hydroxylase and leucoanthocyanidin dioxygenase were up-regulated in the infected leaves (Fig. 3.3) (Hren et al. 2009; Landi and Romanazzi 2011; Dermastia et al. 2015; Paolacci et al. 2017; Rotter et al. 2018). The enzyme activities of phenylalanine ammonia lyase, chalcone synthase/chalcone isomerase, flavanone 3-hydroxylase and polyphenol oxidase were also increased in the infected leaves (Rusjan et al. 2012a). The analysis of the corresponding metabolites showed comparable data (Rusjan et al. 2012b; Rusjan and Mikulic-Petkovsek 2015).

The flavonoid pathway was also affected in the leaves of the cultivars Barbera, Nebbiolo and Blaufränkisch infected with "flavescence dorée" phytoplasmas. A transcriptomic analysis of the genes coding several enzymes involved in this pathway indicated the activation of the anthocyanin accumulation in the infected leaves; these enzymes included chalcone synthase, flavanone 3-hydroxylase, leucoanthocyanidin dioxygenase, UGT-glucose/anthocyanin 3-O-glucosyltransferase, UAGT-transcription factor, anthocyanidine reductase, leucoanthocyanidin reductase, flavonol synthase and FLS-transcription factor (Margaria et al. 2014; Prezelj et al. 2016a). In addition, the expression of the genes that encode naringenin 3-dioxygenase, chalcone synthase, leucoanthocyanidin dioxygenase and flavonoid 3′-monooxygenase was up-regulated upon infection of Mexican lime tree with '*Ca.* P. aurantifolia' (Mardi et al. 2015). Infection of jujube tree with JWB phytoplasmas was also associated with the up-regulation of more than 20 genes and some of their protein products involved in the flavonoid biosynthesis (Ye et al. 2017). The increased transcript levels of the genes involved in the flavonoid biosynthesis have also been demonstrated for paulownia plants infected with PaWB phytoplasmas (Fan et al. 2015c).

3.6 Plant Hormones in the Defence Responses to Phytoplasmas

Although a growing evidence indicates the important roles of the plant hormones in the signalling networks involved in the plant responses to a wide range of pathogens, the underlying molecular mechanisms remain poorly understood (Bari and Jones 2009). The available data on these mechanisms in plants infected with phytoplasmas are still only correlative and are mainly based on comparisons between expression levels of genes involved in hormone metabolism and recent metabolomic data.

In paulownia infected with PaWB phytoplasmas, the differentially expressed genes that are associated with the responses to the *stimuli* of auxins, jasmonic acid, salicylic acid, gibberellins, cytokinins and ethylene indicated the presence of a cross-talk among different hormones and different regulatory mechanisms (Fan et al. 2015c). The transcriptome analysis of jujube infected with JWB phytoplasmas revealed the involvement of the abscissic acid in this interaction, with differential expression of some of the enzymes involved in its synthesis (*i.e.* zeaxanthin, violaxanthin de-epoxidase, abscisic-aldehyde oxidase, phytoene desaturase, and lycopene epsilon cyclase) (Fan et al. 2017). In phytoplasma-infected jujube, 66 genes related to the auxin-activated signalling pathways, 30 to the jasmonic acid signalling pathways and 44 to the salicylic acid signalling pathways were differentially expressed; in addition, a jasmonic acid-induced protein-like was significantly expressed (Ye et al. 2017). In the Mexican lime trees infected with '*Ca.* P. aurantifolia' there was up-regulation of the genes related to the diterpenoid pathway, which leads to biosynthesis and metabolism of the gibberellins. In addition, there was a differential expression of several genes involved in the biosynthesis and signalling of auxins (*i.e.* for indole-3-acetic acid-amido synthetase, SAUR family protein, auxin efflux carrier family protein auxin-responsive protein, and auxin-induced protein), ethylene and brassinosteroids (Mardi et al. 2015).

Jasmonic acid has been shown to have important roles in the plant defence responses against insects and microbial pathogens (Bari and Jones 2009). Trends for increased jasmonic acid and methyl jasmonate levels reported in leaves of tobacco infected with '*Ca.* P. mali' and in grapevine infected with '*Ca.* P. solani' were related to the expression of the genes and protein products involved in the α-linolenic

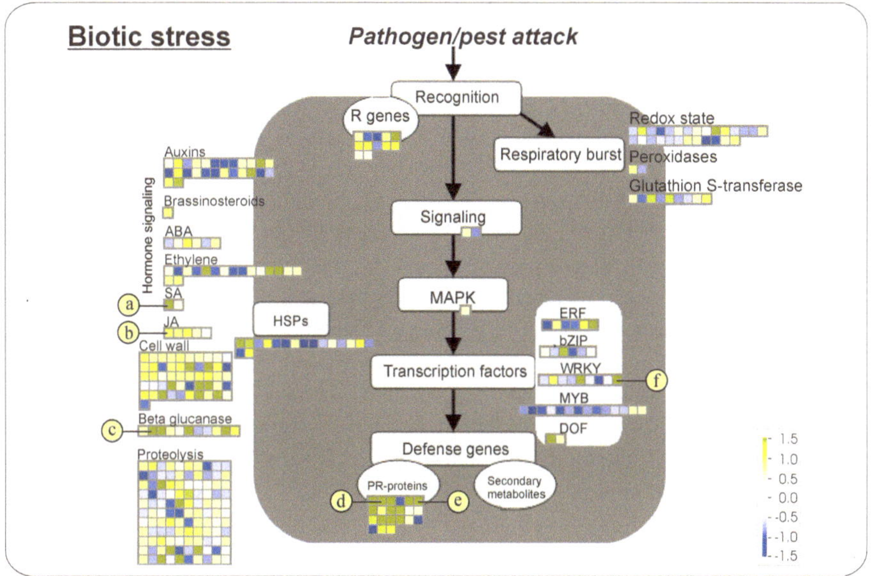

	CloneID	Gene identification
a	Vv_10000965	s-adenosyl-L-methionine:salicylic acid carboxyl methyltransferase
b	Vv_10003710	lipoxygenase
c	Vv_10000389	beta-1,3-glucanase
d	Vv_10000872	thaumatin-like protein
e	Vv_10003617	osmotin-like protein precursor
f	Vv_10013175	WRKY54

Fig. 3.4 A schematic representations of genes involved in biotic stress. Each gene is represented by a coloured square. Yellow colour represents up-regulation and blue colour down-regulation in infected *versus* healthy samples. A few most interesting genes are pointed out. Schemes are slightly adapted visualisations of MapMan pathways. Only differentially expressed genes are presented (adapted from Hren et al. 2009)

acid synthesis pathway (Luge et al. 2014; Paolacci et al. 2017). On the other hand, some of these genes were repressed in the apple trees infected with '*Ca.* P. mali', although they tended to have higher expression levels in plants recovered from the disease, in comparison with diseased plants (Musetti et al. 2013b). Genes that encode the lipoxygenases (LOXs), which are involved in jasmonic acid biosynthesis, were up-regulated in the tomato and grapevine plants infected with '*Ca.* P. solani' (Fig. 3.4) (Hren et al. 2009; Dermastia et al. 2015).

A detailed analysis of four *LOX* genes involved in this interaction showed their differential expression in diseased plants in comparison with those expressed in plants that were recovered from the disease (Paolacci et al. 2017). In tomato infected with '*Ca.* P. solani', the activation of the jasmonic acid-dependent pathway is strain dependent, which is indicated by the expression of the jasmonic acid-related genes

PIN2 and *LoxD* (Ahmad et al. 2013). In *Arabidopsis* plants infected with '*Ca.* P. asteris' strain AY-WB, the binding of the phytoplasma effector SAP11 destabilises CINCINNATA-related TEOSINTE BRANCHED1, CYCLOIDEA, PROLIFERATING CELL FACTORS 1 and 2 transcription factors, which control the plant development and promote the expression of the *LOX* genes (Sugio et al. 2011).

The levels of both jasmonic acid and auxins are decreased in the *Arabidopsis* plants infected with '*Ca.* P. asteris' strain OY. These plants had decreased levels of expression of auxin response factors ARF6 and ARF8, that regulate the floral development in a jasmonic acid-dependent manner, and they thus showed impaired flower development. The same pattern of expression of these genes has been shown after transgenic expression of the phytoplasma virulence peptide TENGU (Minato et al. 2015).

Increased transcript levels for the *hp* gene from the cytokinin signalling pathway has been demonstrated in vein-enriched samples of grapevine infected with '*Ca.* P. solani' and "flavescence dorée" phytoplasmas (Hren et al. 2009; Prezelj et al. 2016a) and also in poinsettia (*Euphorbia pulcherrima* L.) (Nicolaisen and Christensen 2007).

3.7 Possible Roles of Pathogenesis-Related Proteins in the Salicylic Acid-Dependent Host Plant Responses to the Phytoplasma Infection

The source-sink switch that follows pathogen infections in plants is usually associated with coordinated plant defence responses, which enhance the production of secondary metabolites and the expression of defence-related genes (Ehness et al. 1997; Roitsch 1999; Rojas et al. 2014). Some of these defence genes encode several classes of pathogenesis-related (PR) proteins. These occur in multi-gene families and are usually coordinated at the level of their transcription. The PR proteins have been shown to have important roles in plant defences, not only against pathogen infections, but also eliciting acquired resistance (Sudisha et al. 2012). Differences in gene expression and protein levels of these PR proteins in healthy and phytoplasma-infected plants have been demonstrated in apple, jujube, Mexican lime tree, periwinkle, paulownia, grapevine, tomato and garland chrysanthemum (Zhong and Shen 2004; Hren et al. 2009; Ahmad and Eveillard 2011; Landi and Romanazzi 2011; Margaria and Palmano 2011; Santi et al. 2013a, 2013b; Musetti et al. 2013b; Monavarfeshani et al. 2013; Giorno et al. 2013; Liu et al. 2014; Fan et al. 2015c; Prezelj et al. 2016a; Wang et al. 2017; Paolacci et al. 2017; Rotter et al. 2018).

In the leaves infected with phytoplasmas, a significant up-regulation has been demonstrated for the PR protein genes from group 1 (PR-1), group 2 (PR-2; encoding β-1,3-glucanase) and group 5 (PR-5; encoding thaumatin-like and osmotin-PR proteins) (Albertazzi et al. 2009; Hren et al. 2009; Landi and Romanazzi 2011; Santi et al. 2013a, b; Musetti et al. 2013b; Ahmad et al. 2013; Dermastia et al. 2015;

Mardi et al. 2015; Paolacci et al. 2017; Rotter et al. 2018). The expression of these genes is coordinately regulated by the salicylic acid, and they are used as molecular markers for salicylic acid-dependent systemic-acquired resistance signalling. It has been suggested that the infection with '*Ca.* P. solani' induces salicylic acid-dependent systemic-acquired resistance in tomato and grapevine, which delays the phytoplasma multiplication. This was supported by a significant up-regulation of the *PR-1*, *PR-2* and *PR-5* genes in leaf samples of those plants (Hren et al. 2009; Landi and Romanazzi 2011; Santi et al. 2013a, 2013b; Ahmad et al. 2013, 2015; Dermastia et al. 2015; Prezelj et al. 2016a; Paolacci et al. 2017; Rotter et al. 2018). This was accompanied by a significantly increased transcription of the gene that encodes the enzyme S-adenosyl-L-methionine/salicylic acid carboxyl methyltransferase, which is responsible for biosynthesis of methyl salicylate (Hren et al. 2009; Dermastia et al. 2015), by a 26-fold increase in salicylic acid-glucopyranoside (Prezelj et al. 2016b) and a significant increase in free and total salicylate (Paolacci et al. 2017). On the other hand, the abundance of these *PR-1*, *PR-2* and *PR-5* gene transcripts in grapevine plants recovered from the "bois noir" disease associated with the presence of '*Ca.* P. solani' did not differ from the respective transcript levels in the uninfected plants. Moreover, the increases in the levels of the *PR-2* and *PR-5* gene transcripts and salicylic acid-glucopyranoside were less pronounced in the susceptible grapevine cultivar Blaufränkisch infected with the "flavescence dorée" phytoplasmas (Prezelj et al. 2016a). In comparison, in the less susceptible cultivar Nebbiolo, the levels of the PR-5 transcript were ten fold greater (Margaria and Palmano 2011).

These combined data suggest that although the plants react to the phytoplasma presence through a salicylic acid-mediated signalling, the activation of these responses does not confer resistance against the disease (Prezelj et al. 2016a; Paolacci et al. 2017). In addition, it has been shown that during the interactions between grapevines and '*Ca.* P. solani', the activation of the salicylic acid signalling pathway antagonises the jasmonic acid defence responses. However, in plants recovered from the "bois noir" disease associated with this phytoplasma, the jasmonic acid signalling pathway was activated, which was suggested to prevent the subsequent development of the disease symptoms (Paolacci et al. 2017).

3.8 Leafy Flower Transition Induced by Phytoplasmas

An impressive symptom of some phytoplasma-infected plant species is the phyllody, whereby the flower organ development is reprogrammed to the development of leaf-like structures. The homologs of phytoplasma effectors that can induce phyllody symptoms are known as phyllogens (Kitazawa et al. 2017), which include SAP54, a phyllogen from AY-WB strain of '*Ca.* P. asteris' (MacLean et al. 2011, 2014), and PHYL1, a phyllogen from the 'Ca. P. asteris' OY strain (Maejima et al. 2014). It has been shown that phyllogens can induce phyllody in *Arabidopsis thaliana* (MacLean et al. 2014) and other angiosperms (Kitazawa et al. 2017) mediating the degradation of some plant type II MADS-domain transcription factors. While SAP54 interacts with proteins of the RADIATION SENSITIVE23 family that shuttle substrates to the proteasome (MacLean et al. 2014), PHYL1 interacts directly with APETALA1 and SEPALLATA1 type II MADS-domain transcription factors (Maejima et al. 2014, 2015).

3.9 The Role of microRNAs in the Post-Transcriptional Gene Regulation During the Plant Responses to Phytoplasmas

MicroRNAs are among the most important regulatory elements of gene expression in animals and plants. Furthermore, increasing evidence has shown that miRNAs also have crucial roles in the defence responses and the plant adaptation to biotic and abiotic stresses (Sunkar et al. 2012; Borges and Martienssen 2015). Mature miRNAs are single-stranded RNA molecules of approximately 19 nt to 24 nt in length that originate mainly by duplication of pre-existing genes or protein-coding genes. In addition, it appears that transposable elements also contribute to the generation of species-specific miRNAs (Nozawa et al. 2012). Several miRNA families are conserved across the plant species, although in addition, each species has its own specific miRNAs with species-specific functions (Lenz et al. 2011). Studies of differentially expressed miRNAs in Mexican lime trees, grapevines, paulownia and mulberry plants infected with phytoplasmas have revealed their putative targets, which are genes involved in plant morphology, signalling, nutrient homeostasis, stress environmental responses and hormonal metabolism and regulation (Ehya et al. 2013; Fan et al. 2015a, 2015b, 2016; Snyman et al. 2017; Gai et al. 2018).

In the phloem sap of mulberry infected with the mulberry yellow dwarf disease, 86 conserved and 19 novel miRNAs were identified. Among these novel miRNAs, almost 70% are differentially expressed in the infected compared to the healthy plants. Their target genes are involved in metabolic processes, transcription regulation, signalling, stress and environmental responses, development and hormonal metabolism and regulation (Gai et al. 2018). Recent studies have shown potential specific roles for some of these miRNAs that are differentially expressed upon phytoplasma infection. In grafting experiments with mulberry, it has been shown that the miRNA mul-miR482a-5p, differentially expressed in the phloem, can be transported from scion to rootstock, and it was suggested that it reduces the host resistance to biotic stress (Gai et al. 2018). In a transgenic *Arabidopsis* that expressed the *PHY1* gene from the peanut witches' broom phytoplasmas, the degradome profiles showed that miR396 triggers mRNA decay for SHORT VEGETATIVE PHASE (SVP); this indicates that SVP causes a miR396-mediated inhibition of the translation (Yang et al. 2015a).

3.10 Functions of lncRNA During Phytoplasma Infection

Long noncoding RNAs are a group of noncoding RNAs that are >200 nucleotides in length and for which the biological functions are largely unknown. However, recent studies have revealed their interactions with proteins, DNA and RNA, and their regulation of gene expression in chromatin remodelling, nuclear transcription, pre-mRNA splicing and cytoplasmic mRNA translation (Yang et al. 2015b). The whole-genome identification of lncRNAs from paulownia identified the presence of 3,689 lncRNAs. Among these, 112 lncRNAs were associated with PaWB phytoplasma presence. These molecules seem to have roles in the regulation of the genes involved in the hypersensitive responses induced by the reactive oxygen species and the effector-triggered immunity (Fan et al. 2018).

3.11 Concluding Remarks

An explosion in the development of different high-throughput techniques and analyses has led to the unveiling of a large number of previously unknown RNAs and proteins that are involved in phytoplasma interactions with their host plants. Although many of their roles remain unknown to date, or have been extrapolated based on comparisons with other studies, now it is possible to finally start to understand the mechanisms by which phytoplasmas interact with plants.

References

Abbà S, Galetto L, Carle P, Carrère S, Delledonne M, Foissac X, Palmano S, Veratti F, Marzachì C (2014) RNA-Seq profile of "flavescence dorée" phytoplasma in grapevine. *BMC Genomics* **15**, 1088.

Ahmad JN, Eveillard S (2011) Study of the expression of defense related protein genes in "stolbur" C and "stolbur" PO phytoplasma-infected tomato. *Bulletin of Insectology* **64**(Supplement), S159-S160.

Ahmad JN, Renaudin J, Eveillard S (2013) Expression of defence genes in "stolbur" phytoplasma infected tomatoes, and effect of defence stimulators on disease development. *European Journal of Plant Pathology* **139**, 39–51.

Ahmad JN, Renaudin J, Eveillard S (2015) Molecular study of the effect of exogenous phytohormones application in "stolbur" phytoplasma infected tomatoes on disease development. *Phytopathogenic Mollicutes* **5**(1-Supplement), S121-S122.

Albertazzi G, Milc J, Caffagni A, Francia E, Roncaglia E, Ferrari F, Tagliafico E, Stefani E, Pecchioni N (2009) Gene expression in grapevine cultivars in response to "bois noir" phytoplasma infection. *Plant Science* **176**, 792–804.

Ballicora MA, Iglesias AA, Preiss J (2004) ADP-Glucose pyrophosphorylase: a regulatory enzyme for plant starch synthesis. *Photosynthesis Research* **79**, 1–24.

Bantscheff M, Schirle M, Sweetman G, Rick J, Kuster B (2007) Quantitative mass spectrometry in proteomics: a critical review. *Analitical Bioanalitical Chemistry* **389**, 1017–1031.

Bari R, Jones JDG (2009) Role of plant hormones in plant defence responses. *Plant Molecular Biology* **69**, 473–488.

Bertaccini A (2017) Phytoplasmas: dangerous and intriguing bacteria. In: Grapevine yellows diseases and their phytoplasma agents. Eds Dermastia M, Bertaccini A, Constable F and Mehle N. SpringerBriefs in Agriculture. Springer, Switzerland, 1–15 pp.

Bertamini M, Nedunchezhian N (2001a) Decline of photosynthetic pigments, ribulose-1,5-bisphosphate carboxylase and soluble protein contents, nitrate reductase and photosynthetic activities, and changes in tylakoid membrane protein pattern in canopy shade grapevine (*Vitis vinifera* L. cv. Chardonnay) *Photosynthetica* **39**, 529–537.

Bertamini M, Nedunchezhian N (2001b) Effects of phytoplasma ["stolbur"-subgroup ("bois noir"-BN)] on photosynthetic pigments, saccharides, ribulose 1,5-bisphosphate carboxylase, nitrate and nitrite reductases, and photosynthetic activities in field-grown grapevine (*Vitis vinifera* L. cv. Chardonnay). *Photosynthetica* **39**, 119–122.

Bertamini M, Nedunchezhian N, Tomasi F, Grando M. (2002) Phytoplasma ["stolbur"-subgroup ("bois noir"-BN)] infection inhibits photosynthetic pigments, ribulose-1,5-bisphosphate carboxylase and photosynthetic activities in field grown grapevine (*Vitis vinifera* L. cv. Chardonnay) leaves. *Physiological and Molecular Plant Pathology* **61**, 357–366.

Borges F, Martienssen RA (2015) The expanding world of small RNAs in plants. *Nature Revue in Molecular Cell Biology* **16**, 727–741.

Christensen NM, Axelsen KB, Nicolaisen M, Schulz A (2005) Phytoplasmas and their interactions with hosts. *Trends in Plant Science* **10**, 526–535.

Contaldo N, Bertaccini A, Paltrinieri S, Windsor HM, Windsor DG (2012) Axenic culture of plant pathogenic phytoplasmas. *Phytopathologia Mediterranea* **51**, 607–617.

Contaldo N, Satta E, Zambon Y, Paltrinieri S, Bertaccini A (2016) Development and evaluation of different complex media for phytoplasma isolation and growth. *Journal of Microbiological Methods* **127**, 105–110.

Contaldo N, D'Amico G, Paltrinieri S, Diallo HA, Bertaccini A, Arocha Rosete Y (2019) Molecular and biological characterization of phytoplasmas from coconut palms affected by the lethal yellowing disease in Africa. *Microbiological Research* **223–225**, 51–57.

Covington Dunn E, Roitsch T, Dermastia M (2016) Determination of the activity signature of key carbohydrate metabolism enzymes in phenolic-rich grapevine tissues. *Acta Chimica Slovenica* **63**, 757–762.

Croteau R, Kutchan TM, Lewis NG (2000) Natural products (secondary metabolites). In: Biochemistry and Molecular Biology of Plants. Eds Buchanan B, Gruissem W and Jones R. American Society of Plant Physiologists, Rockville, Maryland, United States of America, 1250–1318 pp.

Dermastia M (2017) Interactions between grapevines and grapevine yellows phytoplasmas BN and FD. In: Grapevine yellows diseases and their phytoplasma agents. Eds Dermastia M, Bertaccini A, Constable F and Mehle N. SpringerBriefs in Agriculture. Springer, Switzerland, 47–67 pp.

Dermastia M, Nikolic P, Chersicola M, Gruden K (2015) Transcriptional profiling in infected and recovered grapevine plant responses to '*Candidatus* Phytoplasma solani'. *Phytopathogenic Mollicutes* **5**(1-Supplement), S123-S124.

Ehness R, Ecker M, Godt DE, Roitsch T (1997) Glucose and stress independently regulate source and sink metabolism and defense mechanisms via signal transduction pathways involving protein phosphorylation. *Plant Cell* **9**, 1825–1841.

Ehya F, Monavarfeshani A, Mohseni Fard E, Karimi Farsad L, Khayam Nekouei M, Mardi M, Hosseini Salekdeh G (2013) Phytoplasma-responsive microRNAs modulate hormonal, nutritional, and stress signalling pathways in Mexican lime trees. *Plos One* **8**, e66372.

Endeshaw ST, Murolo S, Romanazzi G, Neri D (2012) Effects of "bois noir" on carbon assimilation, transpiration, stomatal conductance of leaves and yield of grapevine (*Vitis vinifera*) cv. Chardonnay. *Physiology of Plant* **145**, 286–295.

Eom J-S, Chen L-Q, Sosso D, Julius BT, Lin IW, Qu XQ, Braun DM, Frommer WB (2015) SWEETs, transporters for intracellular and intercellular sugar translocation. *Current Opinion on Plant Biology* **25**, 53–62.

Fan G, Dong Y, Deng M, Zhao Z, Niu S, Xu E (2014) Plant-pathogen interaction, circadian rhythm, and hormone-related gene expression provide indicators of phytoplasma infection in *Paulownia fortunei*. *International Journal of Molecular Science* **15**, 23141–23162.

Fan G, Cao X, Niu S, Deng M, Zhao Z, Dong Y (2015a) Transcriptome, microRNA, and degradome analyses of the gene expression of paulownia with phytoplasma. *BMC Genomics* **16**, 896.

Fan G, Niu S, Xu T, Deng M, Zhao Z, Wang Y, Cao L, Wang Z (2015b) Plant–pathogen interaction-related microRNAs and their targets provide indicators of phytoplasma infection in *Paulownia tomentosa* × *Paulownia fortunei*. *Plos One* **10**, e0140590.

Fan G, Xu E, Deng M, Zhao Z, Niu S (2015c) Phenylpropanoid metabolism, hormone biosynthesis and signal transduction-related genes play crucial roles in the resistance of *Paulownia fortunei* to paulownia witches' broom phytoplasma infection. *Genes Genomics* **37**, 913–929.

Fan G, Niu S, Zhao Z, Cao Y (2016) Identification of microRNAs and their targets in *Paulownia fortunei* plants free from phytoplasma pathogen after methyl methane sulfonate treatment. *Biochimic* **127**, 271–280.

Fan G, Cao Y, Wang Z (2018) Regulation of long noncoding RNAs responsive to phytoplasma infection in *Paulownia tomentosa*. *International Journal of Genomics* **2018**, 3174352.

Fan X-P, Liu W, Qiao Y-S, Shang Y-J, Wang G-P, Tian X, Han Y-H, Bertaccini A (2017) Comparative transcriptome analysis of *Ziziphus jujuba* infected by jujube witches' broom phytoplasmas. *Science Horticulturae* **226**, 50–58.

Gai Y-P, Li Y-Q, Guo F-Y, Yuan CZ, Mo YY, Zhang HL, Wang H, Ji XL (2014a) Analysis of phytoplasma-responsive sRNAs provide insight into the pathogenic mechanisms of mulberry yellow dwarf disease. *Science Reporter* **4**, 5378.

Gai YP, Han XJ, Li YQ, Yuan CZ, Mo YY, Guo FY, Liu QX, Ji XL (2014b) Metabolomic analysis reveals the potential metabolites and pathogenesis involved in mulberry yellow dwarf disease. *Plant, Cell and Environment* **37**, 1474–1490.

Gai Y-P, Zhao H-N, Zhao Y-N, Zhu B-S, Yuan S-S, Li S, Guo F-Y, Ji X-L (2018) MiRNA-seq-based profiles of miRNAs in mulberry phloem sap provide insight into the pathogenic mechanisms of mulberry yellow dwarf disease. *Science Reporter* **8**, 812.

Giorno F, Guerriero G, Biagetti M, Ciccotti AM, Baric S (2013) Gene expression and biochemical changes of carbohydrate metabolism in *in vitro* micro-propagated apple plantlets infected by '*Candidatus* Phytoplasma mali'. *Plant Physiology and Biochemistry* **70**, 311–317.

Guthrie JN, Walsh KB, Scott PT, Rasmussen TS (2001) The phytopathology of Australian papaya dieback: a proposed role for the phytoplasma. *Physiological and Molecular Plant Pathology* **58**, 23–30.

Hren M, Nikolić P, Rotter A, Blejec A, Terrier N, Ravnikar M, Dermastia M, Gruden K (2009) "Bois noir" phytoplasma induces significant reprogramming of the leaf transcriptome in the field grown grapevine. *BMC Genomics* **10**, 460.

Jagoueix-Eveillard S, Tarendeau F, Guolter K, Danet J-L, Bové J-M, Garnier M (2001) *Catharanthus roseus* genes regulated differentially by mollicute infections. *Molecular Plant-Microbe Interactions* **14**, 225–233.

Ji X, Gai Y, Zheng C, Mu Z (2009) Comparative proteomic analysis provides new insights into mulberry dwarf responses in mulberry (*Morus alba* L.). *Proteomics* **9**, 5328–5339.

Ji X, Gai Y, Lu B, Zheng C, Mu Z (2010) Shotgun proteomic analysis of mulberry dwarf phytoplasma. *Proteome Science* **8**, 20.

Junqueira A, Bedendo I, Pascholati S (2004) Biochemical changes in corn plants infected by the maize bushy stunt phytoplasma. *Physiological and Molecular Plant Pathology* **65**, 181–185.

Kanehisa M, Goto S, Furumichi M, Tanabe M, Hirakawa M (2010) KEGG for representation and analysis of molecular networks involving diseases and drugs. *Nucleic Acids Research* **38**, D355-D360.

Kitazawa Y, Iwabuchi N, Himeno M, Sasano M, Koinuma H, Nijo T, Tomomitsu T, Yoshida T, Okano Y, Yoshikawa N, Maejima K, Oshima K, Namba S (2017) Phytoplasma-conserved phyllogen proteins induce phyllody across the *Plantae* by degrading floral MADS domain proteins. *Journal of Experimental Botany* **68**, 2799–2811.

Koch K (2004) Sucrose metabolism: regulatory mechanisms and pivotal roles in sugar sensing and plant development. *Current Opinions in Plant Biology* **7**, 235–246.

Landi L, Romanazzi G (2011) Seasonal variation of defense-related gene expression in leaves from "bois noir" affected and recovered grapevines. *Journal of Agricultural Food Chemistry* **59**, 6628–6637.

Lenz D, May P, Walther D (2011) Comparative analysis of miRNAs and their targets across four plant species. *BMC Research Notes* **4**, 483.

Lepka P, Stitt M, Moll E, Seemüller E (1999) Effect of phytoplasmal infection on concentration and translocation of carbohydrates and amino acids in periwinkle and tobacco. *Physiological and Molecular Plant Pathology* **55**, 59–68.

Li M, Feng F, Cheng L (2012) Expression patterns of genes involved in sugar metabolism and accumulation during apple fruit development. *Plos One* **7**, e33055.

Liu L-YD, Tseng H-I, Lin C-P, Lin YY, Huang YH, Huang CK, Chang TH, Lin SS (2014) High-throughput transcriptome analysis for studying the leafy flower transition of *Catharanthus roseus* induced by peanut witches' broom phytoplasma infection. *Plant Cell Physiology* **55**, 942–957.

Luge T, Kube M, Freiwald A, Meierhofer D, Seemüller E, Sauer S (2014) Transcriptomics assisted proteomic analysis of *Nicotiana occidentalis* infected by '*Candidatus* Phytoplasma mali' strain AT. *Proteomics* **14**, 1882–1889.

Machenaud J, Henri R, Dieuaide-Noubhani M, Pracros P, Renaudin J, Eveillard S (2007) Gene expression and enzymatic activity of invertases and sucrose synthase in *Spiroplasma citri* or "stolbur" phytoplasma infected plants. *Bulletin of Insectology* **60**, 219–220.

MacLean AM, Sugio A, Makarova OV, Findlay KC, Grieve VM, Tóth R, Nicolaisen M, Hogenhout SA (2011) Phytoplasma effector SAP54 induces indeterminate leaf-like flower development in *Arabidopsis* plants. *Plant Physiology* **157**, 831–841.

MacLean AM, Orlovskis Z, Kowitwanich K, Zdziarska AM, Angenent GC, Immink RGH, Hogenhout SA (2014) Phytoplasma effector SAP54 hijacks plant reproduction by degrading MADS-box proteins and promotes insect colonization in a RAD23-dependent manner. *Plos Biology* **12**, e1001835.

Maejima K, Iwai R, Himeno M, Komatsu K, Kitazawa Y, Fujita N, Ishikawa K, Fukuoka M, Minato N, Yamaji Y, Oshima K, Namba S (2014) Recognition of floral homeotic MADS domain transcription factors by a phytoplasmal effector, phyllogen, induces phyllody. *Plant Journal* **78**, 541–554.

Maejima K, Kitazawa Y, Tomomitsu T, Yusa A, Neriya Y, Himeno M, Yamaji Y, Oshima K, Namba S (2015) Degradation of class E MADS-domain transcription factors in *Arabidopsis* by a phytoplasmal effector, phyllogen. *Plant Signal Behaviour* **10**, e1042635.

Mardi M, Karimi Farsad L, Gharechahi J, Salekdeh GH (2015) In-depth transcriptome sequencing of Mexican lime trees infected with '*Candidatus* Phytoplasma aurantifolia'. *Plos One* **10**, e0130425.

Margaria P, Palmano S (2011) Response of the *Vitis vinifera* L. cv. Nebbiolo proteome to "flavescence dorée" phytoplasma infection. *Proteomics* **11**, 212–224.

Margaria P, Abbà S, Palmano S (2013) Novel aspects of grapevine response to phytoplasma infection investigated by a proteomic and phospho-proteomic approach with data integration into functional networks. *BMC Genomics* **14**, 38.

Margaria P, Ferrandino A, Caciagli P, Kedrina O, Schubert A, Palmano S (2014) Metabolic and transcript analysis of the flavonoid pathway in diseased and recovered Nebbiolo and Barbera grapevines (*Vitis vinifera* L.) following infection by "flavescence dorée" phytoplasma. *Plant, Cell and Environment* **37**, 2183–2200.

Maust BE, Espadas F, Talavera C, Aguilar M, Santamaría JM, Oropeza C (2003) Changes in carbohydrate metabolism in coconut palms infected with the lethal yellowing phytoplasma. *Phytopathology* **93**, 976–981.

Minato N, Himeno M, Hoshi A, Maejima K, Komatsu K, Takebayashi Y, Kasahara H, Yusa A, Yamaji Y, Oshima K, Kamiya Y, Namba S (2015) The phytoplasmal virulence factor TENGU causes plant sterility by downregulating of the jasmonic acid and auxin pathways. *Science Reports* **4**, 7399.

Monavarfeshani A, Mirzaei M, Sarhadi E, Amirkhani A, Khayam Nekouei M, Haynes PA, Mardi M, Salekdeh GH (2013) Shotgun proteomic analysis of the Mexican lime tree infected with '*Candidatus* Phytoplasma aurantifolia'. *Journal of Proteome Research* **12**, 785–795.

Musetti R, Buxa S V, De Marco F, Loschi A, Polizzotto R, Kogel KH, van Bel AJ (2013a) Phytoplasma-triggered $Ca^{(2+)}$ influx is involved in sieve-tube blockage. *Molecular Plant-Microbe Interactions* **26**, 379–386.

Musetti R, Farhan K, De Marco F, Polizzotto R, Paolacci A, Ciaffi M, Ermacora P, Grisan S, Santi S, Osler R (2013b) Differentially-regulated defence genes in *Malus domestica* during phytoplasma infection and recovery. *European Journal of Plant Pathology* **136**, 13–19.

Nejat N, Cahill DM, Vadamalai G, Ziemann M, Rookes J, Naderali N (2015) Transcriptomics-based analysis using RNA-Seq of the coconut (*Cocos nucifera*) leaf in response to yellow decline phytoplasma infection. *Molecular Genetic Genomics* **290**, 1899–1910.

Nicolaisen M, Christensen NM (2007) Phytoplasma induced changes in gene expression in poinsettia. *Bulletin of Insectology* **60**, 215–216.

Niu S, Fan G, Deng M, Zhao Z, Xu E, Cao L (2016) Discovery of microRNAs and transcript targets related to witches' broom disease in *Paulownia fortunei* by high-throughput sequencing and degradome approach. *Molecular Genetic Genomics* **291**, 181–191.

Nozawa M, Miura S, Nei M (2012) Origins and evolution of microRNA genes in plant species. *Genome Biological Evolution* **4**, 230–239.

Oshima K, Ishii Y, Kakizawa S, Sugawara K, Neriya Y, Misako H, Minato N, Miura C, Shiraishi T, Yamaji Y, Namba S (2011) Dramatic transcriptional changes in an intracellular parasite enable host switching between plant and insect. *Plos One* **6**, e23242.

Paolacci AR, Catarcione G, Ederli L, Zadra C, Pasqualini S, Badiani M, Musetti R, Santi S, Ciaffi M (2017) Jasmonate-mediated defence responses, unlike salicylate-mediated responses, are involved in the recovery of grapevine from bois noir disease. *BMC Plant Biology* **17**, 118.

Prezelj N, Covington E, Roitsch T, Gruden K, Fragner L, Weckwerth W, Chersicola M, Vodopivec M, Dermastia M (2016a) Metabolic consequences of infection of grapevine (*Vitis vinifera* L.) cv. Modra Frankinja with "flavescence dorée" phytoplasma *Frontiers in Plant Science* **7**, 711.

Prezelj N, Fragener L, Weckwerth W, Dermastia M (2016b) Metabolome of grapevine leaf vein-enriched tissue infected with '*Candidatus* Phytoplasma solani'. *Mitteilungen Klosterneubg Rebe und Wein, Obs und Früchteverwertung* **66**, 74–78.

Punelli F, Al Hassan M, Fileccia V, Uva P, Pasquini G, Martinelli F (2016) A microarray analysis highlights the role of tetrapyrrole pathways in grapevine responses to "stolbur" phytoplasma, phloem virus infections and recovered status. *Physiological and Molecular Plant Pathology* **93**, 129–137.

Roitsch T (1999) Source-sink regulation by sugar and stress. *Current Opinions in Plant Biology* **2**, 198–206.

Roitsch T, González MC (2004) Function and regulation of plant invertases: sweet sensations. *Trends in Plant Science* **9**, 606–613.

Rojas CM, Senthil-Kumar M, Tzin V, Mysore KS (2014) Regulation of primary plant metabolism during plant-pathogen interactions and its contribution to plant defense. *Frontieres in Plant Science* **5**, 17.

Rotter A, Camps C, Lohse M, Kappel C, Pilati S, Hren M, Stitt M, Coutos-Thévenot P, Moser C, Usadel B, Delrot S, Gruden K (2009) Gene expression profiling in susceptible interaction of grapevine with its fungal pathogen *Eutypa lata*: extending MapMan ontology for grapevine. *BMC Plant Biology* **9**, 104.

Rotter A, Nikolić P, Turnšek N, Kogovšek P, Blejec A, Gruden K, Dermastia M (2018) Statistical modeling of long-term grapevine response to '*Candidatus* Phytoplasma solani' infection in the field. *European Journal of Plant Pathology* **150**, 653–668.

Rusjan D, Mikulic-Petkovsek M (2015) Phenolic responses in 1-year-old canes of *Vitis vinifera* cv. Chardonnay induced by grapevine yellows ("bois noir"). *Australian Journal of Grape and Wine Research* **21**, 123–134.

Rusjan D, Halbwirth H, Stich K, Gruden K, Fragner L, Weckwerth W, Chersicola M, Vodopivec M, Dermastia M (2012a) Biochemical response of grapevine variety Chardonnay (*Vitis vinifera* L.) to infection with grapevine yellows ("bois noir"). *European Journal of Plant Pathology* **134**, 231–237.

Rusjan D, Veberič R, Mikulič-Petkovšek M (2012b) The response of phenolic compounds in grapes of the variety Chardonnay (*Vitis vinifera* L.) to the infection by phytoplasma "bois noir". *European Journal of Plant Pathology* **133**, 965–974.

Santi S, De Marco F, Polizzotto R, Grisan S, Musetti R (2013a) Recovery from "stolbur" disease in grapevine involves changes in sugar transport and metabolism. *Frontieres in Plant Science* **4**, 171.

Santi S, Grisan S, Pierasco A, De Marco F, Musetti R (2013b) Laser microdissection of grapevine leaf phloem infected by "stolbur" reveals site-specific gene responses associated to sucrose transport and metabolism. *Plant, Cell and Environment* **36**, 343–355.

Snyman MC, Solofoharivelo M-C, Souza-Richards R, Stephan D, Murray S, Burger JT (2017) The use of high-throughput small RNA sequencing reveals differentially expressed microRNAs in response to aster yellows phytoplasma-infection in *Vitis vinifera* cv. Chardonnay. *Plos One* **12**, e0182629.

Sudisha J, Sharathchandra RG, Amruthesh KN, Kumar A, Shetty HS (2012) Pathogenesis related proteins in plant defense response. In: Plant Defence: Biological Control. Springer Netherlands, Dordrecht, The Netherlands, 379–403 pp.

Sugio A, Kingdom HN, MacLean AM, Grieve VM, Hogenhout SA (2011) Phytoplasma protein effector SAP11 enhances insect vector reproduction by manipulating plant development and defense hormone biosynthesis. *Proceedings of the National Academy of Science United States of America* **108**, E1254-E1263.

Sunkar R, Li Y-F, Jagadeeswaran G (2012) Functions of microRNAs in plant stress responses. *Trends in Plant Science* **17**, 196–203.

Thimm O, Bläsing O, Gibon Y, Nagel A, Meyer S, Krüger P, Selbig J, Müller LA, Rhee SY, Stitt M (2004) MAPMAN: a user-driven tool to display genomics data sets onto diagrams of metabolic pathways and other biological processes. *Plant Journal* **37**, 914–939.

Treutter D (2005) Significance of flavonoids in plant resistance and enhancement of their biosynthesis. *Plant Biology* **7**, 581–591.

Wang Z, Liu W, Fan G, Zhai X, Zhao Z, Dong Y, Deng M, Cao Y (2017) Quantitative proteome-level analysis of paulownia witches' broom disease with methyl methane sulfonate assistance reveals diverse metabolic changes during the infection and recovery processes. *PeerJ* **5**, e3495.

Wang H, Ye X, Li J, Tan B, Chen P, Cheng J, Wang W, Zheng X, Feng J (2018) Transcriptome profiling analysis revealed co-regulation of multiple pathways in jujube during infection by '*Candidatus* Phytoplasma ziziphi'. *Gene* **665**, 82–95.

Wei Z, Wang Z, Li X, Zhao Z, Deng M, Dong Y, Cao X, Fan G (2017) Comparative proteomic analysis of *Paulownia fortunei* response to phytoplasma infection with dimethyl sulfate treatment. *International Journal of Genomics* **2017**, 1–11.

Yang C-Y, Huang Y-H, Lin C-P, Lin YY, Hsu HC, Wang CN, Liu LY, Shen BN, Lin SS (2015a) MicroRNA396-targeted SHORT VEGETATIVE PHASE is required to repress flowering and is related to the development of abnormal flower symptoms by the phyllody symptoms1 effector. *Plant Physiology* **168**, 1702–1716.

Yang Y, Wen L, Zhu H (2015b) Unveiling the hidden function of long non-coding RNA by identifying its major partner-protein. *Cell Bioscience* **5**, 59.

Ye X, Wang H, Chen P, Bing Fu, Zhang M, Li J, Zheng X, Tan B, Feng J (2017) Combination of iTRAQ proteomics and RNA-seq transcriptomics reveals multiple levels of regulation in phytoplasma-infected *Ziziphus jujuba* Mill. *Horticultural Research* **4**, 17080.

Zabotin AI, Barysheva TS, Trofimova OI, Lozovaya V, Widholm J (2002) Regulation of callose metabolism in higher plant cells *in vitro*. *Russian Journal of Plant Physiology* **49**, 792–798.

Zhong B-X, Shen Y-W (2004) Accumulation of pathogenesis-related type-5 like proteins in phytoplasma-infected garland chrysanthemum *Chrysanthemum coronarium*. *Acta Biochimica Biophysica Sinica (Shanghai)* **36**, 773–779.

Chapter 4
Plant-Insect Host Switching Mechanism

Kenro Oshima, Kensaku Maejima, and Shigetou Namba

Abstract Phytoplasmas are intracellular parasites of both plants and insects and are spread among plants by insects. How phytoplasmas can adapt to two diverse environments has long been of considerable interest. Transcriptional analysis revealed that phytoplasmas dramatically alter the gene expression in response to their hosts and may use transporters, secreted proteins, and metabolic enzymes in a host-specific manner. Several plasmids cloned from phytoplasmas have been speculated to be involved in insect transmissibility. In addition to plasmids, immunodominant membrane proteins are thought to have some important function for host-parasite interactions. It has been suggested that the protein-protein interaction between immunodominant membrane proteins and insect microfilaments may be correlated with the phytoplasma-transmitting capability of leafhoppers. Further analyses about plant-insect host switching mechanism will contribute to the development of novel methods of pest control for phytoplasma diseases.

Keywords Insect transmissibility · Immunodominant membrane protein · Host switching mechanism

4.1 Introduction

Phytoplasmas are plant pathogenic bacteria of the class *Mollicutes* that were formerly called mycoplasma-like organisms. They are transmitted by insect vectors and infect over 700 plant species that include many economically important crops, fruit trees, and ornamental plants (Lee and Davis 1992; Hogenhout et al. 2008). Infected plants show a wide range of symptoms such as stunting, yellowing, witches' broom (proliferation of shoots), phyllody (formation of leaf-like tissues instead of

K. Oshima (✉)
Department of Clinical Plant Science, Faculty of Bioscience and Applied Chemistry, Hosei University, Tokyo, Japan

K. Maejima · S. Namba
Department of Agricultural and Environmental Biology, Graduate School of Agricultural and Life Sciences, The University of Tokyo, Tokyo, Japan

A. Bertaccini et al. (eds.), *Phytoplasmas: Plant Pathogenic Bacteria - III*,
https://doi.org/10.1007/978-981-13-9632-8_4

flowers), virescence (greening of floral organs), proliferation, purple top (reddening of leaves and stems), and phloem necrosis (Maejima et al. 2014). Phytoplasma infection is often fatal and causes devastating damages to agricultural production around the world (Bertaccini et al. 2014).

Despite the unique features of phytoplasmas have long been of interest to researchers, the difficulty of their culture in media have hindered their molecular and biological characterization. Whole genome sequence has been completed for several phytoplasma strains (Oshima et al. 2004, 2013; Bai et al. 2006; Kube et al. 2008; Tran-Nguyen et al. 2008; Andersen et al. 2013), and this has enabled to better understand some of the molecular mechanisms of virulence and host interaction of a range of phytoplasmas.

4.2 Host-Phytoplasma Interaction

Phytoplasmas, like other members of the class *Mollicutes*, are wall-less pleomorphic bacteria with cell size 200–800 nm (Lee et al. 2000). Their genome sizes are also small, ranging between 530 and 1,350 kbp (Marcone et al. 1999; Lee et al. 2000; Oshima et al. 2013). While the mycoplasmas studied reside intercellularly of their animal hosts, the phytoplasmas are intercellular both in plant hosts and insect vectors (Fig. 4.1). In plants, they were observed exclusively in the nutrient-rich

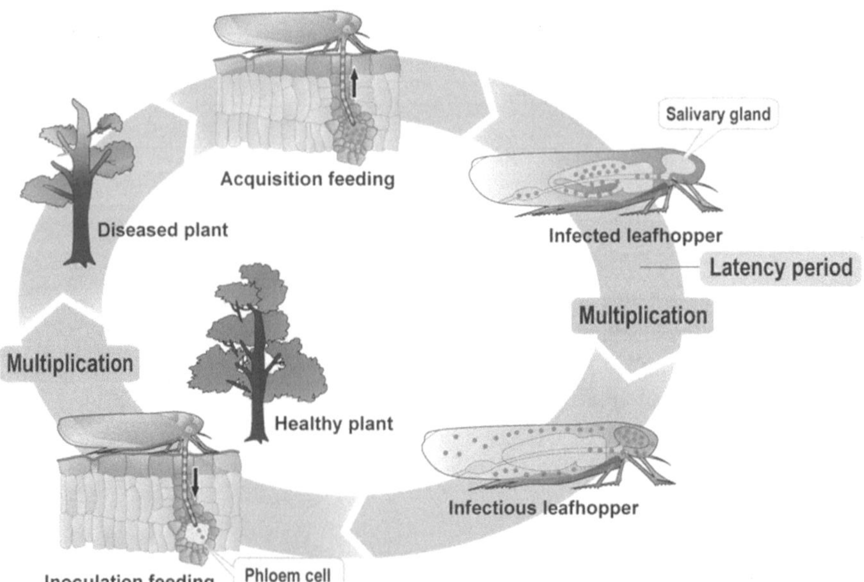

Fig. 4.1 Life cycle of phytoplasmas. Phytoplasma is shown as a dot. Phytoplasmas are unique biologically in that they can parasitize a diverse range of hosts, including plants and insects. Phytoplasmas can reside endocellularly within the plant phloem and feeding insects (leafhoppers) and are spread among plants by the insects

phloem tissue (the transport system for the photoassimilate) through which they spread to the whole plant.

Insect vectors of phytoplasmas are phloem feeders of the order Hemiptera, mostly leafhoppers, planthoppers, and psyllids (Weintraub and Beanland 2006). Insect vectors acquire phytoplasmas with phloem sap during feeding. In the insect, phytoplasmas cross the gut to hemolymph and finally get into salivary gland cells from which they are introduced back into the plants (Webb et al. 1999; Carraro et al. 2001). In plants, symptoms can develop at about 7 days after introduction of the phytoplasmas by the insect vector, but this can take much longer, depending on the phytoplasma and plant species (Carraro et al. 2001). Phytoplasmas have generally a broad plant host range, which depends on the plant feeding behaviour of their insect vectors. While several study showed that phytoplasmas are transmitted at low rates to plant embryos or insect progeny, they depend on both plants and insects for survival and dispersal in nature (Alma et al. 1997; Kawakita et al. 2000; Hanboonsong et al. 2002; Cordova et al. 2003; Khan et al. 2002; Botti and Bertaccini 2006; Tedeschi et al. 2006; Nipah et al. 2007; Calari et al. 2011; Satta et al. 2016, 2017, 2019; Mittelberger et al. 2017).

Beneficial symbiosis has also been observed for other leafhopper-phytoplasma interaction. The longevity and number of offspring of the aster leafhoppers can significantly increase when feeding on infected as compared with healthy plant hosts (Schultz 1973; Beanland et al. 2000). Phytoplasmas can also manipulate plants to become new hosts for leafhoppers that normally do not use these plants (Maramorosch 1958; Purcell 1988).

4.3 Possible Role of Plasmids in Insect Transmission

In the late 1990s, phytoplasma plasmids were cloned, sequenced, and characterized to investigate the relationships among plasmids, pathogenicity, and insect transmissibility (Goodwin et al. 1994; Kuboyama et al. 1998; Rekab et al. 1999; Liefting et al. 2004). In particular, all plasmids have been cloned and completely sequenced in '*Candidatus* Phytoplasma asteris' OY strain (Nishigawa et al. 2003). Each phytoplasma plasmid encodes a replication initiation protein (Rep) involved in rolling-circle replication (Nishigawa et al. 2002a, 2002b), as well as several other unknown proteins. The phytoplasmal Reps are similar not only to the Reps encoded by bacterial plasmids, but also to the Rep of geminivirus (Tsuchizaki et al. 2001) and circovirus (Oshima et al. 2001); therefore, it is of interest to analyze the evolutionary relationships between phytoplasmas and viruses. Although the functions of the other genes in the phytoplasma plasmids are unknown, a spontaneous OY mutant that was not insect transmissible (OY-NIM) (Oshima et al. 2001) was revealed to lack the ability to express ORF3 gene encoded in plasmids, suggesting a relationship between the plasmids and insect transmissibility (Nishigawa et al. 2002a, 2002b). The analysis of promoters on OY-M plasmid revealed that the promoter sequences of ORF3 were absent in the OY-NIM plasmid, implying that OY-NIM have lost its insect transmissibility by the lack of the functional orf3 promoter (Ishii et al. 2009a). Furthermore, during the maintenance of OY-NIM in plant tissue culture for 10 years, the plasmid was gradually lost from OY-NIM as follows: first, the promoter region of *orf3* was lost, followed by the loss of a large region including *orf3*, and finally the plasmid entirely disappeared (Ishii et al. 2009b). These results suggest that the plasmid is not essential for the phytoplasma viability in the plant host. Paradoxically, these results imply that the plasmid may be a key element for phytoplasmas to adapt to insect hosts. The relationships between plasmids and insect transmissibility have been also demonstrated in the research on spiroplasmas (Berho et al. 2006a, 2006b; Breton et al. 2010).

4.4 Protein Secretion System

Since phytoplasmas are cell wall-less and reside inside the host cells, their membrane proteins and secreted proteins function in the cytoplasm of the host plant or insect cell, and are predicted to have some important roles in host-parasite interactions and/or virulence. Phytoplasmas are known to have two secretion systems, the YidC system for the integration of membrane proteins and the Sec system for the integration and secretion of proteins into the host cell cytoplasm. The Sec protein translocation system is essential for cell viability in many bacteria (Economou 1999; Tjalsma et al. 2000). The *Escherichia coli* in the Sec system, is composed of at least 11 proteins and one RNA species, is the best characterized, and among these proteins, SecA, SecY, and SecE are essential for protein translocation and cell viability (Economou 1999), and protein translocation activity can be reconstituted *in vitro* with only these three proteins (Akimaru et al. 1991). In '*Ca.* P. asteris' OY strain (OY), the genes encoding SecA, SecY, and SecE have been identified (Kakizawa et al. 2001, 2004), and SecA expression has been confirmed in phytoplasma-infected plants (Kakizawa et al. 2001; Wei et al. 2004). These three genes have also been identified in other phytoplasma genomes (Bai et al. 2006; Kube et al. 2008; Tran-Nguyen et al. 2008), and *secY* genes have been cloned from a lot of phytoplasma strains (Lee et al. 2006). These results strongly suggest that a functional Sec system is common among phytoplasmas.

4.5 Membrane Proteins of Phytoplasmas

A subset of membrane proteins, usually referred to as immunodominant membrane proteins, constitutes a major portion of the total cellular membrane proteins in most phytoplasmas and it was thought to have some important function for the host-parasite interactions.

	Gene organization	'*Candidatus* Phytoplasma' species
Imp	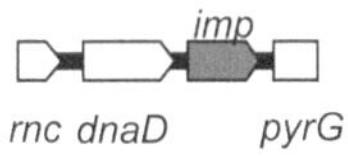 *imp* *rnc dnaD pyrG*	'*Candidatus* Phytoplasma asteris' '*Candidatus* Phytoplasma australiense' '*Candidatus* Phytoplasma solani
IdpA	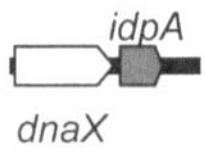 *idpA* *dnaX*	'*Candidatus* Phytoplasma pruni'
Amp	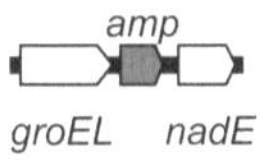 *amp* *groEL nadE*	'*Candidatus* Phytoplasma asteris' '*Candidatus* Phytoplasma australiense' '*Candidatus* Phytoplasma meliae'

Fig. 4.2 Gene organizations of three types of immunodominant membrane proteins of phytoplasma. Immunodominant membrane proteins could be classified into three types, Imp, IdpA, and Amp. Gene organizations around these three immunodominant membrane genes are different each other. Phytoplasma species that has each immunodominant membrane protein is indicated. Rnc, dsRNA-specific ribonuclease; DnaD, chromosome replication initiation protein; PyrG, CTP synthase; DnaX, DNA polymerase III; GroEL, HSP60 family chaperonin; NadE, NAD synthase

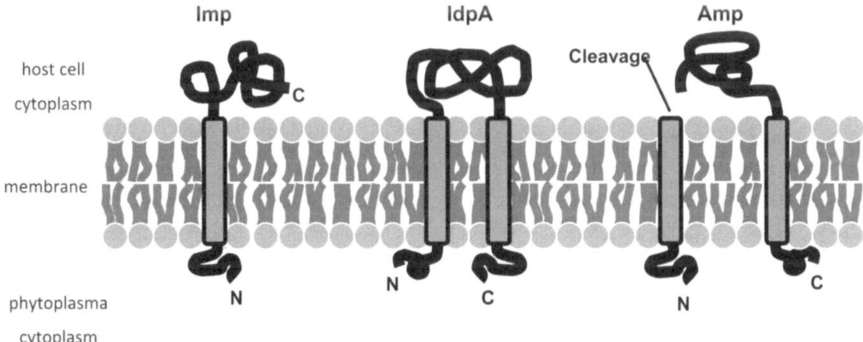

Fig. 4.3 Schematic representation of the hypothetical transmembrane structures of three types of immunodominant membrane proteins. N: N-terminus; C: C-terminus; gray rectangle: transmembrane regions. The N-terminus transmembrane region of Amp is cleaved during protein localization

Immunodominant membrane proteins were classified into three types (Fig. 4.2): immunodominant membrane protein (Imp) (Kakizawa et al. 2009), antigenic membrane protein (Amp) (Kakizawa et al. 2004; Arashida et al. 2008), and immunodominant membrane protein A (IdpA) (Neriya et al. 2011), which are not orthologues (Fig. 4.3). Amp, one of the immunodominant membrane proteins of phytoplasmas (Barbara et al. 2002), has been reported to be a substrate of the Sec system. Amp has a Sec signal sequence at its N-terminus, which is cleaved in OY phytoplasma (Kakizawa et al. 2004), suggesting that the phytoplasma Sec system utilizes recognition and cleavage of a signal sequence, as other bacterial Sec systems. This also suggests that signal sequence prediction programs, such as SignalP (Nielsen et al. 1997) or PSORT (Nakai and Kanehisa 1991), may be applicable to phytoplasma proteins and could be used to identify secretory proteins (Kakizawa et al. 2004).

Cloning amp genes from several phytoplasma strains showed that Amp proteins are under positive selection, and positively selected amino acids can be found in their central hydrophilic domain (Kakizawa et al. 2006). In the *Mollicutes* bacteria, it is reported that mycoplasmas and spiroplasmas have adhesin of binding to the host cell, and phytoplasmas also have an adhesion-like membrane protein named P38 (Neriya et al. 2014). Binding assays indicates that P38 interacts with the insect extract, but not with the plant extract. In addition, the interaction with the insect extract depends on its adhesion-like motif, suggesting that P38 may be a phytoplasma adhesin to the insect host (Neriya et al. 2014).

4.6 Host Specificity of Phytoplasmas

Since mollicute membrane proteins seem to have important roles in the attachment to the host cell and/or in virulence, Amp is a candidate involved in host-phytoplasma interactions. It has been reported that Amp determines insect specificity (Suzuki et al. 2006). Amp of OY phytoplasma was shown to interact with an insect microfilament. OY phytoplasma was localized to the microfilaments of the insect's intestinal tract, and Amp formed a complex with insect proteins involved in this microfilament both *in vitro* and *in vivo*. In addition, the formation of Amp-microfilament complexes was correlated with the phytoplasma-transmitting capability of leafhoppers, suggesting that the interaction plays a major role in determining the phytoplasma transmissibility. The interactions between surface proteins of microbes and host microfilaments have been often reported (Hayward and Koronakis 2002; Cossart et al. 2003); thus, these interactions might be a general system and are important for the successful bacterial infection. Recently, it was found that the Amp of chrysanthemum yellows (CY) phytoplasma also interacts with the insect microfilaments; in addition, it was revealed that ATP synthase alpha and beta subunits of CY phytoplasma could form a complex with Amp and microfilament of insects. Moreover, the formation of these interactions was clearly correlated with the phytoplasma-transmitting capability of leafhoppers (Galetto et al. 2011), suggesting that the formation of Amp and insect microfilament complexes would be common in phytoplasmas and ATP synthase subunits were included in the complexes.

Sequences of Amp were shown to be highly variable (Barbara et al. 2002; Morton et al. 2003; Kakizawa et al. 2004) and under positive selection (Kakizawa et al. 2006), suggesting that mutations in Amp sequences improve the fitness of phytoplasmas, just like many positively selected proteins that were previously reported (Hughes and Nei 1988; Nielsen and Yang 1998; Bishop et al. 2000; Jiggins et al. 2002; Urwin et al. 2002; Andrews and Gojobori 2004). Taken together, sequence variation of Amp might change its target (insect microfilaments) and offer host switching behaviour to phytoplasmas.

The infection of phytoplasmas to insect hosts is composed of several different steps (Nielson 1979). In these steps, there are two major barriers to pass for a successful phytoplasma colonization: the insect intestine and the salivary gland (Markham and Townsend 1979; Purcell et al. 1981; Hogenhout et al. 2008). Although the role of the Amp-microfilament complex remains unclear, it is likely that it may be necessary for the passage through these barriers, especially those in the insect intestine. Also in the case of spiroplasmas, an immunodominant membrane protein (spiralin) plays an important role in the infection to insects. It has been reported that a defective mutant for the spiralin was less effective in its transmissibility (Duret et al. 2003) and that the spiralin binds to the glycoproteins of the insect vector (Killiny et al. 2005). Although molecular details of the function of spiralin are still unknown, it would be a candidate as key component to determine the insect vector specificity. Although no similarity was detected between Amp and spiralin, the immunodominant membrane proteins of phytoplasmas and spiroplasmas would commonly play an important role in their transmission by insect vectors.

4.7 Transcriptional Changes During Host Switching Between Plants and Insects

Since phytoplasmas are intracellular parasites of both plants and insects (Christensen et al. 2005), their ability to adapt to two diverse environments is of considerable interest. Microarray analysis of '*Candidatus* Phytoplasma asteris' strain OY-M revealed that expression of approximately 33% of the genes changes during the host switching between plant and insect, suggesting that the phytoplasmas dramatically alter the gene expression in response to their hosts (Oshima et al. 2011) and may use transporters, secreted proteins, and metabolic enzymes in a host-specific manner.

The genes encoded in the potential mobile unit (PMU) of '*Ca.* P. asteris' strain AY-WB are more expressed in insects than in plants (Toruno et al. 2010), most likely due to an increased production of the extrachromosomal circular type of PMU during the insect infection (Toruno et al. 2010). Differential gene expression between plant and insect hosts has been also reported in '*Ca.* P. asteris' strain OY-M, in which TENGU is over expressed in plant than in insect hosts (Hoshi et al. 2009). As phytoplasmas reside within the host cell, secreted proteins are thought to play crucial roles in the interplay between pathogen and host cell; therefore, the expression of virulence factors might be strictly regulated.

Two types of sigma factors, RpoD and FliA, are encoded in the phytoplasma genomes (Oshima et al. 2004; Bai et al. 2006; Kube et al. 2008; Tran-Nguyen et al. 2008; Andersen et al. 2013). By an *in vitro* transcription assay, it has been demonstrated that RpoD regulates several housekeeping, virulence, and host-phytoplasma interaction genes (Miura et al. 2015). Since hundreds of candidate promoter sequences regulated by RpoD were predicted in the OY-M genome, the transcription start sites have been identified from the genome of OY-M phytoplasma by using RNA-Seq technology (Nijo et al. 2017). From these data, two promoter consensus sequences located upstream the transcription start sites have been predicted. While one was almost identical to the RpoD-dependent consensus promoter sequence, the other was an unidentified novel motif, which might be recognized by another transcription factor such as FliA (Nijo et al. 2017). Further analysis of transcription factors will provide insight into the host-switching mechanism of phytoplasmas.

4.8 Outlook

Since phytoplasmas are unique microorganisms, which can reside intracellularly in different kingdoms, plants and insects, it is of interest how they can adapt to two diverse intracellular environments. Phytoplasmas have long remained the most poorly characterized plant pathogens during a quarter of a century since their discovery. However, recent studies revealed plasmids, membrane proteins, and expression proteins involved in insect transmission. Phytoplasma-related diseases are expected to increase because of global warming and climate change that is advantageous to the cold-sensitive phytoplasma vectors. Therefore, the attempt to develop the pest control of phytoplasmas will become more important in the future. Further analysis of host-switching mechanism of phytoplasmas at the molecular level will provide clues to control these economically important and biologically attractive microorganisms.

References

Akimaru J, Matsuyama S, Tokuda H, Mizushima S (1991) Reconstitution of a protein translocation system containing purified SecY, SecE, and SecA from *Escherichia coli*. *Proceedings of the National Academy of Sciences United States of America* **88**, 6545–6549.

Alma A, Bosco D, Danielli A, Bertaccini A, Vibio M, Arzone A (1997) Identification of phytoplasmas in eggs, nymphs and adults of *Scaphoideus titanus* Ball reared on healthy plants. *Insect Molecular Biology* **6**, 115–121.

Andersen MT, Liefting LW, Havukkala I, Beever RE (2013) Comparison of complete genome sequence of two closely related isolates of '*Candidatus* Phytoplasma australiense' reveals genome plasticity. *BMC Genomics* **14**, 529.

Andrews TD, Gojobori T (2004) Strong positive selection and recombination drive the antigenic variation of the PilE protein of the human pathogen *Neisseria meningitidis*. *Genetics* **166**, 25–32.

Arashida R, Kakizawa S, Ishii Y, Hoshi A, Jung HY, Kagiwada S, Yamaji Y, Oshima K, Namba S (2008). Cloning and characterization of the antigenic membrane protein (Amp) gene and *in situ* detection of Amp from malformed flowers infected with Japanese hydrangea phyllody phytoplasma. *Phytopathology* **98**, 769–775.

Bai X, Zhang J, Ewing A, Miller SA, Jancso Radek A, Shevchenko DV, Tsukerman K, Walunas T, Lapidus A, Campbell JW, Hogenhout SA (2006) Living with genome instability: the adaptation of phytoplasmas to diverse environments of their insect and plant hosts. *Journal of Bacteriology* **188**, 3682–3696.

Barbara DJ, Morton A, Clark MF, Davies DL (2002) Immunodominant membrane proteins from two phytoplasmas in the aster yellows clade (chlorante aster yellows and clover phyllody) are highly divergent in the major hydrophilic region. *Microbiology* **148**, 157–167.

Beanland L, Hoy CW, Miller SA, Nault LR (2000) Influence of aster yellows phytoplasma on the fitness of aster leafhopper (Homoptera: Cicadellidae). *Annals of the Entomological Society of America* **93**, 271–276.

Berho N, Duret S, Danet J-L, Renaudin J (2006a) Plasmid pSci6 from *Spiroplasma citri* GII-3 confers insect transmissibility to the non-transmissible strain *S. citri. Microbiology* **152**, 2703–2716.

Berho N, Duret S, Renaudin J (2006b) Absence of plasmids encoding adhesion-related proteins in non-insect-transmissible strains of *Spiroplasma citri. Microbiology* **152**, 873–886.

Bertaccini A, Duduk B, Paltrinieri S, Contaldo N (2014) Phytoplasmas and phytoplasma diseases: a severe threat to agriculture. *American Journal of Plant Sciences* **5**, 1763.

Bishop JG, Dean AM, Mitchell-Olds T (2000) Rapid evolution in plant chitinases: molecular targets of selection in plant-pathogen coevolution. *Proceedings of the National Academy of Sciences United States of America* **97**, 5322–5327.

Botti S, Bertaccini A (2006) Phytoplasma infection through seed transmission: further observations. 16th International Organization of Mycoplasmology Conference, Cambridge, United Kingdom, 9–14 July, 76, 113.

Breton M, Duret S, Beven L, Dubrana MP, Renaudin J (2010) I-*Sce*I-mediated plasmid deletion and intra-molecular recombination in *Spiroplasma citri*. *Journal of Microbiological Methods* **84**, 216–222.

Calari A, Paltrinieri S, Contaldo N, Sakalieva D, Mori N, Duduk B, Bertaccini A (2011) Molecular evidence of phytoplasmas in winter oilseed rape, tomato and corn seedlings. *Bulletin of Insectology* **64**(Supplement), S157–S158.

Carraro L, Loi N, Ermacora P (2001) Transmission characteristics of the European stone fruit yellows phytoplasma and its vector *Cacopsylla pruni. European Journal of Plant Pathology* **107**, 695–700.

Christensen NM, Axelsen KB, Nicolaisen M, Schulz A (2005) Phytoplasmas and their interactions with hosts. *Trends in Plant Science* **10**, 526–535.

Cordova I, Jones P, Harrison NA, Oropeza C (2003) *In situ* PCR detection of phytoplasma DNA in embryos from coconut palms with lethal yellowing disease. *Molecular Plant Pathology* **4**, 99–108.

Cossart P, Pizarro-Cerda J, Lecuit M (2003) Invasion of mammalian cells by *Listeria monocytogenes*: functional mimicry to subvert cellular functions. *Trends in Cell Biology* **13**, 23–31.

Duret S, Berho N, Danet J-L, Garnier M, Renaudin J (2003) Spiralin is not essential for helicity, motility, or pathogenicity but is required for efficient transmission of *Spiroplasma citri* by its leafhopper vector *Circulifer haematoceps*. *Applied Environmental Microbiology* **69**, 6225–6234.

Economou A (1999) Following the leader: bacterial protein export through the Sec pathway. *Trends in Microbiology* **7**, 315–320.

Galetto L, Bosco D, Balestrini R, Genre A, Fletcher J, Marzachì C (2011) The major antigenic membrane protein of '*Candidatus* Phytoplasma asteris' selectively interacts with ATP synthase and actin of leafhopper vectors. *Plos One* **6**, e22571.

Goodwin PH, Xue BG, Kuske CR, Sears MK (1994) Amplification of plasmid DNA to detect plant-pathogenic mycoplasmalike organisms. *Annals of Applied Biology* **124**, 27–36.

Hanboonsong Y, Choosai C, Panyim S, Damak D (2002) Transovarial transmission of sugarcane white leaf phytoplasma in the insect vector *Matsumuratettix hiroglyphicus* (Matsumura). *Insect Molecular Biology* **11**, 97–103.

Hayward RD, Koronakis V (2002) Direct modulation of the host cell cytoskeleton by *Salmonella* actin-binding proteins. *Trends in Cell Biology* **12**, 15–20.

Hogenhout SA, Oshima K, Ammar E-D, Kakizawa S, Kingdom HN, Namba S (2008) Phytoplasmas: bacteria that manipulate plants and insects. *Molecular Plant Pathology* **9**, 403–423.

Hoshi A, Oshima K, Kakizawa S, Ishii Y, Ozeki J, Hashimoto M, Komatsu K, Kagiwada S, Yamaji Y, Namba S (2009) A unique virulence factor for proliferation and dwarfism in plants identified from a phytopathogenic bacterium. *Proceedings of the National Academy of Sciences United States of America* **106**, 6416–6421.

Hughes AL, Nei M (1988) Pattern of nucleotide substitution at major histocompatibility complex class I loci reveals overdominant selection. *Nature* **335**, 167–170.

Ishii Y, Kakizawa S, Hoshi A, Maejima K, Kagiwada S, Yamaji Y, Oshima K, Namba S (2009a) In the non-insect-transmissible line of onion yellows phytoplasma (OY-NIM), the plasmid-encoded transmembrane protein ORF3 lacks the major promoter region. *Microbiology* **155**, 2058–2067.

Ishii Y, Oshima K, Kakizawa S, Hoshi A, Maejima K, Kagiwada S, Yamaji Y, Namba S (2009b) Process of reductive evolution during 10 years in plasmids of a non-insect-transmissible phytoplasma. *Gene* **446**, 51–57.

Jiggins FM, Hurst GD, Yang Z (2002) Host-symbiont conflicts: positive selection on an outer membrane protein of parasitic but not mutualistic *Rickettsiaceae*. *Molecular Biology Evolution* **19**, 1341–1349.

Kakizawa S, Oshima K, Kuboyama T, Nishigawa H, Jung H-Y, Sawayanagi T, Tsuchizaki T, Miyata S, Ugaki M, Namba S (2001) Cloning and expression analysis of phytoplasma protein translocation genes. *Molecular Plant-Microbe Interactions* **14**, 1043–1050.

Kakizawa S, Oshima K, Nishigawa H, Jung H-Y, Wei W, Suzuki S, Tanaka M, Miyata S, Ugaki M, Namba S (2004) Secretion of immunodominant membrane protein from onion yellows phytoplasma through the Sec protein-translocation system in *Escherichia coli*. *Microbiology* **150**, 135–142.

Kakizawa S, Oshima K, Namba S (2006) Diversity and functional importance of phytoplasma membrane proteins. *Trends in Microbiology* **14**, 254–256.

Kakizawa S, Oshima K, Ishii Y, Hoshi A, Maejima K, Jung H-Y, Yamaji Y, Namba S (2009) Cloning of immunodominant membrane protein genes of phytoplasmas and their *in planta* expression. *FEMS Microbiology Letters* **293**, 92–101.

Kawakita H, Saiki T, Wei W, Mitsuhashi W, Watanabe K, Sato M (2000) Identification of mulberry dwarf phytoplasmas in the genital organs and eggs of leafhopper *Hishimonoides sellatiformis*. *Phytopathology* **90**, 909–914.

Khan AJ, Botti S, Al-Subhi AM, Gundersen-Rindal DE, Bertaccini AF (2002) Molecular identification of a new phytoplasma associated with alfalfa witches' broom in Oman. *Phytopathology* **92**, 1038–1047.

Killiny N, Castroviejo M, Saillard C (2005) *Spiroplasma citri* spiralin acts *in vitro* as a lectin binding to glycoproteins from its insect vector *Circulifer haematoceps*. *Phytopathology* **95**, 541–548.

Kube M, Schneider B, Kuhl H, Dandekar T, Heitmann K, Migdoll AM, Reinhardt R, Seemüller E (2008) The linear chromosome of the plant-pathogenic mycoplasma '*Candidatus* Phytoplasma mali'. *BMC Genomics* **9**, 306.

Kuboyama T, Huang CC, Lu X, Sawayanagi T, Kanazawa T, Kagami T, Matsuda I, Tsuchizaki T, Namba S (1998) A plasmid isolated from phytopathogenic onion yellows phytoplasma and its heterogeneity in the pathogenic phytoplasma mutant. *Molecular Plant-Microbe Interactions* **11**, 1031–1037.

Lee I-M, Davis RE (1992) Mycoplasmas which infect plant and insects. In: Mycoplasmas: Molecular Biology and Pathogenesis. Eds Maniloff J, McElhansey RN, Finch LR, Baseman JB. ASM Press, Washington DC, United States of America, 379–390 pp.

Lee I-M, Davis RE, Gundersen-Rindal DE (2000) Phytoplasma: phytopathogenic mollicutes. *Annual Revue of Microbiology* **54**, 221–255.

Lee I-M, Zhao Y, Bottner KD (2006) *SecY* gene sequence analysis for finer differentiation of diverse strains in the aster yellows phytoplasma group. *Molecular and Cellular Probes* **20**, 87–91.

Liefting LW, Shaw ME, Kirkpatrick BC (2004) Sequence analysis of two plasmids from the phytoplasma beet leafhopper-transmitted virescence agent. *Microbiology* **150**, 1809–1817.

Maejima K, Oshima K, Namba S (2014) Exploring the phytoplasmas, plant pathogenic bacteria. Japanese *Journal of Phytopathology* **80**, 124–133.

Maramorosch K (1958) Beneficial effect of virus diseased plants on nonvector insects. *Tijdschr Plantenziekten* **63**, 383–391.

Marcone C, Neimark H, Ragozzino A, Lauer U, Seemüller E (1999) Chromosome sizes of phytoplasmas composing major phylogenetic groups and subgroups. *Phytopathology* **89**, 805–810.

Markham PJ, Townsend R (1979) Experimental vectors of spiroplasmas. In: Leafhopper Vectors and Plant Disease Agents. Eds Maramorosch K, Harris KF. Academic Press, New York, United States of America, 413–445 pp.

Mittelberger C, Obkircher L, Oettl S, Oppedisano T, Pedrazzoli F, Panassiti B, Kerschbamer C, Anfora G, Janik K (2017) The insect vector vertically transmits the bacterium '*Candidatus* Phytoplasma mali' to its progeny. *Plant Pathology* **66**, 1015–1021.

Miura C, Komatsu K, Maejima K, Nijo T, Kitazawa Y, Tomomitsu T, Yusa A, Himeno M, Oshima K, Namba S (2015) Functional characterization of the principal sigma factor RpoD of phytoplasmas via an *in vitro* transcription assay. *Scientific Reports* **5**, 11893.

Morton A, Davies DL, Blomquist CL, Barbara DJ (2003) Characterization of homologues of the apple proliferation immunodominant membrane protein gene from three related phytoplasmas. *Molecular Plant Pathology* **4**, 109–114.

Nakai K. Kanehisa M (1991) Expert system for predicting protein localization sites in Gram-negative bacteria. *Proteins* **11**, 95–110.

Neriya Y, Sugawara K, Maejima K, Hashimoto M, Komatsu K, Minato N, Miura C, Kakizawa S, Yamaji Y, Oshima K, Namba S (2011) Cloning, expression analysis, and sequence diversity of genes encoding two different immunodominant membrane proteins in poinsettia branch-inducing phytoplasma (PoiBI). *FEMS Microbiology Letters* **324**, 38–47.

Neriya Y, Maejima K, Nijo T, Tomomitsu T, Yusa A, Himeno M, Netsu O, Hamamoto H, Oshima K, Namba S (2014) Onion yellow phytoplasma P38 protein plays a role in adhesion to the hosts. *FEMS Microbiology Letters* **361**, 115–122.

Nielsen H, Engelbrecht J, Brunak S, and von Heijne G (1997) Identification of prokaryotic and eukaryotic signal peptides and prediction of their cleavage sites. *Protein Engeneering* **10**, 1–6.

Nielsen R, Yang Z (1998) Likelihood models for detecting positively selected amino acid sites and applications to the HIV-1 envelope gene. *Genetics* **148**, 929–936.

Nielson MW (1979) Taxonomic relationships of leafhopper vectors of plant pathogens. In: Leafhopper Vectors and Plant Disease Agents. Eds Maramorosch K, Harris KF. Academic Press, New York, United States of America, 3–27 pp.

Nijo T, Neriya Y, Koinuma H, Iwabuchi N, Kitazawa Y, Tanno K, Okano Y, Maejima K, Yamaji Y, Oshima K, Namba S (2017) Genome-wide analysis of the transcription start sites and promoter motifs of phytoplasmas. *DNA Cell Biology* **36**, 1081–1092.

Nipah J, Jones P, Dickinson M (2007) Detection of lethal yellowing phytoplasma in embryos from coconut palms infected with Cape St Paul wilt disease in Ghana. *Plant Pathology* **56**, 777–784.

Nishigawa H, Oshima K, Kakizawa S, Jung H-Y, Kuboyama T, Miyata S, Ugaki M, Namba S. (2002a) Evidence of intermolecular recombination between extrachromosomal DNAs in phytoplasma: a trigger for the biological diversity of phytoplasma? *Microbiology* **148**, 1389–1396.

Nishigawa H, Oshima K, Kakizawa S, Jung H-Y, Kuboyama T, Miyata S, Ugaki M, Namba S (2002b) A plasmid from a non-insect-transmissible line of a phytoplasma lacks two open reading frames that exist in the plasmid from the wild-type line. *Gene* **298**, 195–201.

Nishigawa H, Oshima K, Miyata S, Ugaki M, Namba S (2003) Complete set of extrachromosomal DNAs from three pathogenic lines of onion yellows phytoplasma and use of PCR to differentiate each line. *Journal of General Plant Pathology* **69**, 194–198.

Oshima K, Kakizawa S, Nishigawa H, Kuboyama T, Miyata S, Ugaki M, Namba S (2001) A plasmid of phytoplasma encodes a unique replication protein having both plasmid- and virus-like domains: clue to viral ancestry or result of virus/plasmid recombination? *Virology* **285**, 270–277.

Oshima K, Kakizawa S, Nishigawa H, Jung H-Y, Wei W, Suzuki S, Arashida R, Nakata D, Miyata S, Ugaki M, Namba S (2004) Reductive evolution suggested from the complete genome sequence of a plant-pathogenic phytoplasma. *Nature Genetics* **36**, 27–29.

Oshima K, Ishii Y, Kakizawa S, Sugawara K, Neriya Y, Himeno M, Minato N, Miura C, Shiraishi T, Yamaji Y, Namba S (2011) Dramatic transcriptional changes in an intracellular parasite enable host switching between plant and insect. *Plos One* **6**, e23242.

Oshima K, Maejima K, Namba S (2013) Genomic and evolutionary aspects of phytoplasmas. *Frontiers in Microbiology* **4**, 230.

Purcell AH, Richardson JR, Finlay AH (1981) Multiplication of X-disease agent in a nonvector leafhopper *Macrosteles fascifrons. Annals of Applied Biology* **99**, 283–289.

Purcell AH (1988) Increased survival of *Dalbulus maidis*, a specialist on maize, on non-host plants infected with mollicute plant-pathogens. *Entomologia Experimentalis et Applicata* **46**, 187–196.

Rekab D, Carraro L, Schneider B, Seemüller E., Chen J., Chang CJ, Locci R, Firrao G (1999) Geminivirus-related extrachromosomal DNAs of the X-clade phytoplasmas share high sequence similarity. *Microbiology* **145**, 1453–1459.

Satta E, Contaldo N, Paltrinieri S, Bertaccini A (2016) Biological and molecular proofs of phytoplasma seed transmission in corn. 21th Congress of the IOM, Brisbane, Australia, July 3–7, 61: 65–66.

Satta E, Nanni IM, Contaldo N, Collina M, Poveda JB, Ramírez AS, Bertaccini A (2017) General phytoplasma detection by a q-PCR method using mycoplasma primers. *Molecular and Cellular Probes* **35**, 1–7.

Satta E, Paltrinieri S, Bertaccini A (2019) Phytoplasma transmission by seed. In: Phytoplasmas: Plant Pathogenic Bacteria-II Transmission and Management of Phytoplasma Associated Diseases. Chapter 6. Eds Bertaccini A, Weintraub P, Rao GP, Mori N. Springer, Singapore, 131–147 pp.

Schultz GA (1973) Plant resistance to aster yellows. *Proceedings North Central Branch Entomology Society of America* **28**, 93–99.

Suzuki S, Oshima K, Kakizawa S, Arashida R, Jung HY, Yamaji Y, Nishigawa H, Ugaki M, Namba S (2006) Interaction between the membrane protein of a pathogen and insect microfilament complex determines insect-vector specificity. *Proceedings of the National Academy of Sciences United States of America* **103**, 4252–4257.

Tedeschi R, Ferrato V, Rossi J, Alma A (2006) Possible phytoplasma transovarial transmission in the psyllids *Cacopsylla melanoneura* and *Cacopsylla pruni. Plant Pathology* **55**, 18–24.

Tjalsma H, Bolhuis A, Jongbloed JD, Bron S, van Dijl JM (2000) Signal peptide-dependent protein transport in *Bacillus subtilis*: a genome-based survey of the secretome. *Microbiology and Molecular Biology Revue* **64**, 515–547.

Toruno TY, Seruga-Music M, Simi S, Nicolaisen M, Hogenhout SA (2010) Phytoplasma PMU1 exists as linear chromosomal and circular extrachromosomal elements and has enhanced expression in insect vectors compared with plant hosts. *Molecular Microbiology* **77**, 1406–1415.

Tran-Nguyen LT, Kube M, Schneider B, Reinhardt R, Gibb KS (2008) Comparative genome analysis of '*Candidatus* Phytoplasma australiense' (subgroup tuf-Australia I; rp-A) and '*Ca.* Phytoplasma asteris' strains OY-M and AY-WB. *Journal of Bacteriology* **190**, 3979–3991.

Tsuchizaki T, Nishigawa H, Miyata S, Sawayanagi T, Oshima K, Matsuda I, Kuboyama T, Komoto A, Namba S (2001) *In planta* expression of a protein encoded by the extrachromosomal DNA of a phytoplasma and related to geminivirus replication proteins. *Microbiology* **147**, 507–513.

Urwin R, Holmes EC, Fox AJ, Derrick JP, Maiden MC (2002) Phylogenetic evidence for frequent positive selection and recombination in the meningococcal surface antigen PorB. *Molecular Biology Evolution* **19**, 1686–1694.

Webb D, Bonfiglioli R, Carraro L, Osler R, Symons R (1999) Oligonucleotides as hybridization probes to localize phytoplasmas in host plants and insect vectors. *Phytopathology* **89**, 894–901.

Wei W, Kakizawa S, Jung HY, Suzuki S, Tanaka M, Nishigawa H, Miyata S, Oshima K, Ugaki M, Hibi T, Namba S (2004) An antibody against the SecA membrane protein of one phytoplasma reacts with those of phylogenetically different phytoplasmas. *Phytopathology* **94**, 683–686.

Weintraub PG, Beanland L (2006) Insect vectors of phytoplasmas. *Annual Revue of Entomology* **51**, 91–111.

Chapter 5
Diversity and Functional Importance of Phytoplasma Membrane Proteins

Marika Rossi, Ivana Samarzija, Martina Šeruga-Musić, and Luciana Galetto

Abstract Phytoplasmas are wall-less prokaryotes, associated with hundreds of severe crop diseases worldwide. They inhabit phloem elements and are transmitted by insects of a few hemipteran families. In absence of a cell wall, phytoplasma membrane proteins are in direct contact with insect and plant hosts. The most abundant on their pathogen cells are the immunodominant membrane proteins (IDPs), which have been characterized in several phytoplasma strains, but also other membrane protein families, like variable membrane proteins (Vmps), adhesins, AAA + ATPases, and several transporters, are worth to mention for interactions with hosts and pathogen adaptation to different environments and as molecular markers useful for strain genotyping. Indeed, many of these membrane proteins are under positive selection pressure, and therefore highly variable among the different phytoplasma strains. A review and a schematic summary of the salient literature on phytoplasma membrane proteins are presented. The focuses were the variability of their gene sequences and the molecular characterization of pathogen strains and their functional roles in mediating interactions with plants and insects and in the perception and adaptation to different environments.

Keywords Immunodominant membrane proteins · Variable membrane proteins · Positive selection pressure · Insect transmission specificity · Pathogen-plant interaction · Host switching

---

M. Rossi · L. Galetto (✉)
Istituto per la Protezione Sostenibile delle Piante, CNR, Torino, Italy

I. Samarzija · M. Šeruga-Musić
Department of Biology, Faculty of Science, University of Zagreb, Zagreb, Croatia

A. Bertaccini et al. (eds.), *Phytoplasmas: Plant Pathogenic Bacteria - III*,
https://doi.org/10.1007/978-981-13-9632-8_5

5.1 Overview of Phytoplasma Membrane Proteins

Phytoplasmas are pathogenic bacterial *Mollicutes* inhabiting the plant phloem. They are associated with severe plant diseases in many different species worldwide, with huge economic impacts on several important crops (Maejima et al. 2014; Marcone 2014). These diseases are transmitted by different Hemiptera insect species, mainly leafhoppers, planthoppers and psyllids (Tomkins et al. 2018). Transmission modality is mediated by several factors in a compatible phytoplasma/vector combination (Bosco and D'Amelio 2010). These bacteria possess small genomes, due to a progressive gene reduction, and strongly depend on hosts to survive (Kube et al. 2012; Oshima et al. 2013). Phytoplasmas lack cell wall, and therefore, their membrane is in direct contact with the host cell. Proteins with domains exposed on the outer part of the phytoplasma membrane are not only subjected to selective pressure, and therefore highly variable, but also good candidate in mediating interactions with plants and insect vectors. Moreover, phytoplasmas show a dual life style (in plants and in insects): proteins on bacterial membrane also play crucial roles in perceiving this host switching and in their adaptation to different host environments (Toruño et al. 2010; Oshima et al. 2011; Ishii et al. 2013; Pacifico et al. 2015). Phytoplasmas are possibly able to establish interactions with their hosts, both plants and insects, mainly by secreting effectors responsible for modulation of plant metabolism and producing typical symptoms (Sugio et al. 2011; Tomkins et al. 2018), or buy crossing the gut epithelium and colonizing salivary glands of the insect vectors in order to be transmitted and spread (Rashidi et al. 2015). The present chapter aims at exploring the multifaceted role of phytoplasma membrane proteins, drawing attention to their sequence diversity in relation with evolutionary pressure, their functional interactions with plants and insects, and their involvement in pathogen adaptation to different host environments.

5.2 Immunodominant Membrane Proteins

Three main types highly abundant and immunodominant membrane proteins (IDP) have been identified in phytoplasmas: Amp (antigenic membrane protein), IdpA (immunodominant membrane protein A), and Imp (immunodominant membrane protein) (Kakizawa et al. 2006a). Among all the phytoplasma membrane proteins, the class of immunodominant proteins is certainly the most studied, as far as it concerns either the structure characterization or their functional role in host interaction. The three IDP types identified in phytoplasmas share a central hydrophilic region, with a predicted external orientation, and one or two transmembrane domains, although the overall organization of the different domains is not conserved, as only one N-terminal transmembrane domain is present in Imp, while IdpA and Amp have two of them at the N- and C-terminals, respectively (Fig. 5.1) (Konnerth et al. 2016). The Amp N-terminal transmembrane domain is cleaved and it has no further role as anchor of Amp to the membrane upon insertion of the protein in the plasma membrane (Barbara et al. 2002).

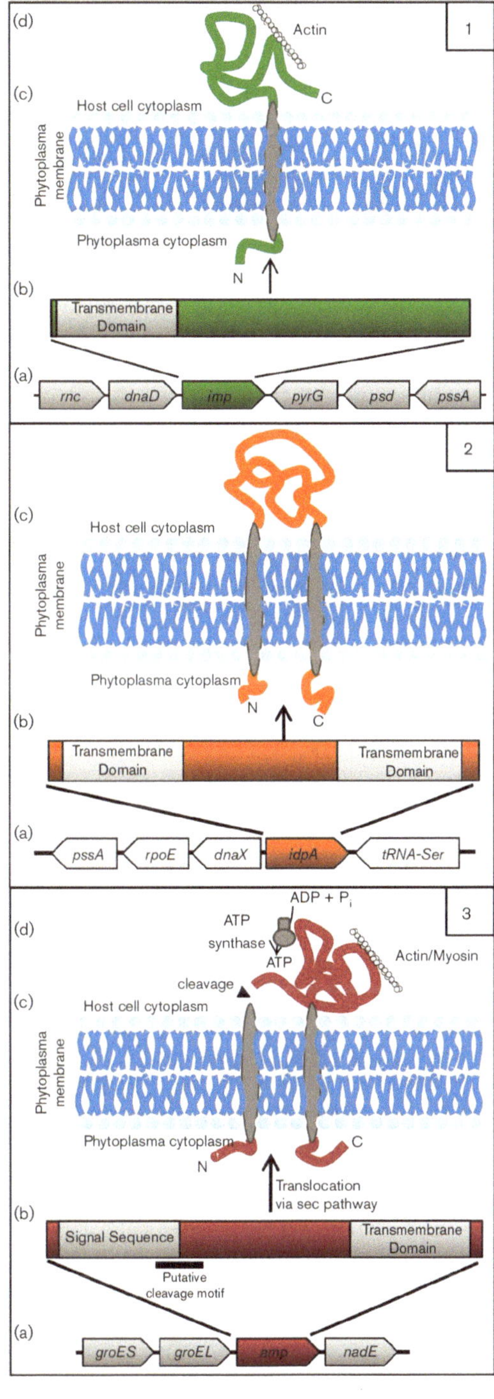

Fig. 5.1 Gene organization and transmembrane structure of phytoplasma IDPs Imp (1), IdpA (2) and Amp (3). (**a**) Localization in phytoplasma genome. (**b**) Putative translation product. (**c**) Putative protein structure. (**d**) Host interaction partners. (Konnerth et al. 2016, Microbiology Reproduced with permission of Microbiology Society)

These three proteins are not all present in phytoplasma genomes sequences so far: the Amp protein is found in '*Candidatus* Phytoplasma solani', namely, Stamp (Fabre et al. 2011), and in different strains of '*Ca.* P. asteris,' among which onion yellows (OY) (Kakizawa et al. 2004), aster yellows, clover phyllody (Barbara et al. 2002), and chrysanthemum yellows (CY) (Galetto et al. 2008), were the most deeply characterized; the IdpA is present in '*Ca.* P. pruni' associated with Western X-disease, WX (Blomquist et al. 2001), and poinsettia witches' broom phytoplasma (Neriya et al. 2011); and the Imp is found in '*Ca.* P. asteris' (Kakizawa et al. 2009), '*Ca.* P. mali' (Berg et al. 1999), '*Ca.* P. prunorum,' and '*Ca.* P. pyri' (Morton et al. 2003), "flavescence dorée" phytoplasma (Abbà et al. 2014) and in different strains of '*Ca.* P. aurantifolia' associated with sweet potato witches' broom (Yu et al. 1998) and with witches' broom disease of lime (WBDL) (Siampour et al. 2013). Several phytoplasmas have genes encoding two types of these proteins, but one of them is generally predominantly expressed. This is the case of OY and WX phytoplasmas, which express on membrane both their major antigenic protein, Amp and IdpA, respectively, as well as Imp (Kakizawa et al. 2006b, 2009). Imp is conserved in many phytoplasmas, and it might thus represent the ancestral IDP (Kakizawa et al. 2009). Moreover, the quantitative ratio between the different immunodominant proteins may vary in the different phytoplasma strains (Neriya et al. 2011). These proteins are highly variable even among closely related strains (Galetto et al. 2008, 2014; Neriya et al. 2011; Siampour et al. 2013) and this variability is higher compared to other metabolic genes or non-coding sequences. Indeed, evolution under strong positive selection has been demonstrated for Amp and Imp (Kakizawa et al. 2006b, 2009; Fabre et al. 2011; Siampour et al. 2013) supporting a possible role of both proteins in determining the phytoplasma fitness.

Given the abundant expression of IDPs on the phytoplasma membranes, there were several attempts to produce antibodies against these antigens. The antibodies were suggested to be used in phytoplasma studies or, ultimately, in suppression of phytoplasma infections in a so-called plantibody-mediated resistance approach (Le Gall et al. 1998; Malembic-Maher et al. 2005; Shahryari et al. 2010, 2013). Although these studies are still only at the beginnings, the results serve as a proof of principle for the mentioned strategy.

5.3 Other Phytoplasma Membrane Proteins

Although the immunodominant and antigenic membrane proteins are the most abundant proteins on the phytoplasma cell surface, there is a number of other studied membrane proteins that are also important or potentially crucial for phytoplasma biology. These include variable membrane proteins (Vmp), proteins of the Sec translocation system and different uptake transporter systems, the recently identified P38 phytoplasma adhesin, AAA$^+$ ATPases and hflB/ftsH proteases, and a subset of membrane proteins encoded on potential mobile units (PMUs) or plasmids.

Phytoplasma *vmp* genes encode membrane proteins such as VmpA, VmpB, and Vmp1 that display similarity to variable surface proteins of animal mycoplasmas. In general, Vmp protein possesses N-terminal signal sequence, large repeated sequences, and a C-terminal transmembrane domain. This organization is shared with some bacterial surface proteins-like adhesins (Renaudin et al. 2015). Among phytoplasmas, the *vmp* genes were first studied in '*Ca*. P. solani' strain PO (Cimerman et al. 2009) and in the "flavescence dorée" phytoplasma (Renaudin et al. 2015). In '*Ca*. P. solani,' it was reported that the *vmp1* gene encoding a protein of 557 amino acids in the *uvrA-vmp1-lig* region might have multiple sequence homologues in the genome (Cimerman et al. 2009). Recently, the "flavescence dorée" phytoplasma VmpA protein has been showed to act as an adhesin, able to interact with insect cells (Arricau-Bouvery et al. 2018). The P38 protein from '*Ca*. P. asteris' is another example of phytoplasma adhesion; it possesses a conserved amino acid sequence known as the *Mollicutes* adhesion motif (MAM) characterized by three β-sheets which has been shown to be important for the interaction between P38 protein and hosts factors (Neriya et al. 2014).

The Sec protein translocation system, essential for the bacterial cell viability (Kakizawa et al. 2001), is a multiprotein cytoplasmic membrane system whose translocase component is constituted from SecY, SecE, and SecA proteins, which are crucial for protein translocation, and SecG protein, which is not essential. SecYEG is a heterotrimeric complex constituting a transmembrane pore. SecA, which is an ATPase, brings secretory proteins to the pore and initiates the protein secretion. Kakizawa and colleagues identified members of the Sec protein translocation system in phytoplasmas (Kakizawa et al. 2001) and suggested that the Sec system is mediating the secretion of onion yellows (OY) phytoplasma Amp immunodominant membrane protein (Kakizawa et al. 2004). SecY, very often used in phytoplasma molecular epidemiology studies, is ubiquitous in living organisms (Kakizawa et al. 2001), and its homologues have been identified in all the three domains of life (*Bacteria*, Archaea, and Eukaryotes).

The AAA^+ ATPases and HflB (synonym FtsH) proteases are other integral membrane phytoplasma proteins in depth studied in '*Ca*. P. mali' (Seemüller et al. 2011, 2013, 2018). These belong to the AAA^+ superfamily, all sharing an ATPase domain and exerting their activity through the energy-dependent remodeling or translocation of macromolecules (Langklotz et al. 2012). Different '*Ca*. P. mali' strains, characterized by a wide range of symptom severity from mild to severe, are clustered in relation with virulence when polymorphisms of these genes are taken into account, suggesting a role for AAA^+ ATPases and HflB/FtsH proteases in phytoplasma virulence and/or suppression of virulence (Seemüller et al. 2013).

Given the parasitic life of phytoplasmas and their dependence on the nutrients that are provided by the host organism, they have a relatively large number of genes encoding membrane transporter systems for the uptake of nutrients. These include *dppABCDF* genes that code the ABC transporter systems responsible for dipeptides/oligopeptides uptake; *artM* the polar amino acids; *potABCD* the spermidine/putrescine; *metNQI* the D-methionine; *znuABC* the zinc/manganese; *citS* the malate/citrate; and, finally, *malEFGK* coding the sugar or the phosphorylated hexose uptake system (Marcone 2014). These proteins are not ubiquitously expressed on the cell surface like the IDPs, but their role in phytoplasma biology is essential. Interestingly, the expression levels of several transporter genes may be differently regulated depending on the phytoplasma life stage, if in plant or in insect hosts. Among others, genes for the mechanosensitive channel, multidrug efflux pumps, and cobalt transporter are upregulated in the plant host, while the zinc, sugar, and oligopeptide transporters are upregulated in the insect host (Oshima et al. 2011).

Frequently, phytoplasma membrane proteins are encoded by PMUs or extrachromosomal elements. Toruño et al. (2010) have shown that '*Ca.* P. asteris' AY-WB PMU1 encodes a series of membrane-targeted proteins. Moreover, depending on the host, PMU1 was shuffling between chromosomal linear (L-PMU1) and extrachromosomal circular (C-PMU1) form. The C-PMU1 copy number and gene expression levels were considerably higher in insects as compared to plants, suggesting that PMU-1 encoding membrane proteins might have an important role in the adaptation of this phytoplasma to the environment or, alternatively, it might provide platform for genetic material exchange with other phytoplasmas (Toruño et al. 2010). A further example of membrane protein encoded in extrachromosomal elements include '*Ca.* P. asteris' ORF3 protein whose coding sequence is located on two '*Ca.* P. asteris' plasmids. ORF3 is expressed by plasmids harbored by strains that are transmissible by an insect vector, but absent in the non-insect-transmissible lines (Ishii et al. 2009).

5.4 Diversity and Sequence Analysis of Membrane Proteins

Amino acid sequence analysis revealed that no similarity is found among the different types of immunodominant membrane proteins previously described. Moreover, the IDPs present in phytoplasmas belonging to the same ribosomal group, although similar in the protein structure, vary significantly on the level of amino acid sequence (Kakizawa et al. 2006a; Konnerth et al. 2016). It was also shown that even phylogenetically close phytoplasmas express different types of major membrane proteins (Neriya et al. 2011).

First investigations into IDP and some other membrane protein gene sequences have showed that they are highly divergent (Barbara et al. 2002), and much more variable than those of the housekeeping genes. Furthermore, for some of them (*e.g.* amp and imp of different '*Ca.* P. asteris' strains; stamp of "stolbur" phytoplasmas; imp of WBDL and of almond witches' broom-associated phytoplasmas), the ratio of nonsynonymous to synonymous substitutions is relatively high, indicating a strong positive selection pressure acting on their sequences (Kakizawa et al. 2006b, 2009; Fabre et al. 2011; Siampour et al. 2013; Quaglino et al. 2015). As positive selection on a protein is an indication of the role in the evolution of an organism, the presence of a high proportion of IDPs is most probably very important for the biology of some phytoplasmas. This is possibly because of their crucial and direct interaction with the host cell factors, especially in the case of insect cells (Kakizawa et al. 2006b, 2009; Fabre et al. 2011). The reason why this interaction is a subject of positive selection acting on the mentioned proteins is currently unknown, but it is presumed that the adaptation to different hosts could have an important role (Fabre et al. 2011).

Due to the high variability of nucleotide sequences, the IDPs and some other membrane protein genes, like *vmp* and *secY*, are frequently used for assessing phytoplasma diversity by genotyping. Examples of IDPs used for differentiating phytoplasma strains include *imp* gene, in '*Ca.* P. aurantifolia' (Siampour et al. 2013; Al-Ghaithi et al. 2018). The *vmp1* gene is strikingly variable in size, with repeated domains of 80 to 84 amino acids in the largest forms of the gene, as well as the variations in 11 nucleotide repeats leading to gene disruption in some of the strains. Hence, this gene is widely used as a diversity and discriminant marker for molecular epidemiology of '*Ca.* P. solani' (Cimerman et al. 2009). Frequently, it is combined with *stamp* gene for the molecular typing of "bois noir" phytoplasma strains (Kostadinovska et al. 2014; Atanasova et al. 2015; Pierro et al. 2018). In the "flavescence dorée"-related phytoplasmas, the variability of the Vmp protein was shown to correlate with the ability to be transmitted by different insect vector (Malembic-Maher et al. 2017). Phylogenetic analysis of different phytoplasma strains across several 16Sr subgroups based on the *secY* gene has greater discriminating power compared to the traditionally used 16S rRNA gene analysis (Lee et al. 2006, 2010). Currently, the *secY* sequence analysis is widely used in multi-locus sequence typing (MLST) of different phytoplasmas (Foissac et al. 2013). In particular, together with *vmp1*, *stamp*, and *tuf* gene analysis, it provides a discriminatory tool for '*Ca.* P. solani' diversity (Murolo and Romanazzi 2015; Plavec et al. 2015). Moreover, together with *imp* and two other non-membrane genes, it was used in genotyping temperate fruit tree phytoplasmas (Danet et al. 2007, 2011), and in combination with the *amp* gene, it was also used in MLST analysis of '*Ca.* P. asteris' strains (Seruga-Musić et al. 2014). "Flavescence dorée" (FD) phytoplasma strains are also distinguished based on the *secY* and several other non-membrane gene sequences. Arnaud et al. (2007) demonstrated the existence of three FD strain clusters in vineyards of southern France and Northern Italy based on MLST encompassing also *secY* gene. Furthermore, this MLST tool was successfully used for genotyping and tracking FD phytoplasmas in other European countries (Plavec et al. 2019). Along with *secY* gene, *secA* was also shown to display the ability to differentiate phytoplasmas and serves as informative alternative molecular marker for phytoplasma detection and differentiation (Hodgetts et al. 2008; Dickinson and Hodgetts 2013).

5.5 Functional Role of Membrane Proteins in the Interaction with Hosts

Table 5.1 summarizes the functional roles assigned to phytoplasma membrane proteins involved in their interaction with hosts, according to literature.

Insect Vectors The propagative transmission modality is mediated by specific interactions between the pathogen and the insect vector, as different barriers in the

Table 5.1 Functional roles of phytoplasma membrane proteins in interactions with insect and plant hosts

Membrane protein	'*Candidatus* Phytoplasma'	Strain	Host	Description	References
Amp	'*Ca.* P. asteris'	OY	Insects	OY Amp specifically interacts *in vitro* with actin and myosin and co-localizes with actin microfilament in the gut tissues of infected insects	Suzuki et al. (2006)
		CY	Insects	CY Amp specifically interacts *in vitro* with actin and ATP synthase and subunits. Vector ATP synthase surface is exposed in the gut epithelium and salivary gland plasma membrane	Galetto et al. (2011)
			Insects	CY Amp is required *in vivo* to cross the midgut epithelium and to colonize the insect salivary glands	Rashidi et al. (2015)
"Stolbur" amp (stamp)	'*Ca.* P. solani'	"Stolbur"	Plants (plantibody-mediated resistance)	Plants producing single-chain variable fragment (scFv) antibody against stamp in the apoplast showed partial resistance to "stolbur," only following infection by grafting and not by insect transmission	Le Gall et al. (1998) and Malembic-Maher et al. (2005)
Imp	'*Ca.* P. aurantifolia'	WBDL	Insects	WBDL Imp specifically *in vitro* interacts with extracts of insect vectors only	Siampour et al. (2011, 2013)
	"Flavescence dorée"	FD	Insects	FD Imp specifically *in vitro* interacts with extracts of insect vectors only	Galetto et al. (2014)
	'*Ca.* P. mali'	PM19	Plants	Extracellular domain of '*Ca.* P. mali' Imp *in vitro* binds the plant actin (both globular and filamentous form) without affecting the cytoskeleton structure. This interaction is likely to mediate the phytoplasma movement within phloem elements	Boonrod et al. (2012)
	'*Ca.* P. aurantifolia'	WBDL	Plants (plantibody-mediated resistance)	Functional scFv antibodies against WBDL Imp, transiently expressed in *N. benthamiana*, accumulated either in the cytosol or in the apoplast	Shahryari et al. (2010, 2013)

(continued)

Table 5.1 (continued)

Membrane protein	'*Candidatus* Phytoplasma'	Strain	Host	Description	References
VmpA	"Flavescence dorée"	FD	Insects	FD VmpA specifically binds to the perimicrovillar membrane of insect midgut and promotes the adhesion to vector cell cultures	Renaudin et al. (2015) and Arricau-Bouvery et al. (2018)
Adhesin P38 (PAM289)	'*Ca.* P. asteris'	OY	Insects	The strong *in vitro* interaction between OY P38 and insect vector extracts is mediated by P38 MAM motif. Its expression is higher in insects than in plants. Its regulation is under control of both RpoD and FliA sigma initiation transcription factors	Neriya et al. (2014), Oshima et al. (2011), and Ishii et al. (2013)
			Plants	OY P38 is expressed in plants, weekly interacts *in vitro* with plant extracts, but P38 MAM-like motif seems not involved	
AAA$^+$ ATPase AP460	'*Ca.* P. mali'	Mild to severe strains	Plants	Phylogenetic analysis of AAA$^+$ ATPase genes shows that different strains cluster distantly according to strain virulence, suggesting a role in determining symptom severity. In AP460, region 1 (N-terminus) has 14 highly conserved substitutions related with symptom attenuation, and region 2 (C-terminus) includes two sites involved in virulence modulation that may affect the ATPase activity	Seemüller et al. (2011, 2013, 2018)

(continued)

Table 5.1 (continued)

Membrane protein	'*Candidatus* Phytoplasma'	Strain	Host	Description	References
MscL	'*Ca*. P. asteris'	OY	Plants	Mechanosensitive channel L (MscL) is significantly more expressed in plant than in insect hosts, treatment of inoculated plants with a MscL inhibitor induces a slower OY multiplication rate	Oshima et al. (2011)
		CY	Plants	MscL is one of the most abundant transcripts both in plants and insects, and its expression in infected plants is significantly higher than in two insect species	Pacifico et al. (2015)
ORF3	'*Ca*. P. asteris'	OY	Insects	The lack of a plasmid encoded ORF3 expression in an OY non-transmissible line could explain the loss of transmissibility	Ishii et al. (2009)

insect body have to be overcome by the bacteria, and once an insect becomes infective, generally it remains infected for the rest of its life (Weintraub and Beanland 2006). The interactions between phytoplasmas and their insect vectors seem to be very specific, but some phytoplasmas could be vectored by different insect species and the same insect species may transmit different phytoplasmas (Lee et al. 1998, 2000).

In vitro studies have been developed to support the presence of interaction between phytoplasma IDPs and insect proteins. The role of '*Ca*. P. asteris' Amp was investigated in different strains (Suzuki et al. 2006; Galetto et al. 2011; Rashidi et al. 2015). In the case of onion yellows (OY) phytoplasma, the formation of a complex between Amp and insect actin microfilaments has been correlated with the phytoplasma transmission capability of leafhoppers, suggesting that the interaction between Amp and insect microfilaments plays a role in phytoplasma transmissibility. Heavy and light chains of insect myosin *in vitro* also interacted with OY Amp, confirming that host the cytoskeleton is involved in mediating the relationship with bacterial cells (Suzuki et al. 2006). Far Western blot and affinity chromatography indicated a specific *in vitro* interaction between the chrysanthemum yellows (CY) phytoplasma Amp and several proteins of the insect vector, among which the actin and the subunits alpha and beta of ATP synthase were identified by MS/MS spectrometry. Moreover, immunofluorescence microscopy also proved that the insect vector ATP synthase beta subunit is present on the microvillar external surface of the gut epithelial cells as well as on the plasma membrane of the salivary gland cells, both crucial organs for the infection process (Galetto et al. 2011). Intriguingly, phytoplasmas lack the functional genes for the production of ATP (F1F0 type ATP synthase) (Oshima et al. 2004), and Amp/ATP synthase interaction may combine specific recognition and host energy exploitation during the colonization process. Five P-type ATPases, presumably involved in generating the membrane potential in phytoplasmas, were identified in the OY genome and three of them were differently regulated in plants or in insect hosts, suggesting that the pathogen may use these P-type ATPases for its adaptation to the two different environments (Oshima et al. 2011). The boost of insect energy metabolism induced in *Euscelidius variegatus* upon '*Ca.* P. asteris' infection (Galetto et al. 2018) could fit within the framework of the exploitation by phytoplasma of the host potential energy. The *in vivo* role of CY Amp in mediating the interactions with the insect vector at two main barriers was also indirectly demonstrated: Amp was required by the phytoplasma to cross the insect gut epithelium and to colonize the salivary glands (Rashidi et al. 2015). Indeed, significant decreases of acquisition and inoculation efficiency were recorded after masking the native Amp on phytoplasma surface with a specific antibody anti-Amp. Confocal microscopy observations showed the co-localization at the gut level of the antibody anti-Amp and the phytoplasma cells, demonstrating that the specific recognition occurred, and supporting the hypothesis that masking native Amp during early stages of gut epithelium colonization, determined a significant reduction in phytoplasma acquisition. Moreover, the internalization assay, optimized to observe the early stages of insect salivary gland penetration by phytoplasmas (Galetto et al. 2019), showed a lower number of phytoplasma cells in the insect salivary glands when the native CY Amp was blocked by its specific antibody. Indeed, the masking of Amp on phytoplasma surface impaired gland entry of pathogen cells, which were rarely detected in glands, only as isolated cells and never packed within intracellular vesicles, as they were instead observed in the absence of anti-Amp antibody (Rashidi et al. 2015).

Some preliminary *in vitro* experiments indicated a role for Imp in the interaction with the insect vectors. A recombinant partial Imp fusion protein of the '*Ca.* P. aurantifolia' strain WBDL was expressed (Siampour et al. 2013) and used in interaction studies with insect proteins (Siampour et al. 2011). Recombinant WBDL Imp specifically interacted in far Western assays with extracts from phytoplasma insect vectors, but not with proteins of non-vector species, suggesting also for Imp a role in determining the insect transmissibility. These results were confirmed by affinity chromatography experiments, despite that the identification by MS/MS spectrometry of isolated vector proteins was not conclusive (Siampour et al. 2011). Similar preliminary results were obtained with "flavescence dorée" Imp, which was expressed as partial recombinant fusion protein (Galetto et al. 2014) and used for *in vitro* interaction studies with vector and non-vector insect species. Also, in this case, Imp seemed to play a promising role in mediating insect transmission specificity (V. Trivellone, personal communication).

The Vmps are generally less abundant than IDPs (Abbà et al. 2014), but also target of a strong selective pressure (Cimerman et al. 2009). In order to determine the role of FD VmpA in the interaction with hosts, a mutant line of *Spiroplasma citri* lacking ScARP adhesins was successfully engineered to express the FD VmpA on plasma membrane (Renaudin et al. 2015). This mutant *S. citri* surface expressing FD VmpA as well as latex beads coated with recombinant VmpA specifically adhered to insect vector cell cultures, demonstrating the role of adhesin for this protein. Moreover, FD VmpA, which was expressed in both midguts and salivary glands of the infected insects, mediated the specific retention of coated latex beads mainly in the perimicrovillar membrane on apical epithelium of anterior midguts dissected from insects artificially fed with beads, confirming its crucial function in the interaction with insect vectors (Arricau-Bouvery et al. 2018).

Adhesins are other membrane proteins involved in pathogen adherence to host cell. In the *Mollicutes*, these proteins are involved in infection processes (Fleury et al. 2002), and a conserved amino acid sequence, namely, *Mollicutes* adhesin motif (MAM), was identified in adhesins P40 of *Mycoplasma agalactiae*, P89 of *S. citri*, and P38 of different phytoplasma strains belonging to the '*Ca.* P. asteris' species (Neriya et al. 2014). Adhesin P38 (PAM289) of OY was fully characterized, predicted to be anchored in plasma membrane, and shown to be expressed in infected plants and insects, as recognized by specific antibody. Interestingly, OY P38 fusion protein *in vitro* interacted strongly with insect vector extracts but weekly with plant protein extracts. Moreover, binding assays with different recombinant P38 constructs (with wild type or mutated/deleted MAM-like motif) demonstrated that the insect interaction was dependent and mediated by this domain (Neriya et al. 2014). Interestingly, P38 mRNA (PAM289) was shown to be more expressed in the insect host than in the plants both by microarray and qRT-PCR approaches (Oshima et al. 2011), consistently with its potential role in interaction with the insect vectors. The expression of gene encoding P38 (PAM289) was controlled by both the sigma initiation transcription factors identified in OY, namely RpoD and FliA. The former is mainly associated with control of phytoplasma gene regulation during insect infection, the latter during phytoplasma infection of both insect and plant hosts (Ishii et al. 2013). Being P38 controlled by both initiation transcription factors, a role in plant infection cannot be excluded for this protein.

Putative transmembrane proteins encoded on phytoplasma plasmids might also have a role in interaction with the insect host. Indeed, a transmembrane protein (ORF3) of '*Ca.* P. asteris' OY strain is preferentially expressed in infected insect vectors, and the lack of such protein expression in the mutant plasmid of a non-insect-transmissible line of the same phytoplasma may explain the loss of transmissibility (Ishii et al. 2009).

Interaction with Plant Hosts As mentioned, the absence of cell wall exposes the phytoplasma membrane proteins to direct interaction with host proteins. These interactions are supposed to be crucial for the efficiency of infections in terms of phytoplasma spread and virulence. Several studies showed the role of the membrane proteins in phytoplasma-insect interaction, but very little is known about the importance of these proteins in terms of phytoplasma-plant interaction. In the case of plants, phytoplasma membrane proteins do not interact with the components of the cellular surface of the host, but directly with the cytosolic component of the sieve tube cells.

It was demonstrated that Imp of '*Ca.* P. mali,' associated with the apple proliferation (AP) disease binds the plant actin both *in vivo* and *in vitro*. Transient expression in *N. benthamiana* of two AP Imp constructs both fused with a GFP reporter domain, one full length (Imp-GFP) and one devoid of the transmembrane domain (Imp$^{\Delta TM}$-GFP), showed that both the chimeric phytoplasma proteins localized along with the actin filaments. The actin-binding site was determined in the C-terminus portion of Imp, downstream the transmembrane domain, in the protein part predicted to be exposed out of the pathogen cell. Moreover, *in vitro* assays demonstrated that Imp$^{\Delta TM}$ bound both globular and filamentous actin without affecting the actin polymerization. Transgenic plant expressing Imp-GFP and Imp$^{\Delta TM}$-GFP did not show any phytoplasma-associated phenotype, and because the interaction between Imp and actin did not induce remodeling of the plant cytoskeleton, the actin-binding capability of Imp seems to be more involved in phytoplasma survival or movement within phloem elements rather than in pathogenesis (Boonrod et al. 2012).

Better characterization of phytoplasma membrane proteins might be a solid starting point to cope with phytoplasma-associated diseases. An attempt in this direction was the expression of the single-chain variable fragment (scFv) antibody directed against the Amp of "stolbur" (stamp) phytoplasma in transgenic tobacco plants (Le Gall et al. 1998; Malembic-Maher et al. 2005). Plants producing scFv in the apoplast showed only partial resistance to "stolbur" phytoplasma and only when the plants were infected by grafting. No resistance was induced when the phytoplasma was transmitted by insects. The authors suggested that the low protection efficiency of scFv in this transgenic line was probably due to the difficulties to reach the cytosol of phloem cells when the protein is produced in the apoplastic space. Unfortunately, the attempt to produce scFv in the cytosol of phloem cells failed since there was no accumulation of the scFv protein in the symplast (Malembic-Maher et al. 2005). More efforts are necessary to improve the stability and the target specificity of antibodies produced in plants and even more to develop strategies to succeed in fight against phytoplasmas. In this direction, a scFv antibody directed against Imp protein of the '*Ca.* P. aurantifolia' WBDL strain was expressed in single plant cell compartments (Shahryari et al. 2010, 2013). Three different constructs, featuring constitutive or phloem-specific promoters in cassettes with or without secretion signals, were transiently expressed in *N. benthamiana*, and the accumulation of functional scFv antibodies was demonstrated either in the cytosol or in the apoplast. Final goal will be the stable expression of this scFv antibody against Imp in lime trees, to suppress WBDL phytoplasma exploiting a plantibody-mediated resistance (Shahryari et al. 2013).

Phytoplasma often induces severe symptoms in leaves, flowers, and fruits, such as stem proliferations, virescence, and phyllody associated with the phytoplasma secretion of effector proteins in the cytoplasm of plant cells (Bai et al. 2009; Hoshi et al. 2009; MacLean et al. 2011; Sugio et al. 2011; Tomkins et al. 2018). Besides these specific symptoms, there are several indirect effects due to the impairment of the sieve tube function such as low productivity, reduced vigor, and plant decline. Symptom severity depends on many factors related to the plant genotype and the sanitary conditions, but also the phytoplasma genotype has a crucial role in determining virulence and pathogenicity. It is the case of the different strains belonging to the '*Ca.* P. mali' that induce a wide range of symptom severity of the AP disease.

Phylogenetic analysis of genes encoding AAA⁺ ATPases and HflB (synonym FtsH) proteases from different '*Ca.* P. mali' strains showed that they cluster distantly according to strain virulence and suggested a role for these proteins in determining the symptom severity (Seemüller et al. 2011, 2013). In bacteria, energy-dependent proteolysis is mediated by AAA⁺ proteases (ATPases associated with various cellular activities), which combine an ATP-driven unfolding activity with the degradation of proteins. Among the AAA⁺ proteases of *Escherichia coli*, FtsH is the only one membrane-anchored and essential for cell viability; it mediates regulation of lipopolysaccharide biosynthesis and plays a role in regulation of the heat shock response (Langklotz et al. 2012). Intriguingly, in four of the fully annotated phytoplasma genomes, 6 to 24 genes were indicated as *hflB* or *ftsH*, and despite some copies turned out to be truncated genes, active transcription was demonstrated for 10 of them in '*Ca* P. mali' strains. Moreover, the enzyme carrying C-tail of several of these '*Ca.* P. mali' proteins was predicted to be extracellularly exposed, oriented toward the phloem sieve tubes, contrary to the inner cytosolic orientation of this domain observed in other bacteria. These aspects suggest the importance of these genes for the phytoplasma biology, even if it is not clear whether they are involved in pathogenicity (Seemüller et al. 2013). The best studied AAA⁺ ATPase is the AP460: two main regions have been characterized in this protein. Region 1, at the N-terminus of the protein sequence, has 14 highly conserved substitutions related with symptom attenuation. The C-terminus part of the protein includes the region 2 which is more variable and shows two highly conserved sites involved in the virulence modulation. Moreover, all the '*Ca.* P. mali' attenuated strains share a group of substitutions next to the ATPase domain that seems to affect ATPase activity (Seemüller et al. 2018).

The mechanisms by which phytoplasma cells can adjust to extremes of temperature, pH, and osmotic pressure are important for their survival in the natural environment, especially as obligate cell-wall parasites, able to survive in such different kinds of environments like insect and plant hosts. The mechanosensitive MscL channel of the well-studied model *E. coli* is involved in this respect in the perception of mechanical stretching of the membrane and plays a fundamental role in protecting the cell from decreases in the environment osmolarity (Sukharev et al. 1997). Interestingly, the expression of OY *mscL* gene is significantly higher when the phytoplasma infects a plant host compared with an insect host, suggesting that the MscL channel may play an important role in the adaptation to the osmotic pressures of the plant-cell environment. This hypothesis was confirmed by the treatment of OY-infected plants with gadolinium chloride, MscL channel inhibitor: phytoplasma multiplication rate was significantly slower in treated plant than in the control, but at four weeks post infection OY loads were similar. Pathogen growth was suppressed by the treatment with the MscL channel inhibitor at the early stage of phytoplasma infection, indicating that the MscL channel may play a crucial role in survival within a plant host cell (Oshima et al. 2011). The importance of MscL channel in the phytoplasma viability was also confirmed by a transcriptional study of CY genes encoding proteins involved in different metabolic pathways. Among 14 analyzed phytoplasma genes, transcripts of the *mscL* gene were the third most abundant ones after those of *amp* and of the effector *tengu* both in plant and in two insect vector species, and the level of *mscL* gene expression was significantly higher in plants than in insects (Pacifico et al. 2015).

5.6 Conclusions and Perspectives

Several lines of evidence indicate that phytoplasma membrane proteins are the key elements in fine network of molecular interaction with hosts. IDPs are surely the most abundant and the most studied, but other protein classes seemed to be crucial in relationship with host cells, such as the cases of Vmps, adhesins, AAA^+ ATPases, and several transporters, which in particular are able to modulate their expression in different kinds of environments and allow pathogen adaptation to plant phloem or insect haemocoel. Both Amp and Imp of phylogenetically distant phytoplasmas interact *in vitro* with plant and insect vector actin. This is a clear signal of the importance of such interaction with a cytoskeleton protein, possibly to guarantee the phytoplasma movement within the host cells. In many cases, the phytoplasma membrane proteins are under a positive selective pressure: this confirms their prominent role in host interaction and environment adaptation and also determines a high variability of their gene sequences, providing molecular tools very useful for genotyping and characterization of different strains. Genes encoding IDPs, Vmps, AAA^+ ATPases, transporters, and in particular the Sec protein translocation system are, together with the 16S ribosomal gene, the most powerful molecular markers used for genetic characterization of these highly divergent pathogens. Phytoplasma membrane proteins are surely a promising starting point for pathogen-host interaction studies, in order to understand the molecular bases of infection mechanisms, to project defense strategies that target those host/pathogen recognition mechanisms, and further develop innovative approaches, such as the expression of plantibody-targeting phytoplasma IDP to obtain phytoplasma-resistant plants. Most of the membrane proteins on phytoplasma cells are still defined as "hypothetical unknown protein" or functionally uncharacterized, and many efforts are needed to fill this gap of knowledge. Therefore, phytoplasma membrane proteins will continue to be a target, in parallel with studies on secreted effectors, the other prominent issue in phytoplasma research field.

References

Abbà S, Galetto L, Carle P, Carrère S, Delledonne M, Foissac X, Palmano S, Veratti F, Marzachì C (2014) RNA-Seq profile of "flavescence dorée" phytoplasma in grapevine. *BMC Genomics* **15**, 1088.

Al-Ghaithi AG, Al-Subhi AM, Al-Mahmooli IH, Al-Sadi AM (2018) Genetic analysis of '*Candidatus* Phytoplasma aurantifolia' associated with witches' broom on acid lime trees. *PeerJ* **5**, e4480.

Arnaud G, Malembic-Maher S, Salar P, Bonnet P, Maixner M, Marcone C, Boudon-Padieu E, Foissac X, (2007) Multilocus sequence typing confirms the close genetic interrelatedness of three distinct "flavescence dorée" phytoplasma strain clusters and group 16SrV phytoplasmas infecting grapevine and alder in Europe. *Applied and Environmental Microbiology* **73**, 4001–4010.

Arricau-Bouvery N, Duret S, Dubrana M-P, Batailler B, Desqué D, Béven L, Danet J-L, Monticone M, Bosco D, Malembic-Maher S, Foissac X (2018) Variable membrane protein A of "flavescence dorée" phytoplasma binds the midgut perimicrovillar membrane of *Euscelidius variegatus* and promotes adhesion to its epithelial cells. *Applied and Environmental Microbiology* **84**, e02487–17.

Atanasova B, Jakovljević M, Spasov D, Jović J, Mitrović M, Toševski I, Cvrković T (2015) The molecular epidemiology of "bois noir" grapevine yellows caused by '*Candidatus* Phytoplasma solani' in the Republic of Macedonia. *European Journal of Plant Pathology* **142**, 759–770.

Bai X, Correa VR, Toruño TY, Ammar E-D, Kamoun S, Hogenhout SA (2009) AY-WB phytoplasma secretes a protein that targets plant cell nuclei. *Molecular Plant-Microbe Interactions* **22**, 18–30.

Barbara DJ, Davies DL, Clark MF, Morton A (2002) Immunodominant membrane proteins from two phytoplasmas in the aster yellows clade (chlorante aster yellows and clover phyllody) are highly divergent in the major hydrophilic region. *Microbiology* **148**, 157–167.

Berg M, Davies DL, Clark MF, Vetten HJ, Maier G, Marcone C, Seemüller E (1999) Isolation of the gene encoding an immunodominant membrane protein of the apple proliferation phytoplasma, and expression and characterization of the gene product. *Microbiology* **145**, 1937–1943.

Blomquist CL, Barbara DJ, Davies DL, Clark MF, Kirkpatrick BC (2001) An immunodominant membrane protein gene from the Western X-disease phytoplasma is distinct from those of other phytoplasmas. *Microbiology* **147**, 571–580.

Boonrod K, Munteanu B, Jarausch B, Jarausch W, Krczal G (2012) An immunodominant membrane protein (Imp) of '*Candidatus* Phytoplasma mali' binds to plant actin. *Molecular Plant-Microbe Interactions* **25**, 889–895.

Bosco D, D'Amelio R (2010) Transmission specificity and competition of multiple phytoplasmas in the insect vector. In: Phytoplasmas: Genomes, Plant Hosts and Vectors. Eds Weintraub PG, Jones P. CABI, Wallingford, United Kingdom, 293–308 pp.

Cimerman A, Pacifico D, Salar P, Marzachì C, Foissac X (2009) Striking diversity of *vmp1*, a variable gene encoding a putative membrane protein of the "stolbur" phytoplasma. *Applied and Environmental Microbiology* **75**, 2951–2957.

Danet J-L, Bonnet P, Jarausch W, Carraro L, Skoric D, Labonne G, Foissac X (2007) *Imp* and *secY*, two new markers for MLST (multilocus sequence typing) in the 16SrX phytoplasma taxonomic group. *Bulletin of Insectology* **60**, 339–340.

Danet J-L, Balakishiyeva G, Cimerman A, Sauvion N, Marie-Jeanne V, Labonne G, Lavina A, Batlle A, Krizanac I, Skoric D, Ermacora P, Ulubas Serçe C, Caglayan K, Jarausch W, Foissac X (2011) Multilocus sequence analysis reveals the genetic diversity of European fruit tree phytoplasmas and supports the existence of inter-species recombination. *Microbiology* **157**, 438–450.

Dickinson M, Hodgetts J (2013) PCR analysis of phytoplasmas based on the *secA* gene. In: Phytoplasma. Methods in Molecular Biology, vol. 938, Eds Dickinson M, Hodgetts J. Humana Press, Totowa, New Jersey, United States of America, 205–215 pp.

Fabre A, Danet J-L, Foissac X (2011) The "stolbur" phytoplasma antigenic membrane protein gene *stamp* is submitted to diversifying positive selection. *Gene* **472**, 37–41.

Fleury B, Bergonier D, Berthelot X, Peterhans E, Frey J, Vilei EM (2002) Characterization of P40, a cytadhesin of *Mycoplasma agalactiae*. *Infection and Immunity* **70**, 5612–5621.

Foissac X, Danet J-L, Malembic-Maher S, Salar P, Šafářová D, Válová P, Navrátil M (2013) Tuf and secY PCR amplification and genotyping of phytoplasmas. In: Phytoplasma. Methods in Molecular Biology, vol 938. Eds Dickinson M, Hodgetts J. Humana Press, Totowa, New Jersey, United States of America, 189–204 pp.

Galetto L, Fletcher J, Bosco D, Turina M, Wayadande A, Marzachì C (2008) Characterization of putative membrane protein genes of the '*Candidatus* Phytoplasma asteris', chrysanthemum yellows isolate. *Canadian Journal of Microbiology* **54**, 341–351.

Galetto L, Bosco D, Balestrini R, Genre A, Fletcher J, Marzachì C (2011) The major antigenic membrane protein of '*Candidatus* Phytoplasma asteris' selectively interacts with ATP synthase and actin of leafhopper vectors. *Plos One* **6**, e22571.

Galetto L, Rashidi M, Yamchi A, Veratti F, Marzachì C (2014) *In vitro* expression of phytoplasma immunodominant membrane proteins. In: Phytoplasmas and Phytoplasma Disease Management: How to Reduce their Economic Impact. Ed Bertaccini A. IPWG - International Phytoplasmologist Working Group, Bologna, Italy, 272–279 pp.

Galetto L, Abbà S, Rossi M, Vallino M, Pesando M, Arricau-Bouvery N, Dubrana MP, Chitarra W, Pegoraro M, Bosco D, Marzachì C (2018) Two phytoplasmas elicit different responses in the insect vector *Euscelidius variegatus* Kirschbaum. *Infection and Immunity* **86**, e00042–18.

Galetto L, Vallino M, Rashidi M, Marzachì C (2019) Immunofluorescence assay to study early events of vector salivary gland colonization by phytoplasmas. In: Phytoplasma - Methods and Protocols. Eds Musetti R, Pagliari L. Springer Science+Business Media, New York, United States of America, 307–317 pp.

Hodgetts J, Boonham N, Mumford R, Harrison N, Dickinson M (2008) Phytoplasma phylogenetics based on analysis of *secA* and 23S rRNA gene sequences for improved resolution of candidate species of '*Candidatus* Phytoplasma'. *International Journal of Systematic and Evolutionary Microbiology* **58**, 1826–1837.

Hoshi A, Oshima K, Kakizawa S, Ishii Y, Ozeki J, Hashimoto M, Komatsu K, Kagiwada S, Yamaji Y, Namba S (2009) A unique virulence factor for proliferation and dwarfism in plants identified from a phytopathogenic bacterium. *Proceedings of the National Academy of Sciences of United States of America*, **106**, 6416–6421.

Ishii Y, Kakizawa S, Hoshi A, Maejima K, Kagiwada S, Yamaji Y, Oshima K, Namba S (2009) In the non-insect-transmissible line of onion yellows phytoplasma (OY-NIM), the plasmid-encoded transmembrane protein ORF3 lacks the major promoter region. *Microbiology* **155**, 2058–2067.

Ishii Y, Kakizawa S, Oshima K (2013) New *ex vivo* reporter assay system reveals that σ factors of an unculturable pathogen control gene regulation involved in the host switching between insects and plants. *MicrobiologyOpen* **2**, 553–565.

Kakizawa S, Oshima K, Kuboyama T, Nishigawa H, Jung H, Sawayanagi T, Tsuchizaki T, Miyata S, Ugaki M, Namba S (2001) Cloning and expression analysis of phytoplasma protein translocation genes. *Molecular Plant-Microbe Interactions* **14**, 1043–50.

Kakizawa S, Oshima K, Nishigawa H, Jung H-Y, Wei W, Suzuki S, Tanaka M, Miyata S, Ugaki M, Namba S (2004) Secretion of immunodominant membrane protein from onion yellows phytoplasma through the Sec protein-translocation system in *Escherichia coli*. *Microbiology* **150**, 135–142.

Kakizawa S, Oshima K, Namba S (2006a) Diversity and functional importance of phytoplasma membrane proteins. *Trends in Microbiology* **14**, 254–256.

Kakizawa S, Oshima K, Jung H-Y, Suzuki S, Nishigawa H, Arashida R, Miyata S-I, Ugaki M, Kishino H, Namba S (2006b) Positive selection acting on a surface membrane protein of the plant-pathogenic phytoplasmas. *Journal of Bacteriology* **188**, 3424–3428.

Kakizawa S, Oshima K, Ishii Y, Hoshi A, Maejima K, Jung H-Y, Yamaji Y, Namba S (2009) Cloning of immunodominant membrane protein genes of phytoplasmas and their *in planta* expression. *FEMS Microbiology Letters* **293**, 92–101.

Konnerth A, Krczal G, Boonrod K (2016) Immunodominant membrane proteins of phytoplasmas. *Microbiology* **162,** 1267–1273.

Kostadinovska E, Quaglino F, Mitrev S, Casati P, Bulgari D, Bianco PA (2014) Multiple gene analyses identify distinct "bois noir" phytoplasma genotypes in the Republic of Macedonia. *Phytopathologia Mediterranea* **53**, 491–501.

Kube M, Mitrovic J, Duduk B, Rabus R, Seemüller E (2012) Current view on phytoplasma genomes and encoded metabolism. *The Scientific World Journal* **2012**, 1–25.

Langklotz S, Baumann U, Narberhaus F (2012) Structure and function of the bacterial AAA protease FtsH. *Biochimica et Biophysica Acta* **1823**, 40–48.

Le Gall F, Bové J-M, Garnier M (1998) Engineering of a single-chain variable-fragment (scFv) antibody specific for the "stolbur" phytoplasma (mollicute) and its expression in *Escherichia coli* and tobacco plants. *Applied and Environmental Microbiology* **64**, 4566–4572.

Lee I-M, Gundersen-Rindal D, Bertaccini A (1998) Phytoplasma: ecology and genomic diversity. *Phytopathology* **88**, 1359–1366.

Lee I-M, Davis RE, Gundersen-Rindal DE (2000) Phytoplasma: phytopathogenic *Mollicutes. Annual Review of Microbiology* **54**, 221–255.

Lee I-M, Zhao Y, Bottner KD (2006) SecY gene sequence analysis for finer differentiation of diverse strains in the aster yellows phytoplasma group. *Molecular and Cellular Probes* **20**, 87–91.

Lee I-M, Bottner-Parker KD, Zhao Y, Davis RE, Harrison NA (2010) Phylogenetic analysis and delineation of phytoplasmas based on *secY* gene sequences. *International Journal of Systematic and Evolutionary Microbiology* **60**, 2887–2897.

MacLean AM, Sugio A, Makarova OV, Findlay KC, Grieve VM, Toth R, Nicolaisen M, Hogenhout SA (2011) Phytoplasma effector SAP54 induces indeterminate leaf-like flower development in *Arabidopsis* plants. *Plant Physiology* **157**, 831–841.

Maejima K, Oshima K, Namba S (2014) Exploring the phytoplasmas, plant pathogenic bacteria. *Journal of General Plant Pathology* **80**, 210–221.

Malembic-Maher S, Gall FL, Danet J-L, Borne FD de, Bové J-M, Garnier-Semancik M (2005) Transformation of tobacco plants for single-chain antibody expression via apoplastic and symplasmic routes, and analysis of their susceptibility to "stolbur" phytoplasma infection. *Plant Science* **168**, 349–358.

Malembic-Maher S, Desqué D, Khalil D, Salar P, Danet J-L, Bergey B, Duret S, Beven L, Arricau-Bouvery N, Jović L, Krnjajić S, Angelini E, Filippin L, Ember I, Kölber M, Della Bartola M, Materazzi A, Lang F, Jarausch B, Maixner M, Foissac X (2017) When a palearctic bacterium meets a nearctic insect vector: genetic and ecological insights into the emergence of the grapevine "flavescence dorée" epidemics in Europe. IOBC-WPRS Meeting, 15–20 October, Riva Del Garda (Verona), Italy, 211–213.

Marcone C (2014) Molecular biology and pathogenicity of phytoplasmas. *Annals of Applied Biology* **165**, 199–221.

Morton A, Davies DL, Blomquist CL, Barbara DJ (2003) Characterization of homologues of the apple proliferation immunodominant membrane protein gene from three related phytoplasmas. *Molecular Plant Pathology* **4**, 109–114.

Murolo S, Romanazzi G (2015) In-vineyard population structure of '*Candidatus* Phytoplasma solani' using multilocus sequence typing analysis. *Infection, Genetics and Evolution* **31**, 221–230.

Neriya Y, Sugawara K, Maejima K, Hashimoto M, Komatsu K, Minato N, Miura C, Kakizawa S, Yamaji Y, Oshima K, Namba S (2011) Cloning, expression analysis, and sequence diversity of genes encoding two different immunodominant membrane proteins in poinsettia branch-inducing phytoplasma (PoiBI). *FEMS Microbiology Letters* **324**, 38–47.

Neriya Y, Maejima K, Nijo T, Tomomitsu T, Yusa A, Himeno M, Netsu O, Hamamoto H, Oshima K, Namba S (2014) Onion yellow phytoplasma P38 protein plays a role in adhesion to the hosts. *FEMS Microbiology Letters* **361**, 115–122.

Oshima K, Kakizawa S, Nishigawa H, Jung H-Y, Wei W, Suzuki S, Arashida R, Nakata D, Miyata S, Ugaki M, Namba S (2004) Reductive evolution suggested from the complete genome sequence of a plant-pathogenic phytoplasma. *Nature Genetics* **36**, 27–29.

Oshima K, Ishii Y, Kakizawa S, Sugawara K, Neriya Y, Himeno M, Minato N, Miura C, Shiraishi T, Yamaji Y, Namba S (2011) Dramatic transcriptional changes in an intracellular parasite enable host switching between plant and insect. *Plos One* **6**, e23242.

Oshima K, Maejima K, Namba S (2013) Genomic and evolutionary aspects of phytoplasmas. *Frontiers in Microbiology* **4**, 230.

Pacifico D, Galetto L, Rashidi M, Abbà S, Palmano S, Firrao G, Bosco D, Marzachì C (2015) Decreasing global transcript levels over time suggest that phytoplasma cells enter stationary phase during plant and insect colonization. *Applied and Environmental Microbiology* **81**, 2591–2602.

Pierro R, Passera A, Panattoni A, Casati P, Luvisi A, Rizzo D, Bianco PA, Quaglino F, Materazzi A (2018) Molecular typing of "bois noir" phytoplasma strains in the Chianti Classico area (Tuscany, Central Italy) and their association with symptom severity in *Vitis vinifera* Sangiovese. *Phytopathology* **108**, 362–373.

Plavec J, Križanac I, Budinšćak Ž, Škorić D, Šeruga-Musić M (2015) A case study of FD and BN phytoplasma variability in Croatia: multigene sequence analysis approach. *European Journal of Plant Pathology* **142**, 591–601.

Plavec J, Budinšćak Ž, Križanac I, Škorić D, Foissac X, Šeruga-Musić M (2019) Multi-locus sequence typing reveals the presence of three distinct "flavescence dorée" phytoplasma genetic clusters in Croatian vineyards. *Plant Pathology* **68**, 18–30.

Quaglino F, Kube M, Jawhari M, Abou-Jawdah Y, Siewert C, Choueiri E, Sobh H, Casati P, Tedeschi R, Molino Lova M, Alma A, Bianco PA (2015) '*Candidatus* Phytoplasma phoenicium' associated with almond witches' broom disease: from draft genome to genetic diversity among strain populations. *BMC Microbiology* **15**, 148.

Rashidi M, Galetto L, Bosco D, Bulgarelli A, Vallino M, Veratti F, Marzachì C (2015) Role of the major antigenic membrane protein in phytoplasma transmission by two insect vector species. *BMC Microbiology* **15**, 193.

Renaudin J, Béven L, Batailler B, Duret S, Desqué D, Arricau-Bouvery N, Malembic-Maher S, Foissac X (2015) Heterologous expression and processing of the "flavescence dorée" phytoplasma variable membrane protein VmpA in *Spiroplasma citri*. *BMC Microbiology* **15**, 82.

Seemüller E, Kampmann M, Kiss E, Schneider B (2011) HflB gene-based phytopathogenic classification of '*Candidatus* Phytoplasma mali' strains and evidence that strain composition determines virulence in multiply infected apple trees. *Molecular Plant-Microbe Interactions* **24**, 1258–1266.

Seemüller E, Sule S, Kube M, Jelkmann W, Schneider B (2013) The AAA$^+$ ATPases and HflB/FtsH proteases of '*Candidatus* Phytoplasma mali': phylogenetic diversity, membrane topology, and relationship to strain virulence. *Molecular Plant-Microbe Interactions* **26**, 367–376.

Seemüller E, Zikeli K, Furch ACU, Wensing A, Jelkmann W (2018) Virulence of '*Candidatus* Phytoplasma mali' strains is closely linked to conserved substitutions in AAA$^+$ ATPase AP460 and their supposed effect on enzyme function. *European Journal of Plant Pathology* **150**, 701–711.

Seruga-Musić M, Duc Nguyen H, Cerni S, Mamula Đ, Oshima K, Skorić D (2014) Multilocus sequence analysis of '*Candidatus* Phytoplasma asteris' strain and the genome analysis of *Turnip mosaic virus* co-infecting oilseed rape. *Journal of Applied Microbiology* **117**, 774–785.

Shahryari F, Safarnejad MR, Shams-Bakhsh M, Jouzani GRS (2010) Toward immunomodulation of witches' broom disease of lime (WBDL) by targeting immunodominant membrane protein (IMP) of '*Candidatus* Phytoplasma aurantifolia'. *Communications in Agricultural and Applied Biological Sciences* **75**, 789–795.

Shahryari F, Safarnejad MR, Shams-Bakhsh M, Schillberg S, Nölke G (2013) Generation and expression in plants of a single-chain variable fragment antibody against the immunodominant membrane protein of '*Candidatus* Phytoplasma aurantifolia'. *Journal of Microbiology and Biotechnology* **23**, 1047–1054.

Siampour M, Galetto L, Bosco D, Izadpanah K, Marzachì C (2011) *In vitro* interactions between immunodominant membrane protein of lime witches' broom phytoplasma and leafhopper vector proteins. *Bulletin of Insectology* **64**(Supplement), S149-S150.

Siampour M, Izadpanah K, Galetto L, Salehi M, Marzachí C (2013) Molecular characterization, phylogenetic comparison and serological relationship of the Imp protein of several '*Candidatus* Phytoplasma aurantifolia' strains. *Plant Pathology* **62**, 452–459.

Sugio A, MacLean AM, Kingdom HN, Grieve VM, Manimekalai R, Hogenhout SA (2011) Diverse targets of phytoplasma effectors: from plant development to defense against insects. *Annual Review of Phytopathology* **49**, 175–195.

Sukharev SI, Blount P, Martinac B, Kung C (1997) Mechanosensitive channels of *Escherichia coli*: the *mscL* gene, protein, and activities. *Annual Review of Physiology* **59**, 633–657.

Suzuki S, Oshima K, Kakizawa S, Arashida R, Jung H-Y, Yamaji Y, Nishigawa H, Ugaki M, Namba S (2006) Interaction between the membrane protein of a pathogen and insect microfilament complex determines insect-vector specificity. *Proceedings of the National Academy of Sciences, United States of America* **103**, 4252–4257.

Tomkins M, Kliot A, Marée AF, Hogenhout SA (2018) A multi-layered mechanistic modelling approach to understand how effector genes extend beyond phytoplasma to modulate plant hosts, insect vectors and the environment. *Current Opinion in Plant Biology* **44**, 39–48.

Toruño TY, Seruga-Musić M, Simi S, Nicolaisen M, Hogenhout SA (2010) Phytoplasma PMU1 exists as linear chromosomal and circular extrachromosomal elements and has enhanced expression in insect vectors compared with plant hosts: PMUs are extrachromosomal elements. *Molecular Microbiology* **77**, 1406–1415.

Weintraub PG, Beanland L (2006) Insect vectors of phytoplasmas. *Annual Review of Entomology* **51**, 91–111.

Yu YL, Yeh KW, Lin CP (1998) An antigenic protein gene of a phytoplasma associated with sweet potato witches' broom. *Microbiology* **144**, 1257–1262.

Chapter 6
Phytoplasma Cultivation

Nicoletta Contaldo and Assunta Bertaccini

Abstract The possibility to grow phytoplasmas in complex media as the mycoplasmas was ruled out for more than 40 years due to the inconsistency of the first isolation trials. The use of micropropagated infected periwinkle shoots first and tissues infected directly from the fields on artificial media was recently confirmed. Isolation and cultivation together with the first biochemical and biological characterization of some phytoplasmas were consistently obtained. Phytoplasmas were successfully grown from grapevine, cassava and coconut palm tissues severely infected and in some cases also from asymptomatic tissues from trees or plants growing in severely infected areas. Phytoplasmas from diverse ribosomal groups were isolated in particular from cassava with frog skin the detected prokaryotes in culture were molecularly identified as belonging to the 16SrIII group as in the original plants. Aster yellows and 'stolbur' phytoplasmas, group 16SrI and 16SrXII, respectively, were consistently grown from diverse host plants such as grapevine and coconut palms. Seed transmission in corn of aster yellows phytoplasmas was also confirmed by isolation from seedlings of viable cells. The growth and biochemical and biological characterization of these bacteria is therefore the most important recent step in the study of phytoplasmas that will allow to improve their knowledge and to carry out focused management in the field in order to reduce their impact on cultivated and wild crops worldwide.

Keywords Isolation · Phytoplasmas · Cultivation · Identification · Biochemical properties

N. Contaldo (✉) · A. Bertaccini
Department of Agricultural and Food Sciences, *Alma Mater Studiorum* – University of Bologna, Bologna, Italy

A. Bertaccini et al. (eds.), *Phytoplasmas: Plant Pathogenic Bacteria - III*,
https://doi.org/10.1007/978-981-13-9632-8_6

6.1 Introduction

Doi and coworkers in 1967 discovered by electron microscopy observations pleomorphic, wall-less bacteria in the phloem of many plant species affected by yellows-type diseases previously believed to be caused by viruses (Kunkel 1926, 1931, 1955; Nasu et al. 1967; Mc Coy et al. 1989; Lee and Davis 1992; Maramorosch 2011). These microorganisms named mycoplasma-like (MLO) were later renamed phytoplasmas since they were shown to be clearly differentiable from mycoplasmas, recognized pathogens for human and animals since long time (Nocard and Roux 1898), for their ribosomal RNA gene (Lim and Sears 1989) and are now classified based on molecular analyses on the 16S ribosomal gene at the level of the '*Candidatus*' genus. The disappearance of symptoms after antibiotic (*i.e.* tetracycline) treatments provided additional evidence to support their pathogenic role in the diseased plants (Ishiie et al. 1967). After the discovery of MLO, the plant pathologists took up the challenge to culture these organisms in growth media. In the early 1970s, reports on MLO cultivation were published (Lombardo and Pignattelli 1970; Lin et al. 1970; Giannotti and Vago 1971; Ghosh et al. 1975; Skripal et al. 1984), but none of the described methodology proved to be repeatable, and scientists start to consider phytoplasmas as unculturable microorganisms (Maramorosch 2011). Trials were carried out using media able to support the mycoplasma or the spiroplasma growth (Skripal and Malinivskaya 1984; A. Bertaccini, 1990, unpublished; Poghosyan and Lebsky 2004) until phylogenetic data on mollicutes (Gasparich et al. 2004) clearly linked the phytoplasmas to one of the mycoplasma groups and mycoplasma-based media were tested for phytoplasma cultivation (Fig. 6.1) with the successful isolation of 16SrXII-A phytoplasmas from micropropagated periwinkle shoots (Bertaccini et al. 1992, 2010).

By continuation of these trials, phytoplasma cultivation in artificial media was achieved (Contaldo et al. 2012) using diverse phytoplasmas in micropropagated

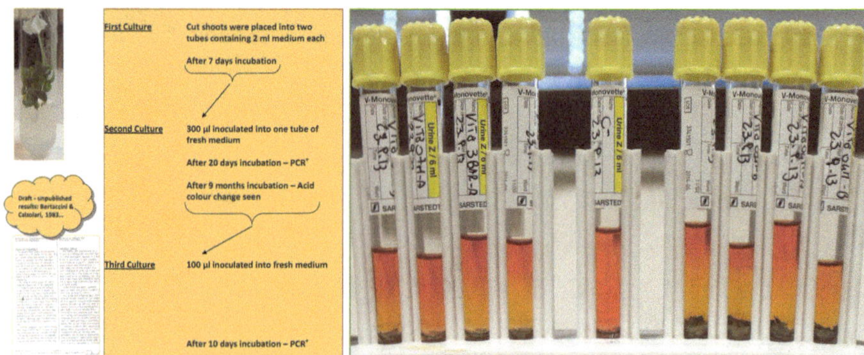

Fig. 6.1 From left: scheme used for first phytoplasma isolation from micropropagated periwinkles in collection and isolation tubes containing infected plant materials; the tube labeled C-contains healthy plant materials

infected periwinkle shoots such as *Chrysanthemum* yellows (strain CY-TO from Italy, 16SrI-B), tomato big bud (strain TBB from Australia, 16SrII-D) and "stolbur" (strains CH-1 from Italian grapevine and STOL, from Serbian pepper). These results were confirmed by the isolation soon after of the potato witches' broom (strain PWB from the USA, 16SrVI-A), lime witches' broom (strain WBDL from Oman, 16SrII-B) and *Picris echioides* yellows (strain PEY from Italy, 16SrIX-C) phytoplasmas (Contaldo et al. 2013, 2014a). More recently, by using the same media, phytoplasmas were isolated and grown from cassava infected by the frog skin disease associated with 16SrIII-L strain employing fragments of roots, petioles, stems, leaves and embryos as a source for phytoplasma isolation (Alvarez et al. 2009, 2010; Rao et al. 2018).

6.2 Isolation Media Development

Attempts to cultivate phytoplasmas have been performed using artificial media similar to those used for spiroplasma growth (Davis et al. 1972; Saglio et al. 1973; Müller et al. 1975; Elmendorf 1977; Whitcomb and Tully 1975; Jones et al. 1977), but the experiments were not repeatable, and the results were therefore not accepted by the scientific community. To explain the difficulties in cultivating phytoplasmas, two hypotheses have been formulated: the developed media lack essential elements and phytoplasmas are host cell-dependent. The first growth in artificial media was obtained when infected micropropagated periwinkle shoots were used as source and the isolation was carried out on available commercial media with composition not disclosed by the provider (Mycoplasma Experience Ltd) (Contaldo et al. 2012). In order to achieve the cultivation from naturally diseased plant materials, media supporting the growth from different plants sources and with flexible and modifiable composition were specifically designed (Contaldo et al. 2016a). These complex media (CB) are based on TSB (Oxoid, UK) pH 7.3 ± 0.2 which essentially contains tryptone and soy peptone, and enriched with a supplement containing 20 ml of sterile horse serum (Oxoid, SR0035), and 25 µg/ml of ampicillin (Sigma, A9393) and 50 µg/ml of nystatin (Sigma, N6261) both 0.22 µm filter sterilized, 10 ml of autoclaved yeast extract (25% w/v) and 0.005% of phenol red for each 80 ml of medium. These media were used to support the isolation and growth of phytoplasmas from infected field-collected grapevine samples. Comparative trials showed that the new developed CB medium could support the phytoplasma growth in the same manner as the previous Piv media. In spite of the relatively long time required for incubation in liquid medium (from 2 to 15 days), colony growth on agar usually occurs within 2–5 days (Fig. 6.2) as for the majority of cultured bacteria.

The composition of solid medium, CBs, was mainly TSB, NaCl 20 g/l, agar No. 3 12 g/l (Oxoid, LP0013), ampicillin and nystatin as before. One of the key points that was shown to be important within phytoplasma isolation from plant sources was the optimization of the growing conditions, by the use of plate incubation in a specific 2.5 l anaerobic jar (Oxoid, AG0025) in a microaerophilic atmosphere using

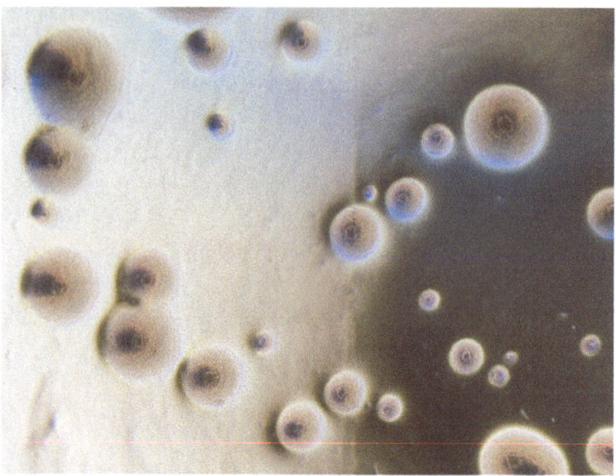

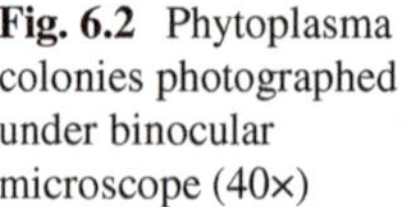

Fig. 6.2 Phytoplasma colonies photographed under binocular microscope (40×)

CampyGen sachets (Oxoid, CN0025) (Contaldo et al. 2014b, c, 2016a). The results obtained from phytoplasma isolation and cultivation from field-infected grapevine samples were confirmed by the phytoplasma isolation from coconut palms infected by Côte d'Ivoire lethal yellowing (CILY) (Fig. 6.3) (Contaldo et al. 2019). The samples employed for the phytoplasma isolation and culturing were representative of the various stages of the disease (Arocha Rosete et al. 2017), and leaves, inflorescences and trunk boring were the tissues employed for the isolation.

While the leaves and the inflorescences were used as described (Contaldo et al. 2016a), a change in the isolation procedure was employed for the trunk boring samples. In fact, approximately 100 mg from each plant were placed in two separate sterile mortars and mixed with 1 ml of CBl using a sterile pestle. Immediately after, 100 μl of the mixtures were transferred into monovette tubes and incubated (Contaldo et al. 2019).

The use of commercial TSB in liquid and solid media formulation for phytoplasma isolation and subsequent culture supports also the growth of bacterial endophytes present in the plant source materials (Fig. 6.4). Different combinations of antibiotics, carbon sources and NaCl concentration were therefore evaluated by visual observation of the degree of turbidity of the liquid media, due to the presence of contaminating microorganisms and/or symbionts inside the isolation tissue, and by the ability to form colonies in agar (Fig. 6.5) (Contaldo et al. 2018a).

The colony purification was performed by gentle filtration of the entire liquid culture through 0.8 μm membrane filters (Whatman, Maidstone, UK) to avoid plugging, followed by $10^{-3}/10^{-4}$ serial dilutions in CBl. Both the undiluted filtrate and the filtrate dilutions were cultured on solid CB plates and incubated at 25 ± 1 °C. This process, including the filtration, is repeated three times from single descending colonies (Fig. 6.6).

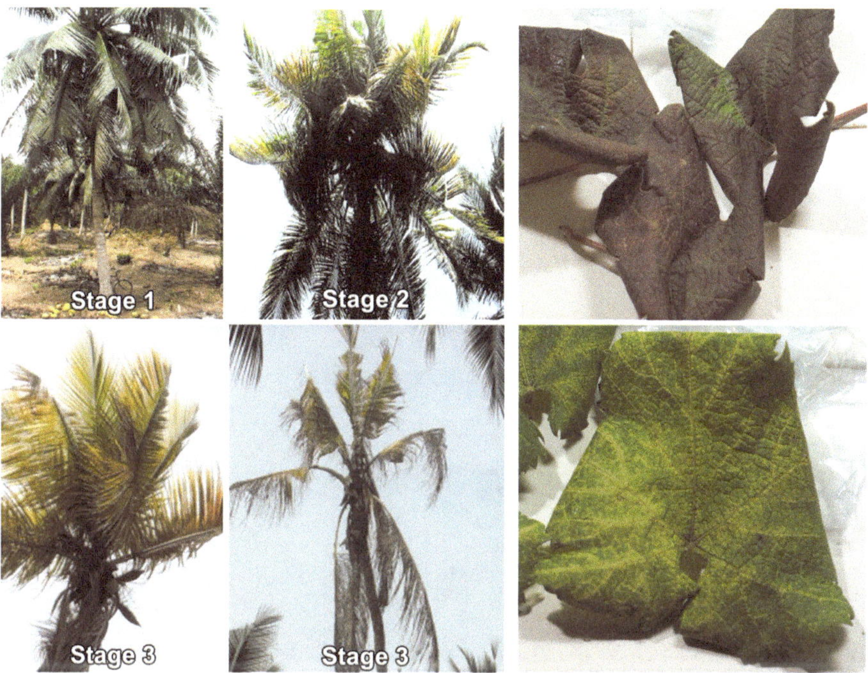

Fig. 6.3 On the left CILY symptoms on coconut palms at different CILY disease stages in Côte d'Ivoire (kindly provided by Y. Arocha-Rosete) and on the right grapevine leaves infected with grapevine yellows phytoplasmas (GY) used for phytoplasma isolation: top a red grapevine variety and bottom a white variety

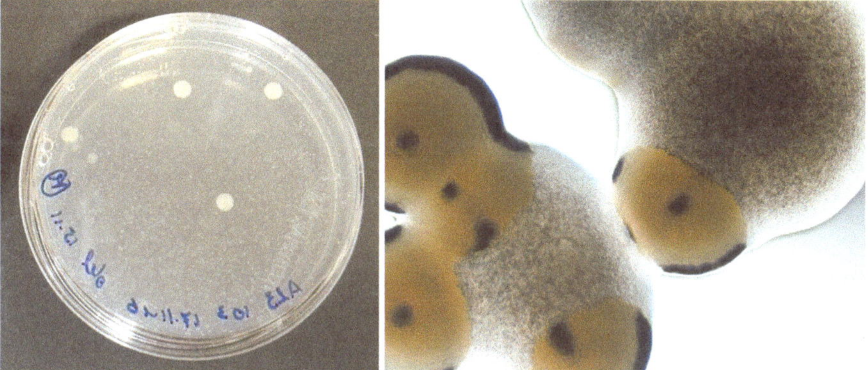

Fig. 6.4 From the left endophytes and phytoplasma colonies in a plate and (right) photographed under a bifocal optical microscope (40 X) two days after plating

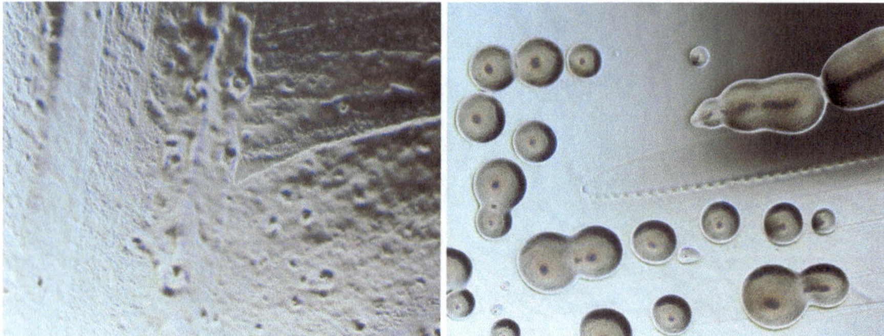

Fig. 6.5 Close-up: two Petri dishes on the left with colonies treated with tetracycline and on the right treated with penicillin

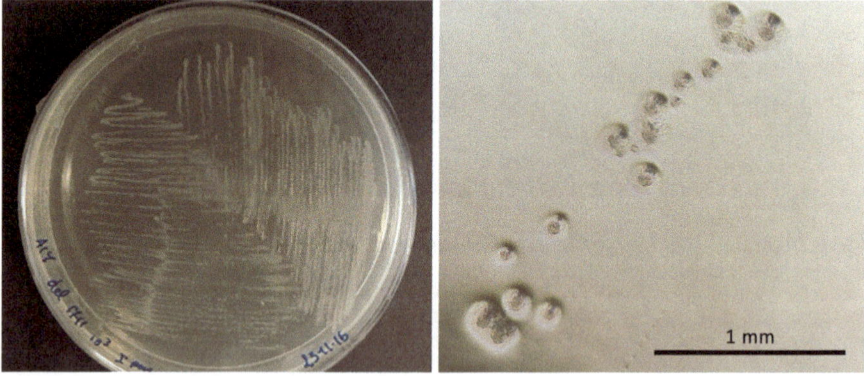

Fig. 6.6 On the left: Petri dish containing phytoplasma colonies derived from the 10^{-3} liquid media dilutions showing the purification by streaking onto plate. On the right: phytoplasma colonies after the purification procedure photographed with binocular microscope

6.3 Phytoplasma Identification in Liquid and Solid Media

One of the main challenges with phytoplasma cultivation is to define a rapid, repeatable and reliable detection method from both solid and liquid cultures. The DNAs extracted from both media were amplifiable on 16Sr gene only after nested PCR with universal primers and only scattered amplification was obtained on other non-ribosomal genes.

This was the case of amplicons obtained on *ef-Tu* gene on aster yellows (AY) and "stolbur" (STOL) isolates from periwinkle shoots (Contaldo et al. 2012) and of STOL colonies grown on agar media after isolation from field-collected grapevine that tested positive on stamp gene (Fig. 6.7) (Contaldo et al. 2016b). The RFLP profiles and their sequences obtained after direct sequencing in both directions with primers used for amplification are however useful to confirm the phytoplasma pres-

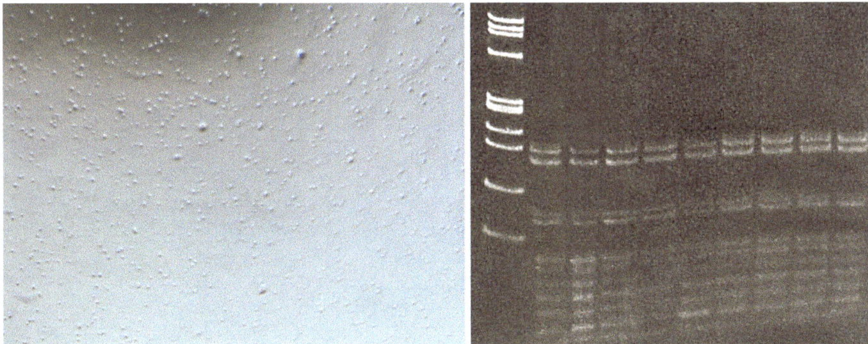

Fig. 6.7 From the left "stolbur" phytoplasma colonies in agar plate two days after inoculation under bifocal microscope (40 X) and left polyacrylamide (6.7%) gel showing RFLP profiles of the *stamp* gene amplicons of the same colonies digested with *Tru1*I. M, Marker DNA phiX174 *Hind*III digested

ence and identity (Fig. 6.8) (Contaldo et al. 2012, 2015a, b). The difficulty in amplifying nonribosomal genes from these isolates is probably due to both low DNA concentration (number of cells) and the presence of mixed bacterial and/or phytoplasma population in the media.

In the cassava frog skin isolation, the phytoplasma identity was verified using a quantitative PCR technique specific for the isolated phytoplasma ('*Candidatus* Phytoplasma pruni'-related). Furthermore, the pathogenicity of five isolates containing phytoplasma DNAs was proved using stem injection in *in vitro* cassava shoots: severe CFSD and typical root symptoms were observed 6 months after inoculation (Alvarez et al. 2017). A preliminary qPCR analysis using universal primers designed on the 16Sr gene (Saccardo et al. 2012) was recently set (Contaldo et al. 2019) to verify the phytoplasma presence and titre in two culture suspensions obtained from CILY-infected coconut palms; at the inoculation time (T0) and after 18 h (T18), the methodology was able to demonstrate the increase of the phytoplasma titre.

6.4 Biological and Nutritional Properties of Selected Phytoplasma Strains

Some biological and nutritional properties of '*Ca.* P. asteris' strains isolated from grapevine were determined. They were not surviving with sucrose as source of carbon and were very well differentiable from the *Acholeplasma laidlawii* strain used as control for the arginine hydrolysis ability (Contaldo et al. 2018a).

The same results were confirmed in '*Ca.* P. solani' (16SrXII-A) and '*Ca.* P. palmicola' (16SrXXII-B) isolates from coconut tissues. For these two latter strains, the antibiotic susceptibility was also evaluated (Fig. 6.9). The two strains

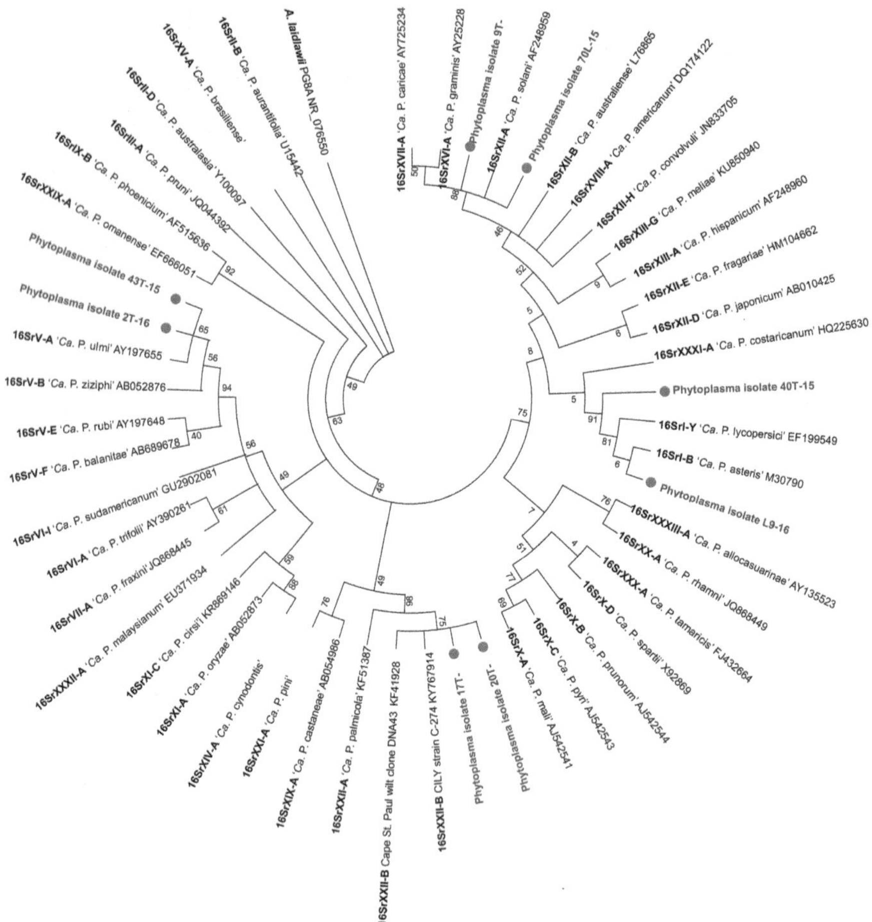

Fig. 6.8 Phylogenetic tree showing the evolutionary relationships of reference phytoplasmas and phytoplasma strains isolated from coconut palms infected by CILY disease (in green) using the neighbour-joining method in MEGA6

studied, indeed, revealed the maximum susceptibility to tobramycin, followed by polymyxin and tetracycline; an intermediate susceptibility to 5-fluorouracil was also observed, but they resulted completely resistant to cephalexin hydrate and rifampicin (Contaldo et al. 2019).

The biochemical characterization of the '*Ca.* P asteris', '*Ca.* P. solani', and '*Ca.* P. palmicola' isolates revealed that they shared metabolic features with *Mycoplasma* species group III (glucose positive/arginine positive), which are represented by several mycoplasma species (*i.e. Mycoplasma fermentans*) (Brown et al. 2007; Freundt et al. 1979).

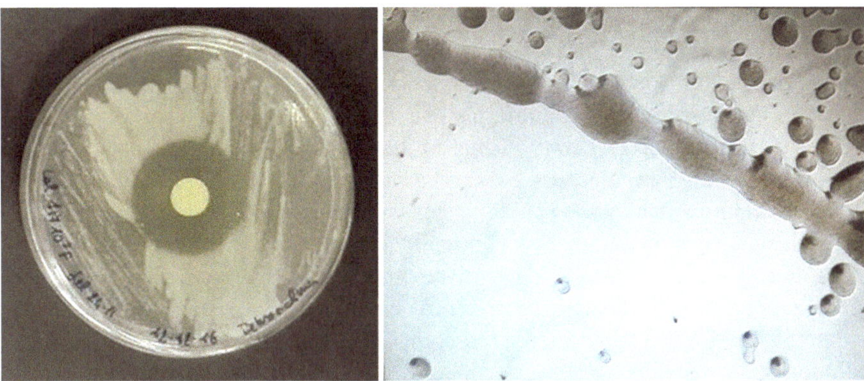

Fig. 6.9 Phytoplasma colonies treated with tetracycline with the standard disc diffusion method: on the left inhibition halo produced by the antibiotic on the colonies and on the right colonies inside and outside the inhibition halo photographed under binocular microscope (40 X)

6.5 Mixed Phytoplasma Infection

The isolation of phytoplasmas from naturally infected grapevine and coconut plants highlighted how starting from samples in which one phytoplasma was detected by PCR could provide colonies in which different phytoplasmas are detected (Contaldo et al. 2018b, 2019). These results suggest that the medium employed is not phytoplasma specific and support the growth of phytoplasmas that are present in the endobiome of the plants at concentration that are below the routine threshold of molecular PCR detection.

The comparison between the phytoplasma subgroups found in the plant material prior to culture, and those found in the liquid and solid media, indicates the frequent presence of phytoplasmas in the tissues used for isolation that are not the same as those isolated. These somewhat contradictory results on phytoplasma identification obtained from plant material, liquid and solid cultures, confirm the common presence of a mixed phytoplasma infection. While it is very likely that the prevalent strain in the plant material is the one determined by the molecular analyses in plants, the medium used in these isolation trials appears to favour phytoplasmas enclosed in specific ribosomal groups. Moreover, also the differences in the identity of the phytoplasma that were detected from testing the DNA extracts from the CB liquid cultures and plate colonies further indicate the presence of mixtures of phytoplasmas. The presence of multiple phytoplasmas in a single plant host may modulate the expression of disease symptoms, as it was suggested regarding the presence of multiple apple proliferation phytoplasma strains in apple trees or for other phytoplasma-associated diseases (Seemüller and Schneider 2007; Rid et al. 2016). On the other hand, the medium-dependent growth kinetics for '*Ca.* P. solani' (16SrXII-A) and '*Ca.* P. palmicola' (16SrXXII-B) isolates from coconut indicated their different performances in the used medium. This phenomenon was clearly evidenced by the longer growth timeframe for the 16SrXII-A isolate. The 16SrXXII-B phytoplasma corresponds to the prevalent subgroup associated with CILY in Côte d'Ivoire; however, this isolate was shown to multiply less efficiently than the other in the media used. The shorter survival time in the artificial medium for this isolate, may be associated with the presence of a limiting factor in the cell-free medium, clearly demonstrating that the various phytoplasma isolates exhibit different performances in the same growth medium. This is possibly related to different biological behaviours, and indicates that very likely diverse phytoplasmas possess different plant colonization abilities. Further biochemical comparative studies may lead to the identification of new chemicals or bio-compounds which may improve the media composition to support a better growth performance and survival rate of other phytoplasmas or phytoplasma strains.

6.6 Insect Transmissibility from Colonies

Trials were carried out to verify if the phytoplasmas detected in the colonies are able to be acquired by insects. Aster yellows colonies from grapevine (isolates Gl1is and Gl2is) were used to investigate the acquisition ability of the leafhopper *Scaphoideus titanus* Ball that was demonstrated to transmit aster yellows phytoplasma through the eggs (transovarial transmission) (Danielli et al. 1996; Alma et al. 1997) and is able to acquire and transmit this phytoplasmas to plants under controlled conditions (Alma et al. 2001) since in nature it transmits only "flavescence dorée" phytoplasmas (Mori et al. 2002). Preliminary results showed that the different *S. titanus* nymph stages that were feeding on solid medium with and without phytoplasma colonies showed a good capacity of survival with mean survival time (LT50) of 4.4 ± 0.13 and 5.5 ± 0.35 days for nymph instars and adults, respectively (Fig. 6.10).

Insects fed for 24 hours on phytoplasma-inoculated solid medium showed a percentage of phytoplasma acquisition ranging from 55% to 80%. The presence of

Fig. 6.10 Steps for the insect acquisition on plates containing phytoplasma colonies

aster yellows (AY) in the insects was detected up to 15 days post acquisition. The 16S ribosomal gene of phytoplasma detected in leafhopper at 7 days post inoculation (AY-St1) was 99% identical to the ones detected in the culture used for insect feeding. The alignment of these amplicons with aster yellows strains OY-M (16SrI-B) and AY-WB (16SrI-A) shows the presence of four SNPs between isolate Gl-1is and strain AY-St1; two of these SNPs are in common between AY-St1 and strain AY-WB, while the other two SNPs are specific for strain AY-St1 (Fig. 6.11). The nested PCR amplification of 30 *S. titanus* specimens not used for acquisition trials and maintained on healthy grapevine resulted in 28 negative specimens and 2 positive to specific 16SrV group primers, "flavescence dorée" possible phytoplasmas (Angelini et al. 2018), indicating that no 16SrI infection was present in the *S. titanus* rearing. Some of the healthy grapevine plants employed for insect feeding during the experiment were tested after 10 months and resulted positive to '*Ca*. P. asteris' presence.

6.7 Viability of Seed Transmitted Phytoplasmas

Phytoplasma seed transmission is still a poorly investigated topic in spite of the growing evidences of its presence in several agricultural relevant species. In order to better understand the mechanisms behind it and to verify the viability of phytoplasmas in the progenies, phytoplasma isolation trials were conducted from plantlets derived from symptomatic and phytoplasma-infected corn seeds (Satta et al. 2016, 2019). Preliminary trials were carried out on nine seedlings, six of which, positive to aster yellows and "stolbur" phytoplasmas, were used as sources for isolation trials. From three plants the isolation allowed to detect aster yellows and "stolbur" phytoplasma DNAs after chloroform/phenol extraction from 1 ml of liquid medium from both isolation tubes and from tubes obtained after serial dilution. Further tests on 79 seedlings resulted in 17 of them positive for aster yellows and "stolbur" phytoplasmas. Isolation from three of these plants after 30 days from germination resulted positive for phytoplasma DNA in tubes deriving after 2–10 serial dilution in fresh medium. Reisolation carried out at 90 days from germination confirmed these results from one plant. Plating carried out together with DNA extraction from liquid medium produced colonies of different sizes and shapes. The same type of colonies was obtained from plating tubes maintained for 7 months at 25°C. Single colonies were picked and transferred in broth for purification steps at several times, and small colonies were obtained resulting to consistently contain aster yellows DNAs; moreover this type of colony growth was observed for at least three subsequent passages in liquid/solid media carried out every 5 days. These preliminary results are indicating the viability of phytoplasmas isolated from corn seedlings.

```
                 545        555        565        575        585        595
AY-St1     CAAGCGTTAT CCGGAATTAT TGGGCGTAAA GGGTGCGTAG GCTGTTAAAT AAGTCTATGG
Gl-1is     CAAGCGTTAT CCGGAATTAT TGGGCGTAAA GGGTGCGTAG GCGGTTAAAT AAGTTTATGG
OY-M       CAAGCGTTAT CCGGAATTAT TGGGCGTAAA GGGTGCGTAG GCGGTTAAAT AAGTTTATGG
AYWB       CAAGCGTTAT CCGGAATTAT TGGGCGTAAA GGGTGCGTAG GCTGTTAAAT AAGTTTATGG

                 605        615        625        635        645        655
AY-St1     TCTAAGTGCA ATGCTCAACA TTGTGATGCT ATAAAAACTG TTTAGCTAGA GTAAGATAGA
Gl-1is     TCTAAGTGCA ATGCTCAACA TTGTGATGCT ATAAAAACTG TTTAGCTAGA GTAAGATAGA
OY-M       TCTAAGTGCA ATGCTCAACA TTGTGATGCT ATAAAAACTG TTTAGCTAGA GTAAGATAGA
AYWB       TCTAAGTGCA ATGCTCAACA TTGTGATGCT ATAAAAACTG TTTAGCTAGA GTAAGATAGA

                 665        675        685        695        705        715
AY-St1     GGCAAGTGGA ATTCCATGTG TAGTGGTAAA ATGCGTAAAT ATATGGAGGA ACACCAGTAG
Gl-1is     GGCAAGTGGA ATTCCATGTG TAGTGGTAAA ATGCGTAAAT ATATGGAGGA ACACCAGTAG
OY-M       GGCAAGTGGA ATTCCATGTG TAGTGGTAAA ATGCGTAAAT ATATGGAGGA ACACCAGTAG
AYWB       GGCAAGTGGA ATTCCATGTG TAGTGGTAAA ATGCGTAAAT ATATGGAGGA ACACCAGTAG

                 725        735        745        755        765        775
AY-St1     CGAAGGCGGC TTGCTGGGTC TTTACTGACG CTGAGGCACG AAAGCGTGGG GAGCAAACAG
Gl-1is     CGAAGGCGGC TTGCTGGGTC TTTACTGACG CTGAGGCACG AAAGCGTGGG GAGCAAACAG
OY-M       CGAAGGCGGC TTGCTGGGTC TTTACTGACG CTGAGGCACG AAAGCGTGGG GAGCAAACAG
AYWB       CGAAGGCGGC TTGCTGGGTC TTTACTGACG CTGAGGCACG AAAGCGTGGG GAGCAAACAG

                 785        795        805        815        825        835
AY-St1     GATTAGATAC CCTGGTAGTC CACGCCGTAA ACGATGAGTA CTAAACGTTG GGTAAAACCA
Gl-1is     GATTAGATAC CCTGGTAGTC CACGCCGTAA ACGATGAGTA CTAAACGTTG GGTAAAACCA
OY-M       GATTAGATAC CCTGGTAGTC CACGCCGTAA ACGATGAGTA CTAAACGTTG GGTAAAACCA
AYWB       GATTAGATAC CCTGGTAGTC CACGCCGTAA ACGATGAGTA CTAAACGTTG GGTAAAACCA

                 845        855        865        875        885        895
AY-St1     GTGTTGAAGT TAACACATTA AGTACTCCGC CTGAGTAGTA CGTACGCAAG TATGAAACTT
Gl-1is     GTGTTGAAGT TAACACATTA AGTACTCCGC CTGAGTAGTA CGTACGCAAG TATGAAACTT
OY-M       GTGTTGAAGT TAACACATTA AGTACTCCGC CTGAGTAGTA CGTACGCAAG TATGAAACTT
AYWB       GTGTTGAAGT TAACACATTA AGTACTCCGC CTGAGTAGTA CGTACGCAAG TATGAAACTT

                 905        915        925        935        945        955
AY-St1     AAAGGAATTG ACGGGACTCC GCACAGGCGG TGGATCATGT TGTTTAATTC GAAGGTACCC
Gl-1is     AAAGGAATTG ACGGGACTCC GCACAAGCGG TGGATCATGT TGTTTAATTC GAAGGTACCC
OY-M       AAAGGAATTG ACGGGACTCC GCACAAGCGG TGGATCATGT TGTTTAATTC GAAGGTACCC
AYWB       AAAGGAATTG ACGGGACTCC GCACAAGCGG TGGATCATGT TGTTTAATTC GAAGGTACCC

                 965        975        985        995       1005       1015
AY-St1     GAAAAACCTC ACCAGGTCTT GACATGCTTC TGCAAAGCTG TAGAAACACA GTGGAGGTTA
Gl-1is     GAAAAACCTC ACCAGGTCTT GACATGCTTC TGCAAAGCTG TAGAAACACA GTGGAGGTTA
OY-M       GAAAAACCTC ACCAGGTCTT GACATGCTTC TGCAAAGCTG TAGAAACACA GTGGAGGTTA
AYWB       GAAAAACCTC ACCAGGTCTT GACATGCTTC TGCAAAGCTG TAGAAACACA GTGGAGGTTA

                1025       1035       1045       1055       1065       1075
AY-St1     TCAGTTGCAC AGGTGGTGCA TGGTTGTCGT CAGCTCGTGT CGTGAGATGT TGGGTTAAGT
Gl-1is     TCAGTTGCAC AGGTGGTGCA TGGTTGTCGT CAGCTCGTGT CGTGAGATGT TGGGTTAAGT
OY-M       TCAGTTGCAC AGGTGGTGCA TGGTTGTCGT CAGCTCGTGT CGTGAGATGT TGGGTTAAGT
AYWB       TCAGTTGCAC AGGTGGTGCA TGGTTGTCGT CAGCTCGTGT CGTGAGATGT TGGGTTAAGT

                1085       1095       1105       1115       1125       1135
AY-St1     CCCGCAACGA GCGCAACCCT TATTGTTAGT TGCCAGCACG TAATGGTGGG GACTTTAGCA
Gl-1is     CCCGCAACGA GCGCAACCCT TATTGTTAGT TACCAGCACG TAATGGTGGG GACTTTAGCA
OY-M       CCCGCAACGA GCGCAACCCT TATTGTTAGT TACCAGCACG TAATGGTGGG GACTTTAGCA
AYWB       CCCGCAACGA GCGCAACCCT TATTGTTAGT TGCCAGCACG TAATGGTGGG GACTTTAGCA
```

Fig. 6.11 Alignment of aster yellows strains: Gl-1is, GenBank Accession Number (Acc. No.) KP890829; AY-St1, Acc. No. KP890830; aster yellows strains OY-M (16SrI-B, Acc. No. NC_005303); and AY-WB (16SrI-A, Acc. No. NC_007716). The four SNPs are between isolate Gl-1is and strain AY-St1; two of these are in common between AY-St1 and AY-WB (red squares), while the other two are specific for strain AY-St1 (green squares)

6.8 Conclusions

The phytoplasma cultivation was an important breakthrough in the study of their biology since, despite a reduced genome size in comparison to their ancestors, their retain an independent metabolism that allows them to survive outside the host species very likely with restricted ability to grow in artificial media. This versatility is a unique property among plant-inhabiting microbes cultured up to now, shared only with some animal- or plant-infecting viruses and with a few other microorganisms such as the causal agent of malaria. The phytoplasma isolation and biological characterization are relevant for disease management and containment measures to contrast their epidemics.

References

Alma A, Bosco D, Danielli A, Bertaccini A, Vibio M, Arzone A (1997) Identification of phytoplasmas in eggs, nymphs of *Scaphoideus titanus* Ball reared on healthy plants. *Insect Molecular Biology* **2**, 115–121.

Alma A, Palermo S, Boccardo G, Conti M (2001) Transmission of chrysanthemum yellows, a subgroup 16SrI-B phytoplasma, to grapevine by four leafhopper species. *Journal of Plant Pathology* **83**, 181–187.

Alvarez E, Mejía JF, Llano GA, Loke JB, Calari A, Duduk B, Bertaccini A (2009) Detection and molecular characterization of a phytoplasma associated with frogskin disease in Cassava. *Plant Disease* **93**, 1139–1145.

Alvarez E, Mejía JF, Pardo JM (2010) Development of a real-time PCR assay, to detect and quantify a 16SrIII-L phytoplasma associated with cassava frogskin disease (CFSD). *Phytopathology* **100**, S5.

Alvarez E, Betancourth C, Muñoz J (2017) Pathogenicity of a 16SrIII-L phytoplasma associated with frogskin disease of cassava (*Manihot esculenta* Crantz) in Colombia. *Phytopathology* **107**, S2.5.

Angelini E, Constable F, Duduk B, Fiore N, Quaglino F, Bertaccini A (2018) Grapevine phytoplasmas. In: Phytoplasmas: Plant Pathogenic Bacteria-I. Characterization and Epidemiology of Phytoplasma-Associated Diseases. Eds Rao GP, Bertaccini A, Fiore N, Liefting L. Springer, Singapore, 123–151 pp.

Arocha Rosete Y, Atta Diallo H, Konan Konan JL, Yankey N, Saleh M, Pilet F, Contaldo N, Paltrinieri S, Bertaccini A, Scott JA (2017) Improving sampling and detection for the coconut lethal yellowing phytoplasma in coconut-growing villages of Grand-Lahou, Côte d'Ivoire. *Annals of Applied Biology* **170**, 333–347.

Bertaccini A, Davis RE, Lee I-M (1992) *In vitro* micropropagation for maintenance of mycoplasmalike organisms in infected plant tissues. *HortScience* **27**, 1041–1043.

Bertaccini A, Contaldo N, Calari A, Paltrinieri S, Windsor HM, Windsor DG (2010) Preliminary results of axenic growth of phytoplasmas from micropropagated infected periwinkle shoots. 18th Congress of the International Organization for Mycoplasmology, Chianciano Terme, Italy, July 11–16 147, 153.

Brown DR, Whitcomb RF, Bradbury JM (2007) Revised minimal standards for description of new species of the class *Mollicutes* (division *Tenericutes*). *International Journal of Systematic and Evolutionary Microbiology* **57**, 2703–2719.

Contaldo N, Bertaccini A, Paltrinieri S, Windsor HM, Windsor DG (2012) Axenic culture of plant pathogenic phytoplasmas. *Phytopathologia Mediterranea* **51**, 607–617.

Contaldo N, Bertaccini A, Paltrinieri S, Windsor DG, Windsor HM (2013) Cultivation of several phytoplasmas from a micropropagated plant collection. *Petria* **23**, 13–18.

Contaldo N, Windsor DG, Windsor HM, Paltrinieri S, Satta E, Bertaccini A (2014a) Phytoplasma cultivation from a micropropagated plant collection. IOM 2014, June 1-6, Blumenau, Brazil 33, 40.

Contaldo N, Satta E, Bertaccini A, Windsor DG (2014b) Methods for isolation by culture, and subsequent molecular identification, of phytoplasmas from plants sourced in the field. IOM 2014, June 1-6, Blumenau, Brazil 106, 56.

Contaldo N, Satta E, Zambon Y, Canel A, Paltrinieri S, Bertaccini A (2014c) Morphological and molecular identification of phytoplasmas in culture from plants sourced in the field. *Journal of Plant Pathology* **96**, S.4.39.

Contaldo N, Satta E, Paltrinieri S, Gherardi M, Laurita R, Stancampiano A, Zambon Y, Colombo V, Bertaccini A (2015a) Interaction between cold atmospheric plasma and phytoplasmas in micropropagated infected periwinkle shoots. *Phytopathogenic Mollicutes* **5**(1-Supplement), S127-S128.

Contaldo N, Zambon Y, Satta E, Paltrinieri S, Bertaccini A (2015b) Cultivation in chemically defined media of phytoplasmas from field-infected grapevine plants showing yellows symptoms. 18th Congress of the International Council for the Study of Virus and Virus-like Diseases of the Grapevine. September 7-11, Ankara, Turkey, 121–123.

Contaldo N, Satta E, Zambon Y, Paltrinieri S, Bertaccini A (2016a) Development and evaluation of different complex media for phytoplasma isolation and growth. *Journal of Microbiological Methods* **127**, 105–110.

Contaldo N, Zambon Y, Paltrinieri S, Mori N, Mitrovic J, Duduk B, Bertaccini A (2016b) Characterization of '*Candidatus* Phytoplasma solani' strains from grapevines, *Hyalesthes obsoletus*, reference strains in periwinkle and in colonies. *Mitteilungen Klosterneuburg* **66**, 63–69.

Contaldo N., D. Dolanc, A. Bertaccini, M. Dermastia. 2018a. Isolation of diverse phytoplasmas from symptomatic grapevine samples. "Bois noir" Fifth Workshop, Liubljana September 18 and 19, a16.

Contaldo N., G. D'Amico, Y. Zambon, A. Bertaccini. 2018b. '*Candidatus* Phytoplasma asteris' isolated from grapevine: preliminary metabolic features. 22th Congress of the International Organization for Mycoplasmology (IOM) July 9–12, Portsmouth, NH, USA 54, 40.

Contaldo N, D'Amico G, Paltrinieri S, Diallo HA, Bertaccini A, Arocha Rosete Y (2019) Molecular and biological characterization of phytoplasmas from coconut palms affected by the lethal yellowing disease in Africa. *Microbiological Research* **223–225**, 51–57.

Danielli A, Bertaccini A, Bosco D, Alma A, Vibio M, Arzone A (1996) May evidence of 16SrI-group-related phytoplasmas in eggs, nymphs and adults of *Scaphoideus titanus* Ball suggest their transovarial transmission? *IOM Letters* **4**, 190–191.

Davis RE, Worley JF, Whitcomb RF, Ishijima T, Steere RL (1972) Helical filaments produced by a mycoplasma-like organism associated with corn stunt disease. *Science* **176**, 521–523.

Doi YM, Teranaka M, Yora K, Asuyama H (1967) Mycoplasma or PLT-group-like microorganisms found in the phloem elements of plants infected with mulberry dwarf, potato witches' broom, aster yellows, or paulownia witches' broom. *Annals of the Phytopathological Society of Japan* **33**, 259–266.

Elmendorf E (1977) Newly discovered prokaryotic plant pathogens. A bibliography. R 2901 Research Bulletin College Agriculture University of Wisconsin, Madison, United States of America, 1–85 pp.

Freundt EA, Erno H, Lemcke RM (1979) Identification of mycoplasma. In: Methods in Microbiology. Ed Bergan N, Academic Press, London, United Kingdom, 396 pp.

Gasparich GE, Whitcomb RF, Dodge D, French FE, Glass J, Williamson DL (2004) The genus *Spiroplasma* and its non-helical descendants: phylogenetic classification, correlation with phenotype and roots of the *Mycoplasma mycoides* clade. *International Journal of Systematic and Evolutionary Microbiology* **54**, 893–918.

Ghosh SK, Raychaudhuri SP, Chenulu VV, Varma A (1975) Isolation, cultivation and characterization of mycoplasma-like organisms from plants. *Proceedings of the Indian National Science Academy* **41B**, 362–366.

Giannotti J, Vago C (1971) Role des mycoplasmes dans l'étiologie de la phyllodie du trèfle: culture et transmission expérimentale de la maladie. *Physiologie Vegetale* **9**, 541–553.

Jones AL, Whitcomb RF, Williamson DL, Coan ME (1977) Comparative growth and isolation of spiroplasmas in media beside on insect tissue culture formulation. *Phytopathology* **67**, 738–746.

Ishiie T, Doi Y, Yora K, Asuyama H (1967) Suppressive effects of antibiotics of tetracycline group on symptom development of mulberry dwarf disease. *Annals of the Phytopathological Society of Japan* **33**, 267–275.

Kunkel LO (1926) Studies on aster yellows. *American Journal of Botany* **23**, 646–705.

Kunkel LO (1931) Celery yellows of California not identical with the aster yellows of New York. *Boyce Thompson Institute* **4**, 405–414.

Kunkel LO (1955) Cross protection between strains of aster yellow-type viruses. *Advances in Virus Research* **3**, 251–273.

Lee I-M, Davis RE (1992) Mycoplasmas which infect plants and insects. In: Mycoplasmas: Molecular Biology and Pathogenesis. Eds Maniloff J, McElhansey RN, Finch LR, Baseman JB. American Society for Microbiology. Washington DC, United States of America, 379–390 pp.

Lim PO, Sears BB (1989) 16S rRNA sequence indicates that plant-pathogenic mycoplasmalike organisms are evolutionarily distinct from animal mycoplasmas. *Journal of Bacteriology* **171**, 5901–5906.

Lin SC, Lee CS, Chin RJ (1970) Isolation and cultivation of, and inoculation with a mycoplasma causing white leaf disease of sugarcane. *Phytopathology* **60**, 795–797.

Lombardo G, Pignattelli P (1970) Cultivation in a cell-free medium of a mycoplasma-like organism from *Vinca rosea* with phyllody symptoms on the flowers. *Annals of Microbiology* **20**, 83–88.

Maramorosch K (2011) Historical reminiscences of phytoplasma discovery. *Bulletin of Insectology* **64**(Supplement), S5-S8.

Mc Coy RE, Caudwell A, Chang CJ, Chen TA, Chiykowskyi LN, Cousin MT, Dale de Leeuw GTN, Golino DA, Hackett KJ, Kirkptrick BC, Marwitz R, Petzold H, Shina RH, Sugiura M, Whitcomb RF, Yang IL, Zhu BM, Seemüller E (1989) Plant diseases associated with mycoplasmalike organisms. In: The Mycoplasmas, volume 5. Eds Whitcomb RF, Tully JG. Academic Press, New York, United States of America, 545–640 pp.

Mori N, Martini M, Bressan A, Guadagnini M, Girolami V, Bertaccini A (2002) Experimental transmission by *Scaphoideus titanus* Ball of two molecularly distinct "flavescence dorée" type phytoplasmas. *Vitis* **41**, 99–102.

Müller HM, Kleinhempel H, Rabitschuch AW, Spaar D, Müller JH (1975) Mykoplasmen in der Phytopathologie, Bibliographisches Verzeichnis der Literatur fur die Jahre 1967–1974. Institut fur Landwritschaftliche Information und Dokumentation, Berlin.

Nasu S, Sugiura M, Wakimoto T, Iida TT (1967) On the pathogen of rice yellow dwarf virus. *Annals of Phytopathological Society Japan* **33**, 343–344.

Nocard E, Roux ER (1898) Le microbe de la péripneumonie. *Annales de l'Institute Pasteur* **12**, 240–262.

Poghosyan A, Lebsky V (2004) Aislamiento y estudio ultrastructural de tres cepas de fitoplasmas causantes de enfermedades tipo "stolbur" en solanaceae. *Fitopatologia Colombiana* **8**, 21–30.

Rao GP, Bertaccini A, Fiore N, Liefting L (2018) Phytoplasmas: Plant Pathogenic Bacteria-I. Characterization and Epidemiology of Phytoplasma-Associated Diseases. Springer, Singapore, 345 pp.

Rid M, Mesca C, Ayasse M, Gross J (2016) Apple proliferation phytoplasma influences the pattern of plant volatiles emitted depending on pathogen virulence. *Frontieres in Ecological Evolution* **3**, 152.

Saccardo F, Martini M, Palmano S, Ermacora P, Scortichini M, Loi N, Firrao G (2012) Genome drafts of four phytoplasma strains of the ribosomal group 16SrIII. *Microbiology* **158**, 2805–2814.

Saglio P, L'Hospital M, Lafleche D, Dupont G, Bové J-M, Tully JG, Freundt EA (1973) *Spiroplasma citri* gen. and sp. n.: a mycoplasma-like organism associated with "stubborn" disease of citrus. *International Journal of Systematic Bacteriology* **23**, 191–204.

Satta E, Contaldo N, Paltrinieri S, Bertaccini A (2016) Biological and molecular proof of phytoplasma seed transmission in corn. 21th Congress of the IOM, Brisbane, Australia, July 3-7 61, 65–66.

Satta E, Paltrinieri S, Bertaccini A (2019) Phytoplasma transmission by seed. In: Phytoplasmas: Plant Pathogenic Bacteria-II. Transmission and Management of Phytoplasma Associated Diseases. Chapter 6. Eds. Bertaccini A, Weintraub PG, Rao GP, Springer, Singapore 131–147 pp.

Seemüller E, Schneider B (2007) Differences in virulence and genomic features of strains of '*Candidatus* Phytoplasma mali', the apple proliferation agent. *Phytopathology* **97**, 964–970.

Skripal IG, Malinivskaya LP (1984) Medium SMIMB-72 for isolation and cultivation of phytopathogenic mycoplasmas. *Microbiologiceskii Zhurnal* **46**, 71–75.

Skripal IG, Malinovskaya LP, Onishchenko AN (1984) A method for isolation of mycoplasmas from plants affected by yellows diseases. *Microbiologiceskii Zhurnal* **46**, 93–96.

Whitcomb RF, Tully GJ (1975) The Mycoplasmas III. The Academic Press, New York, United States of America, 351 pp.

Chapter 7
Molecular and Serological Approaches in Detection of Phytoplasmas in Plants and Insects

Assunta Bertaccini, Nicola Fiore, Alan Zamorano, Ajay Kumar Tiwari, and Govind Pratap Rao

Abstract The impact of phytoplasmas in agriculture has become serious, and early diagnosis is the best option to prevent the disease spread. Very often the symptoms-based diagnostics is not sufficient or able to discriminate among the diverse phytoplasmas. Until the early 1980s, the phytoplasma presence in diseased plants was detected by transmission electron microscopy observation, and DAPI staining that was developed to detect the pathogen under fluorescent microscopy. Enzyme-linked immunosorbent assay (ELISA) was rarely used since the antisera were developed only for a few phytoplasma-associated diseases. Around 1990, advances in molecular biology enabled direct detection of phytoplasma DNA by hybridization and polymerase chain reaction technologies. PCR amplification of the 16S rRNA genes of phytoplasmas has become the key in phytoplasma disease detection, and now, several variants of PCR like nested and quantitative PCR, microarrays, and NGS are used for detection of phytoplasmas in both plants and insects. The approach using RFLP analyses and or sequencing of PCR-amplified 16S rDNA fragments provides a simple, reliable, and rapid mean for differentiation and identification of known phytoplasma strains. In this chapter up-to-date accounts of developments in serological and molecular approaches for the phytoplasma identification in plants and insect vectors are summarized.

A. Bertaccini (✉)
Department of Agricultural and Food Sciences, *Alma Mater Studiorum* – University of Bologna, Bologna, Italy

N. Fiore · A. Zamorano
Faculty of Agricultural Sciences, Department of Plant Protection, University of Chile, Santiago, Chile

A. K. Tiwari
Sugarcane Research Institute, Shahjahanpur, Uttar Pradesh, India

G. P. Rao
Division of Plant Pathology, Indian Agricultural Research Institute, New Delhi, India

A. Bertaccini et al. (eds.), *Phytoplasmas: Plant Pathogenic Bacteria - III*,
https://doi.org/10.1007/978-981-13-9632-8_7

Keywords Serological detection · PCR assays · Nested PCR · Quantitative PCR · RFLP · 16S rDNA

7.1 Introduction

The development of a reliable method for the detection of phytoplasmas in infected tissues is still one of the biggest challenges for the study of these elusive plant pathogens. The long-time absence of a proper culture media for these bacteria and the present difficulties observed in the cultivation of most of the species of the genus '*Candidatus* Phytoplasma' have a direct impact in the available information of molecular targets for the development of improved detection methods. This information is of high concern regarding the impact of these diseases in agriculture worldwide, because the control of this pathogen is based mainly in the prevention of the infection, due to the absence of a direct field control (Bertaccini 2007).

Up to now, 34 ribosomal groups and more than 200 subgroups have been identified that are able to colonize more than 700 species of plants. All these groups were defined on the bases of 16S rRNA gene sequences, although other conserved genes have been used to strengthen, cross confirm, and/or better detail this classification (Martini et al. 2007a; Makarova et al. 2012).

Before the development of molecular techniques, the detection of phytoplasma diseases was difficult also due to the fact that they were not yet cultured. The phytoplasma identification relied for more than two decades on diagnostic techniques such as observation of symptoms, insect or dodder/graft transmission to host plants, and DAPI staining together with electron microscopy observation of ultrathin sections of the phloem tissue. Phytoplasma strains were also differentiated and identified by their biological properties, such as the similarity in symptoms induced in infected plants, plant host, and insect vector ranges (Chiykowski 1991; Errampalli et al. 1991; Lee and Davis 1992), but this was laborious and time-consuming, and often the results were inconsistent. In many cases, the identities of the insect vectors remained unknown, further complicating the identification based on biological criteria. Serological diagnostic techniques for the detection of phytoplasmas began to emerge in the 1980s when polyclonal and monoclonal antisera were produced and tested also for the phytoplasma detection using fast and handling methods such as ELISA and immunofluorescence (Lee et al. 1993a; Chen et al. 1993, 1994).

However, in the last 20 years, the applications of DNA-based methods allowed to distinguish different molecular clusters inside these prokaryotes mainly by dot and Southern blot hybridization and PCR technology. Dot and Southern blot hybridization assays were also used in phytoplasma detection for some years; however, both are currently completely replaced by PCR assays. More recently, different variants of PCR (nested PCR, quantitative PCR, digital droplet PCR) techniques have been developed that have been shown to be effective methods of phytoplasma detection within the plants and insects (Christensen et al. 2004; Bertaccini et al. 2014). PCR using universal primers are now routine assays for the detection of phytoplasmas and several universal and many phytoplasma group-specific primers have been designed for their detection (Deng and Hiruki 1991; Lee et al. 1993b, c, 2000; Lorenz et al. 1995; Smart et al. 1996; Gundersen and Lee 1996). Nested PCR assay, designed to increase both sensitivity and specificity, was performed by using group-specific primers and is therefore capable of detection of dual or multiple phytoplasmas present in the infected tissues in case of mixed infection (Lee et al. 1994, 1995). The majority of primers used in PCR are derived from 16S ribosomal gene, which, due to the conservation and the presence also of variable regions, is ideal for phylogenetic analysis that is the basis for the classification (Lee et al. 1993a, 1998). However the amplification with group-specific primers is rarely allowing phytoplasma identification, and other techniques such as restriction fragment length polymorphism (RFLP) and sequence analysis are necessary to reach a suitable classification (Lee et al. 1998; IRPCM 2004). The development of a reliable detection method is necessary to minimize the impact of these pathogens in agriculture worldwide. Thus, molecular tools based in genetic sequences are still the most recommended methods to detect the phytoplasmas. Molecular detection of phytoplasmas involves sampling of the tissues, extraction of DNA, selection of gene-specific primers, PCR assays, RFLP, or sequencing and sequence analysis. An essential prerequisite is the maintenance of phytoplasma strains in suitable hosts or under appropriate conditions (frozen or lyophilized). In order to obtain the best performance of a detection method, three main steps must be considered: the nucleic acid extraction, the DNA target selection, and the detection method.

7.2 DNA Extraction from Plants and Insects

The quality of the DNA extraction is the first high-relevance step that must be considered in phytoplasma detection. Since phytoplasmas reside almost exclusively in sieve tubes, the starting material for DNA extraction should include as much phloem tissue as possible. Different extraction protocols are applied, depending on the tissue involved and the required sensitivity of the test. Plants contain also many secondary metabolites, such as polyphenols and polysaccharides, which may inhibit the polymerase reaction, and they are significantly increased in phytoplasma-infected plants (Choi et al. 2004). PCR inhibition can be overcome by diluting the DNA extracts as reported in DNA extracts from mature papaya leaves, apricot leaf veins, and apple tree roots (Padovan and Gibb 2001; Heinrich et al. 2001; Brzin et al. 2003). It is therefore essential to remove inhibitory substances during the DNA extraction steps.

The availability of DNA-based methods has greatly improved the detection of phytoplasma presence in plant and insect hosts (Kirkpatrick et al. 1987, 1995;

Kollar et al. 1990; Lee et al. 1991; Ahrens and Seemüller 1992). The polymerase chain reaction (PCR) technology is the widely used method of choice for phytoplasma detection and requires a quality template DNA extracted from diseased plants and insect vectors of suitable quality and concentration. Several phytoplasma DNA enrichment procedures were developed by different laboratories, but none of these protocols was universally adopted (Jiang and Chen 1987; Garcia-Chapa et al. 2004). Moreover the amount of phytoplasma DNA in total DNA extracted from infected plants/insects can be further increased using appropriate phytoplasma enrichment procedures (Kirkpatrick et al. 1987; Ahrens and Seemüller 1992; Prince et al. 1993) mainly based on differential centrifugations. Most host nuclear and chloroplast DNA is eliminated during a low-speed centrifugation step, whereas most polysaccharides, phenolic compounds, and other enzyme-inhibiting molecules are discarded in the supernatant following a high-speed centrifugation step. The resulting phytoplasma-enriched pellet is then processed using a high-salt cetyl trimethyl ammonium bromide (CTAB)-based buffer followed by chloroform/isoamyl alcohol extraction prior to isopropanol precipitation. This procedure is effective in producing highly purified DNA from fresh tissues from a wide variety of herbaceous and woody plants. A method for the enrichment of phytoplasma DNA was also developed by Prabu et al. (2008), in which differential filtration methods used in concentrating phytoplasmas were adopted. The major advantage of the phytoplasma enrichment procedure is that a substantial proportion of the extracted DNA is originating from phytoplasmas. The other procedures are based on the CTAB extraction method described by Doyle and Doyle (1990), which involve treatments with CTAB-based buffer to lyse cells and purify DNA followed by deproteination and recovery of DNA. These protocols work well for extracting total DNA from fresh, frozen, or lyophilized tissues from a wide variety of plant hosts as well as insect vectors. Because a few manipulations are required, they are faster and easier to perform than the phytoplasma enrichment methods. In addition, they result in very high yields and provide DNA that is less pure, but of suitable quality for use with standard molecular techniques including PCR assays. Commercially available kits have also been employed for DNA extraction from diseased plants for detection of phytoplasmas by PCR (Green et al. 1999) and DNA extraction techniques from insects were also developed for quality phytoplasma DNA extractions (Zhang et al. 1998; Bosco et al. 2002; Palmano 2001; Mandrioli 2008).

7.3 Target DNA Selection

Phytoplasma differentiation is routinely based on amplification of 16S rRNA gene sequences and allowed to distinguish 34 ribosomal groups (16Sr) and more than 200 subgroups indicated by capital letters (Bertaccini and Lee 2018). The use of additional genetic markers could enhance the discrimination resolving power for finer phytoplasma strain distinction. Since the first phytoplasma-specific primers were

Table 7.1 List of universal primers for amplification of 16S rRNA gene of phytoplasmas

Name	Sequence 5′-3′	Gene	Reference
P1	AAG AAT TTG ATC CTG GCT CAG GAT T	16Sr	Deng and Hiruki (1991)
P7	CGT CCT TCA TCG GCT CTT	23Sr	Schneider et al. (1995)
R16F2n	GAA ACG ACT GCT AAG ACT GG	16Sr	Gundersen and Lee (1996)
R16R2	TGA CGG GCG GTG TGT ACA AAC CCC G	16Sr	Lee et al. (1995)
F1	AAG ACG AGG ATA ACA GTT GG	16Sr	Davis and Lee (1993)
B6	TAG TGC CAA GGC ATC CAC TGT G	16Sr	Padovan et al. (1995)
R16mF2	CAT GCA AGT CGA ACG GA	16Sr	Gundersen and Lee (1996)
R16mR2	CTT AAC CCC AAT CAT CGA	16Sr	
P1A	AAC GCT GGC GGC GCG CCT AAT AC	16Sr	Lee et al. (2003)
P7A	CCT TCA TCG GCT CTT AGT GC	23Sr	
U5	CGG CAA TGG AGG AAA CT	16Sr	Lorenz et al. (1995)
U3	TTC AGC TAC TCT TTG TAA CA	16Sr	
M1(758f)	GTC TTT ACT GAC GC	16Sr	Gibbs et al. (1995)
M2(1232r)	CTT CAG CTA CCC TTT GTA AC	16Sr	
R0	GAA TAC CTT GTT ACG ACT TAA CCC C	16Sr	Lee et al. (1995)
P3	GGA TGG ATC ACC TCC TT	16Sr	Schneider et al. (1995)
P4	GAA GTC TGC AAC TCG ACT TC	16Sr	
P5	CGG CAA TGG AGG AAA CT	16Sr	
16Sr-SR	GGT CTG TCA AAA CTG AAG ATG	IS	Lee et al. (2006)
Pc399	AAC GCC GCG TGA ACG ATG AA	16Sr	Skrzeczkowski et al. (2001)
Pc1694	ATC AGG CGT GTG CTC TAA CC	IS	
PA2f	GCC CCG GCT AAC TAT GTG C	16Sr	Heinrich et al. (2001)
PA2r	TTG GTG GGC CTA AAT GGA CTC	IS	
SN910601	GTT TGA TCC TGG CTC AGG ATT	16Sr	Namba et al. (1993)
SN910502	AAC CCC GAG AAC GTA TTC ACC	16Sr	
1F7	AGT GCT TAA CAC TGT CCT GCT A	16Sr	Manimekalai et al. (2010)
7R3	TTG TAG CCC AGA TCA TAA GGG GCA	16Sr	
3Fwd	ACC TGC CTT TAA GAC GAG GA	16Sr	
3rev	AAA GGA GGT GAT CCA TCC CCA CCT	16Sr	
7R2	GAC AAG GGT TGC GCT CGT TTT	16Sr	
5Rev	ACC CCG AGA ACG TAT TCA CCG CGA	16Sr	

designed in 1991 (Deng and Hiruki), several primer pairs have been developed and validated both for general and group-specific phytoplasma detection (Tables 7.1 and 7.2). However the selected conserved regions of 16S rRNA gene are also conserved in other *Mollicutes*. Thus, the efficiency of the process is led by the use of at least one more specific primer per round of amplification. During the last years, a constant increase of the available genetic information on phytoplasma strains has led to the identification of several potential molecular markers: despite this, the 16S rRNA gene is the most used marker for the development of new protocols, not only since it is the basis for classification, but also because there is a great number of strains that can be aligned and compared for the identification of conserved regions after

Table 7.2 List of group-specific primers for amplification of 16S rRNA gene of phytoplasmas

Name	Ribosomal group (16Sr) amplified	Sequence 5′-3′	Reference
R16(I)F1	-I, -II, -IX, -XII, -XV	TAA AAG ACC TAG CAA TAG G	Lee et al. (1994)
R16(I)R1	-I, -II, -IX, -XII, -XV	CAATCCGAACTAAGACTCT	
R16(III)F2	-III	AAGAGTGGAAAAACTCCC	
R16(III)R1	-III	TTC GAA CTG AGA TTG A	
LY16Sf	-IV	CAT GCA AGT CGA ACG GAA ATC	Harrison et al. (2002)
LY16Sr	-IV	GCT TAC GCA GTT AGG CTG TC	
R16(V)F1	-V	TTA AAA GAC CTT CTT CGG	Lee et al. (1994)
R16(V)R1	-V	TTC AAT CCG TAC TGA GAC TAC C	
R16(X)F1	-X	GACC CGC AAG TAT GCT GAG AGA TG	Lee et al. (1995)
R16(X)R1	-X	CAA TCC GAA CTG AGA GTC T	
fO1	-X	CGG AAA CTT TTA GTT TCA GT	Lorenz et al. (1995)
rO1	-X	AAG TGC CCA ACT AAA TGA T	

the universal detection with primers on this gene. The studies developed using a multilocus approach must be therefore based first on this gene for the determination of the phytoplasma taxon among those described (Lee et al. 1998; IRPCM 2004).

Searching the Best DNA Target In order to find a universal tool for phytoplasma detection, an exhaustive search among the genetic information available for these pathogens must be performed, identifying the mostly conserved genes that may be the perfect target for this purpose. In the cellular context, there are several molecules involved in life-or-death processes that must conserve their structure along the evolutionary chain. Among them, ribosomes are considered one of the most important (if not the most important) due to their utility in the interpretation of the genetic code, reading the information carried by the mRNA and assembling the amino acids delivered by transfer RNA. In an evolutionary perspective, the whole process must be extremely conserved, regarding the relevance of the final products, the proteins, which are involved in most of the processes carried out in the living cells. Prokaryotic ribosomes are composed of 65% rRNA and 35% ribosomal proteins that are strongly assembled in a single structure. The main skeleton of this structure is the ribosomal RNA that serve as a scaffold for the proteins that catalyse the generation of the amino acid chain. Thus, the conservation of its sequence is a guarantee for the correct translation and, in consequence, the survival of the cell.

In the 1980s, Carl Woese demonstrated that 16S rRNA was the perfect target for studying the bacterial evolution. Its presence in every prokaryotic cell and the presence of highly conserved domains that interact with ribosomal proteins are two of the most important characteristics that support this statement (Lane et al. 1985). In approximately 1,500 bp length of a full 16S rRNA molecule, there are nine variable regions that allow to obtain a good classification of bacterial families (Ramazzotti and Bacci 2018). Targeting all or some of these regions is useful for phylogenetic and taxonomic studies, but, when the objective is the identification of a single short sequence suitable for genus-specific primers, the selection gets more complicated.

Since the first phytoplasma-specific primers were developed in 1991 (Deng and Hiruki 1991), several primer pairs have been tested for the same purpose (Table 7.1); however, the selected conserved regions of 16Sr RNA gene are also highly conserved in other *Mollicutes*. Figures 7.1, 7.2, and 7.3 show partial alignments of 16S-23S regions highlighting the matching regions of three primer pairs that have been found among the most useful set of primers. The P1/P7 primer pairs are the external primers that amplify the longest fragment including the almost entire sequence of 16S ribosomal RNA gene, 16S–23S ribosomal RNA intergenic spacer, the complete tRNA-Ile gene, and a small portion of the 23S ribosomal RNA gene. P1 matches a highly conserved region in the beginning of the 16S ribosomal gene, being present in most of the *Mollicutes* species. In the other extreme of the gene, P7 primer shows

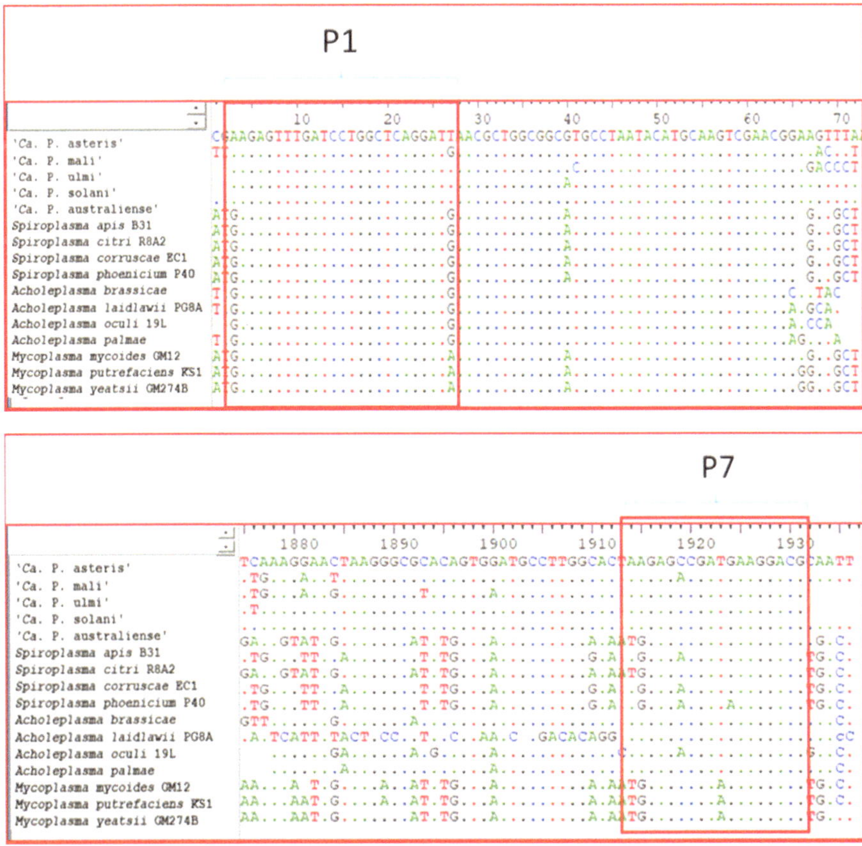

Fig. 7.1 Partial alignments of 16S-23S rRNA genes of *Mollicutes* species, highlighting genomic regions of matching with phytoplasma PCR primers P1/P7

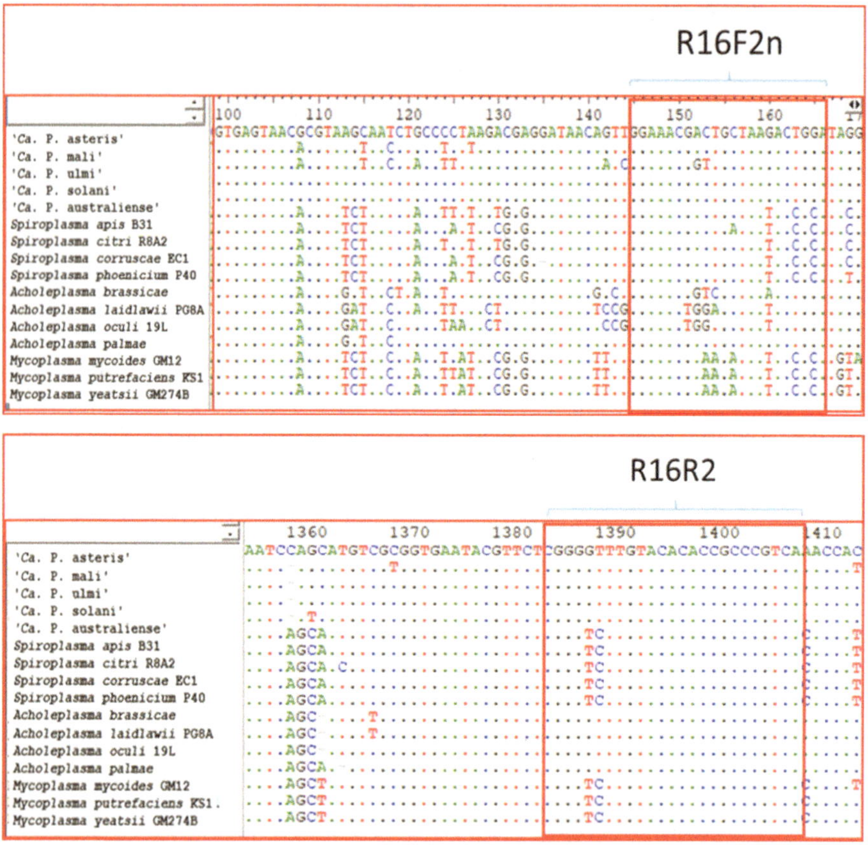

Fig. 7.2 Partial alignments of 16S-23S rRNA genes of *Mollicutes* species, highlighting genomic regions of matching with phytoplasma PCR primers R16F2n/R16R2

high conservation among phytoplasmas, but when it is compared against the same sequence of other *Mollicutes*, two or three nucleotide polymorphisms are present in the 3′ region, depending on the species. This generates a high specificity of the PCR for phytoplasma amplification (Fig. 7.1). The following primer pair that is widely used for phytoplasma detection is R16F2n/R16R2, being also the most useful for obtaining the amplification product mainly used for phytoplasma classification. The resulting 1,244 bp amplicon is used for phylogeny studies, and also for rapid identification in association with RFLP techniques (Zhao et al. 2009). In the same way than P1/P7, this primer pair lays its specificity in one primer. The alignment of partial 16S ribosomal gene regions shows a highly conserved region among *Mollicutes* in the R16R2 matching region, but in the R16F2n matching region, there is a low matching with other potential contaminant species (Fig. 7.2). Finally, the primer pair 758f/1232r (= M1/M2) shows a higher specificity for phytoplasma sequences in both matching regions, being an excellent candidate for phytoplasma-

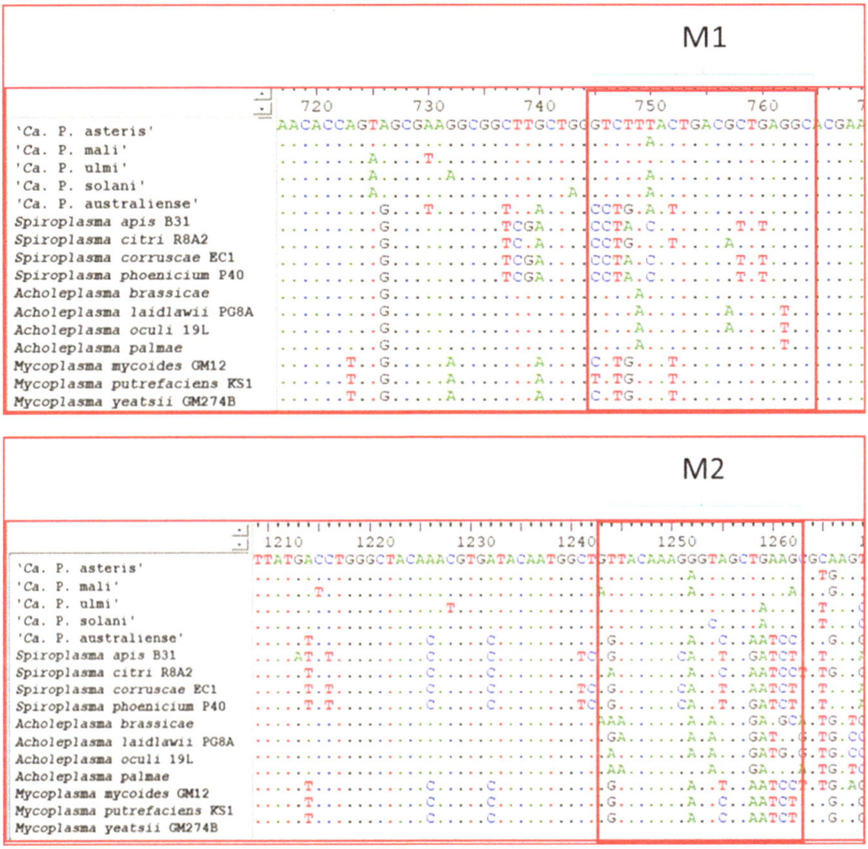

Fig. 7.3 Partial alignments of 16S-23S rRNA genes of *Mollicutes* species, highlighting genomic regions of matching with phytoplasma PCR primers M1/M2

specific detection (Fig. 7.3). Up to now, the combination of these three primer pairs results quite convenient in order to detect and characterize the phytoplasmas in different crops and insect vectors or potential vectors.

7.4 PCR-Based Techniques

Until these days, several phytoplasma group-specific detection protocols have been developed using PCR and quantitative PCR techniques, mostly based on the 16S rRNA gene sequence. PCR-based techniques, followed by a nested amplification, are the preferred protocol for amplifying the 16S rRNA gene, but the relatively high possibility of false positives is an important issue that must be considered in this diagnostic procedure. The use of a complementary technique such as RFLP or sequencing of the amplified gene must be added for the phytoplasma identification. Another marker used is the spacer region (SR) separating the 16S from the 23S rRNA gene of phytoplasmas. Phylogenetic analysis of the entire 16S-23S-SR (Gibbs et al. 1998; Kenyon et al. 1998; Khan et al. 2002) or variable regions flanking the tRNA-ile gene (Kirkpatrick et al. 1994; Schneider et al. 1995) is in some cases useful to clarify the phylogenetic position of phytoplasmas. Sometimes PCR is sufficient for the amplification of 16S ribosomal phytoplasma DNA, but often nested PCR is essential to get phytoplasma amplification because of the low phytoplasma titre especially in infected woody plants or for the presence of inhibitor of the PCR reaction or for the competition with the 16S rDNA from endophytic bacteria (A. Bertaccini, unpublished).

The restriction fragment length polymorphism (RFLP) analysis of PCR-derived amplicons is the fast method to carry out phytoplasma identification as well as typing for epidemiological studies (Schneider et al. 1993). It can be performed by cutting DNA using restriction endonucleases (wet RFLP) and/or by using sequence information combined with computer analysis (*in silico* RFLP). The genes and/or genome fragments amplified by PCR can also be used for subsequent sequencing and phylogenetic analysis.

Restriction Fragment Length Polymorphism Analysis Since the collective RFLP pattern characteristic of each phytoplasma is unique (Lee et al. 1998), the PCR/RFLP analyses on 16Sr RNA gene ideally allow detection and differentiation of all phytoplasmas. This system allocates the up-to-now worldwide detected phytoplasmas in 34 groups and more than 200 subgroups (Bertaccini and Lee 2018). Moreover, this system is more flexible for epidemiological studies than the use of the '*Candidatus*' taxa designation (IRPCM 2004) adopted until now for 43 phytoplasmas. At present, RFLP analysis of PCR-amplified rDNA is the routine method used for differentiation and classification of phytoplasmas, and can be improved further by using a large fragment comprising the entire 16S rRNA gene and 16S-23S rDNA spacer region. However, the 16S rRNA gene does not always seem sufficiently variable to allow distinction of phytoplasma strains that differ in plant host or vector specificity. RFLP analysis of randomly cloned chromosomal DNA fragments has also been employed, mainly to differentiate closely related phytoplasmas (Daire et al. 1997; Botti and Bertaccini 2003).

So far phytoplasmas from more than a thousand plant species have been sequenced and are available in the GenBank; interestingly many of the sequences are deposited in the last one decade, and availability of these high-quality sequences makes it easier to classify the phytoplasma through *in silico* RFLP. Streamlined computer-assisted RFLP analysis for rapid identification and classification of phytoplasmas was developed (Wei et al. 2007; Zhao et al. 2009). This work resulted in an expanded classification scheme in which ten possible new phytoplasma 16S rRNA groups and numerous subgroups were identified (Lee et al. 2007) and are available in the *i*PhyClassifier, a web-based research tool for quick identification of phytoplasmas. Based on RFLP pattern similarity coefficient scores, it gives immediate suggestions on group and subgroup classification status of the phytoplasma strain under study. The *i*PhyClassifier aligns the query sequence with that of strains in the database reported as '*Candidatus* Phytoplasma' species, calculates sequence similarity scores, and assigns the phytoplasmas under study into respective '*Ca.* Phytoplasma' species according to the guidelines (IRPCM 2004). Additional functions include the delineation of phytoplasma ribosomal groups and subgroups.

Several universal or group-specific primers designed on the 16Sr RNA sequence were developed (Table 7.1). They can be used in different combinations in direct, nested, or semi-nested systems for routine detection of phytoplasmas as well as for identification of new phytoplasmas (Bertaccini et al. 1992; Lee et al. 1995; Lorenz et al. 1995; Namba et al. 1993). Phytoplasma differentiation is routinely based on 16S rRNA gene sequences, which is carried out by RFLP analysis of PCR-amplified DNA sequences using 17 endonuclease restriction enzymes (Lee et al. 1998). It is also possible to sequence the PCR or nested PCR products and then to use the aligned sequences of at least 1,200 bp (amplicons obtained with the R16F2n/R2 primer combination) for phytoplasma identification *in silico* (Wei et al. 2007, 2008; Cai et al. 2008; Kakizawa et al. 2014; Zhao et al. 2009). The 16S rDNA amplicons obtained with the listed primers must be sequenced at least in both senses with or without previous cloning. The aligned consensus sequences obtained can be loaded directly in the *i*PhyClassifier (https://plantpathology.ba.ars.usda.gov/cgibin/resource/iphyclassifier.cgi), in pDRaw (http://www.acaclone.com/), or in other systems such as the barcode system of the EPPO-Q-bank (https://www.eppo.int/RESOURCES/eppo_databases/eppo_q_bank) to generate *in silico* profiles or provide preliminary phytoplasma classification. The *i*PhyClassifier can tentatively provide '*Candidatus* species', ribosomal groups and subgroups, and identity coefficient and provides *in silico* RFLP pictures comparing the target sequence to those available in the classifier. However all the data also must be confirmed by wet RFLP with appropriate enzymes. The barcode system allows matching the sequence to those of reference available in the EPPO-Q-bank and corresponding to phytoplasma strains maintained in collection in periwinkle (http://www.ipwgnet.org/collection).

Cloning and sequencing PCR products The cloning of the amplified product followed by its sequencing is another complementary technique that can be used for phytoplasma identification. Amplified products are purified by gel elution/purification and sequenced directly or cloned prior to sequencing. For cloning, DNA fragments are ligated into plasmid vector, *e.g.* pGEM-T and recombinant plasmid are used to transform *Escherichia coli* strains. A number not lower than 5 selected recombinant clones must then be screened for phytoplasma rDNA inserts. Plasmid DNA is then purified, and sequencing of both strands is performed by Sanger method. Primers for sequencing PCR products are the same as for PCR amplification, whereas the standard primers are used for sequencing the cloned fragments. The sequences are then compared with known phytoplasma sequences in GenBank using the Basic Local Alignment Search Tool (BLASTn) (http://www.ncbi.nlm.nih.gov/blast).

7.5 Further Molecular Phytoplasma Detection Methods

Terminal Restriction Fragment Length Polymorphism T-RFLP is a fingerprinting technique which combines both detection and identification in a single method, with the added benefit of inbuilt controls which removes the risk of false-negative results and in addition highlights potential false-positive results. It has also been noted that the standard universal primers can amplify products from closely related bacteria such as *Bacillus* spp., leading to occasional false positives and misdiagnosis (Harrison et al. 2002). Due to this other genes have been studied as potential diagnostic targets such as the *rp*, *tuf*, *secY*, and *secA* genes. Nevertheless whilst these can be less prone to false-positive results, most of these targets do not give universal detection of phytoplasmas. The technique involves combination of PCR amplification with one fluorescent-labelled primer, restriction enzyme digestion with one or more enzymes, and separation of the fragments on an automated DNA sequencer through high-resolution electrophoresis to allow the detection of the fluorescent dye-labelled products, called terminal restriction fragments (TRFs). Once the TRFs have been determined by interrogation of the output electropherogram (or T-RFLP profiles), these are compared to a table of expected TRFs which allows the 16Sr group to be identified. T-RFLP differs from typical RFLP analysis in that only the terminal fragment is detected due to the presence of the fluorescent dye. The other (nonterminal) fragments have no dye and are therefore not detected as in the standard RFLP where the pattern of all of the fragments is visualized. A primary advantage of T-RFLP is that in a single reaction one set of primer amplifies both the plant and phytoplasma DNA and TRFs are generated which represent both. This means that easy and rapid confirmation of the absence of false negatives is achieved confirming the quality of the DNA and the absence of inhibitors. The combination of all these factors (universal detection, determination of false negatives and false positives, sensitivity) makes T-RFLP a tool that is particularly suited for situations such as higher throughput screening, for example, of an outbreak site, where large numbers of samples require screening and confirmation of the 16Sr phytoplasma group. In some situations this may be more appropriate than quantitative PCR as it identifies the phytoplasma in one step. The qPCR is also a universal assay which however requires downstream identification to the group level or the use of group-specific primers which fail to detect infections with other phytoplasmas. T-RFLP also has the ability to detect mixed infections that involve phytoplasmas from different ribosomal groups since they can be detected in the same assay.

Quantitative PCR The qPCR also known as real-time PCR offers several advantages. It is a fast, sensitive, and reliable detection technique amenable to high throughput. Two fluorescent chemistries are available, intercalating dyes or hybridization probes. Intercalating dyes are relatively less expensive than TaqMan® hybridization probes, but the TaqMan® hybridization probes are most commonly used for phytoplasma detection. The qPCR may be designed for universal detection of phytoplasmas, for group- or subgroup-specific detection, or for simultaneous detection of several phytoplasma strains. It may also be used for relative or absolute quantification of target DNA in host plants and in insect vectors. Therefore, qPCR plays an important role in phytoplasma detection as well as in host-pathogen interaction and in epidemiological studies. The technology represents an advanced variant of the polymerase chain reaction, in which the accumulation of amplified fragments of nucleic acid is measured as the increase of fluorescent signal during or after each reaction cycle. Two types of detection chemistry are available: (i) intercalating dyes binding nonspecifically to double-stranded DNA that is generated during PCR and (ii) sequence-specific fluorogenic hybridization probes, which employ the principle of fluorescence resonance energy transfer (FRET) (Bustin and Nolan 2004). Numerous qPCR protocols have so far been described for phytoplasma detection, including qualitative assays for the detection of their presence in host plants and/or insect vectors (Baric and Dallavia 2004; Galetto et al. 2005; Crosslin et al. 2010; Pelletier et al. 2009; Herath et al. 2010; Petrzik et al. 2011; Minguzzi et al. 2016) and quantitative approaches for the determination of the pathogen titre (Bianco et al. 2004; Torres et al. 2005; Hren et al. 2007; Aldaghi et al. 2007; Baric et al. 2011; Jawhari et al. 2015; Linck et al. 2017). The most commonly applied procedure for assessing the plant pathogen load is the absolute qPCR, which depends on external standard curves to which the amplification signal of the target nucleic acid is related. A melt curve analysis is typically performed after the qPCR run to verify the specificity of the test and to check the presence of primer dimers or unspecific amplicons based on the melting temperature difference of the amplicons produced. Each amplicon has an expected melting temperature (Tm), the temperature at which 50% dissociations of dsDNA to ssDNA occur. The melting temperature curve may be transformed by the thermocycler software into melting peaks. Slight differences in amplicon sequence, sometimes of even one base, may be reflected by a change in the Tm peak. This allows multiplexing, *i.e.* detection of more than one pathogen in a single run, when their respective amplicon melting temperatures have distinct peaks in the melt curve analysis (Anniballi et al. 2012; Satta et al. 2017). Absolute quantification provides the advantage of assessing the exact number of pathogen DNA copies in the infected host tissues. However, it must be considered that the accuracy of quantification depends on the quality of the standards employed.

The qPCR-based techniques can improve both the sensitivity and the specificity of the assay. The sensitivity is increased by the use of fluorescent molecules that can be detected by the corresponding real-time machine. The use of a single-step reaction decreases the risk of contaminations avoiding false positives. The specificity is improved by the use of TaqMan-based probes that can be enhanced by the use of specific modifications like BHQplus (Christensen et al. 2004; Jawhari et al. 2015). The qPCR detection is based upon the measurement of emitted fluorescence during the PCR amplification. SYBR® Green is a commonly used intercalating agent; however, it binds nonspecifically to all amplicons and thereby potentially compromises the accuracy of quantification and may even lead to false positives. Fluorogenic hybridization probes, such as TaqMan® probes, ensure a higher degree of specificity as they require not only hybridization of primers, but also the TaqMan® probe to bind the template. As phytoplasma titre can vary significantly during the season and in different plant tissues, quantification of phytoplasmas can be measured utilizing this technique. Quantification is an important parameter in screening plants for susceptibility to phytoplasmas. Universal phytoplasma qPCR-based detection systems have been developed (Christensen et al. 2004; Marzachí and Bosco 2005; Hodgetts et al. 2009). Galetto et al. (2005) developed qPCR assays for the specific detection of "flavescence dorée", "bois noir", and apple proliferation phytoplasmas using SYBR® Green and an assay for the detection of 16SrV, 16SrX, and 16SrXII phytoplasma groups using TaqMan®. Angelini et al. (2007) published qPCR assays for phytoplasmas of "flavescence dorée", "bois noir", and aster yellows detection. New qPCR protocols were developed to detect the agents of apple proliferation, European stone fruit yellows, and pear decline disease agents are based on SYBR® Green chemistry and allowed a detection of phytoplasmas either in the 16SrX group (3) or specific for '*Ca*. P. mali' (Jarausch et al. 2004; Baric and Dallavia 2004; Galetto et al. 2005; Nikolić et al. 2010; Monti et al. 2013) and '*Ca*. P. prunorum' (Martini et al. 2007b; Yvon et al. 2009). For the detection of '*Ca*. P. prunorum', the TaqMan® chemistry was also applied (Pignatta et al. 2008). Babini et al. (2008) have reported the specific detection of '*Ca*. P. pyri' by a multiplex qPCR assay.

Reverse Transcription PCR The PCR-based assays can target either DNA or RNA. As the transcriptome is context-dependent, *i.e.* the mRNA content varies with physiology or development of organism, the choice for the RT-PCR method must be focused on a constitutively expressed gene, such as the ribosomal gene that has the additional advantage of the high copy number present in a living cell, in contrast to the two copies of the 16S rRNA gene present in one phytoplasma genome (Schneider and Seemüller 1994), resulting in an increased amount of the initial template for amplification. Comparison between a diagnostic method that uses nested PCR (on phytoplasma DNA-enriched preparations) and RT-PCR (using crude sap) has been performed, and statistical analyses demonstrated substantial agreement between the two sets of results (Margaria et al. 2007). Furthermore, adapting the RT-PCR to the quantitative TaqMan® system noticeably increases the sensitivity of the assay and has been successfully applied to the detection of the "flavescence dorée" phytoplasma (Margaria et al. 2009).

Microarrays This technology offers a generic assay that can potentially detect and differentiate all phytoplasmas (Nicolaisen et al. 2013). DNA microarray is a powerful tool for identification and differentiation of phytoplasma strains, and detection is based on oligonucleotide probes spotted on glass slides, allowing one sample to be analysed on each slide. However, formats for higher throughput have also been developed, such as multiple arrays on each slide (Nicolaisen and Bertaccini 2007) and arrays in microtitre plates. The main technology for microarray manufacture is robotic spotting of pre-synthesized oligonucleotide probes on glass slides with special surfaces to allow irreversible binding of probes to the surface (Dufva 2005). The key factor in microarray development is the probe design that starts with identification of the most suitable target region. Probes of different lengths can be used on microarrays depending on the application. Short probes (15–25 nucleotides) exhibit reduced sensitivity, but show better discrimination of minor sequence differences, whereas longer probes (>50 nucleotides) show increased sensitivity, but do not discriminate between target sequences with only a few sequences of difference. Generally a pre-microarray amplification step, such as PCR, is required to enhance sensitivity, for phytoplasma detection. To allow detection of the hybridization event, the PCR product must be labelled. Several approaches are available to label PCR products with Cy3 or other dyes (Zhang et al. 2005); however, it has been noted that this may decrease PCR efficiency. Alternatively, products may be biotin-labelled during PCR, followed by incubation with Cy3-streptavidin after hybridization. Post-PCR labelling procedures include random-primed labelling of the PCR product with Cy3-coupled nucleotides (Nicolaisen et al. 2013). Franke-Whittle et al. (2006) compared different labelling methods and found that, in their hands, labelling with a Cy3-coupled forward primer gave the best results. Only a limited number of probes were designed so far, and more probes could be designed for each 16S ribosomal group to improve reliability. Other probes designed for specific purposes such as identification of closely related strains in specific 16S ribosomal groups or '*Candidatus* Phytoplasma' species could be designed from other gene regions as long as it is possible to design flanking primers for PCR amplification.

DNA Barcoding This is an identification method based on comparison of a short DNA sequence with known sequences from a database, and it has been developed for European quarantine phytoplasma identification. DNA barcoding is a technique where short DNA sequences (DNA barcodes) are used for species identification (Hebert et al. 2003; Makarova et al. 2012). Briefly, DNA from an unidentified species is extracted, amplified with a set of generic primers, sequenced, and finally the sequence is compared with the sequences from a database of identified strains to assign the sample to a taxon. The main advantage of this method is that no morphological traits are needed for identification, but a limitation is the risk of contamination resulting in false identifications and a variable detection limit, which depends on the DNA concentration in the preparations. Since phytoplasma DNA is present as a minor component in a plant/insect DNA background and phytoplasmas often coexist with other bacteria, the primers should not amplify plant DNA or unrelated bacterial DNA. Among many regions tested, two barcodes, based on the *tuf* and 16S rRNA genes, proved to be suitable and were selected as DNA barcodes for phytoplasmas (Makarova et al. 2012; Contaldo et al. 2015, 2018).

Next generation sequencing (NGS) Nowadays, the advancement of next-generation sequencing technologies notably increased the available information regarding phytoplasma genomic content, however a study developed by Firrao et al. (2013), based on complete and draft genomic sequences of several phytoplasmas, still validated the importance of 16S rRNA gene in the phytoplasma classification, even when a multilocus analysis of 107 shared genes was performed. Besides the phylogenetic focus, the great amount of information obtained by these methods also give the chance of identifying new conserved genomic regions that can be selected as target for sensitive and specific phytoplasma detection. Several genomic comparison tools as Artemis or Mauve allow the identification of conserved regions where the researchers have to focus for the selection of the appropriate amplification target. Other online tools like RAST (http://rast.nmpdr.org) allow the annotation and subsequent genomic comparison of the genomes. This helps to identify several housekeeping genes, align them, and determine the most conserved for nucleotide and amino acid identity. No matter the size of the gene, if the purpose is the improvement of detection procedures, the use of the NGS information together with techniques like qPCR and LAMP can help increasing the detection specificity in a smaller fragment size (less than 200 nucleotides). This takes more relevance when performing the same procedure showed with universal PCR primers with previously developed universal qPCR primers, the matching with other mollicute sequences is high enough to presume nonspecific amplifications, very important point especially when the detection of phytoplasmas is performed in insect vectors.

Loop-Mediated Isothermal Amplification The LAMP method has been widely used due to its high efficiency, specificity, and simplicity (Notomi et al. 2000). It requires two long outer primers and two short inner primers that recognize six specific sequences in the target DNA. These inner primers contain sense and antisense sequences in the DNA which hybridize the target sequence and initiate the DNA synthesis. Next, the outer primer carries out a strand displacement DNA synthesis and produces a single-stranded DNA which works as a template for the second inner and outer primers producing a DNA molecule with a loop structure (Nagamine et al. 2002). The unremitting cycling reaction accumulates products with repeated sequences of target DNA of different sizes. The LAMP method is considered superior to the PCR and microarray-based methods due to its cost-effectiveness, high specificity, better sensitivity, and convenient procedure (conducted at constant temperature without the need for a thermal cycler) and evaluation method. LAMP is approximately 10–100 times more sensitive than PCR and can amplify the original amount 109–1,010 times in 45–60 minutes, significantly faster than PCR (Nagamine et al. 2002; Li et al. 2007; Bhat et al. 2013). When two loop primers are used, the reaction sensitivity is usually increased tenfold and the reaction time reduced to 30 minutes (Li and Ling 2014). LAMP amplicons can be easily visualized by colour indicators, by the turbidity of magnesium pyrophosphate formed during the reaction (precipitate), or by agarose gel electrophoresis (Goto et al. 2009). LAMP assays have been successfully used to detect phytoplasmas infecting papaya, cassava, grapevines, palm trees, napier grass, potatoes, coconut, periwinkle, and some insect hosts (Tomlinson et al. 2010; Bekele et al. 2011; Obura et al. 2011; Hodgetts et al. 2011; Sugawara et al. 2012; Kogovsek et al. 2015; Dickinson 2015; Vu et al. 2016). The LAMP assays can be developed in the laboratory for subsequent application in the field (Chapter 9), where the simplicity of isothermal amplification makes LAMP a suitable method for rapid detection of phytoplasmas with levels of sensitivity and specificity approaching those of more complex and time-consuming methods.

Recombinase Polymerase Amplification The RPA is another rapid, isothermal amplification method with high specificity and sensitivity and does not require an initial heating step to denature the target DNA as it relies on an enzymatic activity to separate the dsDNA in order to assist primer binding to the target sequences (Yan et al. 2014; Londono et al. 2016; Lobato and O'Sullivan 2018). In RPA, the isothermal amplification of specific DNA fragments is achieved by the combination of enzymes and proteins, *i.e.* recombinase single-stranded binding protein (SSB), and strand-displacing polymerase, used at a constant low temperature. The displaced strand is stabilized by SSB, and the polymerase initiates synthesis. RPA products can be visualized in gel after purification, although fluorescence and/or hybridization, have also been reported. The reaction begins with the integration of a recombinase protein with the primers prior to their annealing to specific sequences in the DNA target (Zhang et al. 2014). Following the primer annealing, the recombinase dissociates from the primers and leaves their 3′ accessible to the DNA polymerase to initiate the amplification. This creates a D-loop which is stabilized by a single-stranded binding protein (SSB) to keep the DNA open as a DNA polymerase with strand displacement activity, continues the amplification. Using RPA, billions of DNA copies can be generated efficiently in 60 minutes with an incubation temperature between 37°C and 42°C (Piepenburg et al. 2006; Yan et al. 2014). The low incubation temperature and short reaction time make RPA a suitable assay for use in point-of-care diagnostic applications. Furthermore, primer design is simple without consideration of annealing temperature as they form a complex with the recombinase to target the homologous sequences. RPA is highly sensitive with a detection limit as low as 6.25 fg of genomic DNA input with a specificity >95% (Boyle et al. 2014). RPA assay has been successfully reported for the specific detection of '*Ca.* P. mali' (Valasevich and Schneider 2017) and '*Ca.* P. pruni' from crude sap of sweet cherry tissues (Villamor and Eastwell 2019). The assays amplify a fragment of the *imp* gene, and amplimers were detected either by fluorescence in real-time mode (TwistAmp®exo assay), using a fluorophore-labelled probe, or by direct visualization employing a lateral flow device (TwistAmp®nfo assay/Milenia®HybriDetect). In comparison with a TaqMan real-time PCR assay based on the same target gene, the RPA assays were equally sensitive, but results are obtained faster. Simplified nucleic acid extraction procedures from plant tissue with Tris- and CTAB-based buffers revealed that crude Tris–DNA extracts are a suitable source for RPA tests. The assays are suitable for high-throughput screening of plant material and diagnostic and can be therefore combined with a simplified DNA extraction procedure.

Digital Droplet PCR The advent of digital droplet PCR (ddPCR) platforms is enabling the expansion of this technology for research and diagnostic applications. The distinctive feature of ddPCR is the separation of the reaction mixture into thousands to millions of droplets followed by detection of the amplification. The distribution of target sequences is according to the Poisson distribution allowing accurate and absolute quantification of the target, and omits the need to use reference materials increasing the accuracy of quantification at low target concentrations compared to qPCR. The ddPCR has also shown higher resilience to inhibitors in a number of different types of samples. This assay was used for absolute quantification of "flavescence dorée" phytoplasmas targeting the *secY* gene (Mehle et al. 2014). A similar TaqMan-based assay was developed to detect phytoplasmas infecting palms in Florida. When compared with real-time and nested PCR assays, ddPCR was capable of detecting the phytoplasmas at lower concentrations and belonging to diverse phytoplasma groups (Bahder et al. 2018).

7.6 Serological Methods

Serological detection is a convenient and economical method which allows examination of many samples in a short time. Serological tools have been used with success also to detect different phytoplasmas in leafhopper vectors or potential vectors, by immunofluorescence (Lherminier et al. 1990), immunosorbent electron microscopy (Sinha 1979; Sinha and Benhamou 1983), dot blot or ELISA (Boudon-Padieu et al. 1989). Polyclonal and monoclonal antibodies have been produced in the time against numerous phytoplasmas (Table 7.3).

The relative sensitivities of polyclonal antibodies produced against several phytoplasmas such as "stolbur", aster yellows, peanut witches' broom, and phytoplasmas associated with faba bean, sesame phyllody, apple proliferation, and sandal spike were detected using immunosorbent assays or indirect ELISA procedure (Hobbs et al. 1987; Bellardi et al. 1992; Saeed et al. 1993; Berg et al. 1999; Lin and Chen 1986; Thomas and Balasundaran 2001; Loi et al. 2002). A specific monoclonal antibody against phytoplasmas associated with rice yellow dwarf (RYD) was produced and specifically detected the phytoplasma by ELISA, immunofluorescent staining, and tissue-blotting techniques. The antibody recognized two polypeptides, 16 kDa and 41 kDa, of RYD-phytoplasma by Western blotting (Chang et al. 1995). In other approaches, tissue blotting with direct or indirect antigen detection was also used for specific phytoplasma detection (Lin and Chen 1985). The Bermuda grass and sugarcane white leaf phytoplasmas were purified from infected tissues and used for polyclonal antibody production; this antiserum was successfully employed as specific tool for the phytoplasma detection in plate-trapped antigen enzyme-linked immunosorbent assay (PTA-ELISA), dot immunoblotting assay (DIBA), and tissue print immuno assay (TPIA), and no cross-reaction between this antiserum and other tested phytoplasmas was reported (Biabani et al. 2009; Biabani Khankahdani and Ghasemi 2011). China-tree decline-infected material in Argentina was used to produce a polyclonal antiserum employed in immuno-dot blot assays discriminating between infected and healthy plants and was able to bind to phytoplasma-like structures as observed by immuno-electron microscopy and immunogold labelling (Gomez et al. 1996).

Immuno-capture PCR assay, in which the phytoplasma of interest is first selectively captured by specific antibody adsorbed on microtitre plates and then the phytoplasma DNA is released and amplified using specific or universal

Table 7.3 Antisera produced and employed for phytoplasma detection

Phytoplasma strain	16Sr classification (country)	Antiserum type	Methodology	Literature
Aster yellows	16SrI (United States of America)	Polyclonal and monoclonal	Immunodiffusion	Sinha and Benhamou (1983), Lin and Chen (1985), (1986), Lee et al. (1993b), Errampalli and Fletcher (1993), Jiang et al. (1988)
Primula yellows	16SrI (United Kingdom)		ELISA, fluorescence microscopy, Western blots	Clark et al. (1989)
Onion yellows	16SrI-B (Japan)	SecA monoclonal	Western blot and histochemical staining	Kakizawa et al. (2001), Wei et al. (2004)
Sandal spike	16SrI-B (India)	Polyclonal		Thomas and Balasundaran (2001)
Alfalfa witches' broom	16SrII (Iran)	Polyclonal	DIBA	Salehi et al. (2011)
Peanut witches' broom	16SrII (India)		ELISA	Hobbs et al. (1987)
Lime witches' broom	16SrII-B (Iran)	Monoclonal	Western blot	Siampour et al. (2013)
		Imp monoclonal	DAS-ELISA and DIBA	Shahryari et al. (2011, 2013)
		Polyclonal		Mirzai et al. (2009)
Faba bean phyllody	16SrII-C (Sudan)		ELISA, dot blot	Saeed et al. (1993)
Sweet potato witches' broom	16SrII-D (Taiwan)	Monoclonal	ELISA, immunofluorescence, tissue blotting	Shen and Lin (1993, 1994)
		Amp monoclonal	Western blot	Yu et al. (1998)
Western X-disease	16SrIII-A (United States of America)	Polyclonal	Western blot, ELISA	Blomquist et al. (2001)
Clover phyllody	16SrIII-B (Italy)	Monoclonal	Immunofluorescence, dot blot	Chen et al. (1994)
"Flavescence dorée"	16SrV-C (France)	Polyclonal	ELISA	Seddas et al. (1993)
Apple proliferation	16SrX-A (United Kindgom, Germany and Italy)		Immunocapture PCR	Rajan and Clark (1985)
		Monoclonal	ELISA, Western blot	Loi et al. (2002)
		Polyclonal, Imp monoclonal		Berg et al. (1999)

(continued)

Table 7.3 (continued)

Phytoplasma strain	16Sr classification (country)	Antiserum type	Methodology	Literature
European stone fruit yellows	16SrX-B (Hungary)	Imp monoclonal		Mergenthaler et al. (2001)
Napier grass stunt	16SrXI (Kenya)	Monoclonal		Wambua et al. (2017)
Sugarcane grassy shoot and white leaf	16SrXI (Thailand, India)	Polyclonal	ELISA	Sarindu and Clark (1993), Viswanathan (1997)
Sugarcane white leaf	16SrXI (Iran)		PTA-ELISA, DIBA, TPIA	Biabani et al. (2009)
Rice yellow dwarf	16SrXI-B (Taiwan)	Monoclonal	ELISA, immunofluorescence, tissue blotting	Chang et al. (1995)
Italian periwinkle green petals	16SrXII-A (Italy)	Polyclonal	Agar immunodiffusion	Bellardi et al. (1992),
"Stolbur"	16SrXII-A (France)	Monoclonal	ELISA, tissue blotting	Fos et al. (1992)
Grapevine yellows	16SrXII-A	Monoclonal	Immunofluorescence, dot blot	Chen et al. (1993)
Japanese hydrangea phyllody	16SrXII-D (Japan)	Amp	*In situ* Western blot	Arashida et al. (2008)
China-tree decline	16SrXIII (Argentina)	Polyclonal	Dot blot, immuno-electron microscopy, and immunogold labelling	Gomez et al. (1996)
Bermudagrass white leaf	16SrXIV (Iran)		PTA-ELISA, DIBA, TPIA	Biabani Khankahdani and Ghasemi (2011)

primers, was also used as an alternative method to increase detection sensitivity (Rajan and Clark 1985).

Whilst the serological method could help in avoiding the lengthy extraction procedures to prepare target DNAs, the traditional approach to prepare antibody against phytoplasmas results in antisera with relatively low titre, contamination with plant-derived immunogens, and occurrence of cross-reactions with healthy crude extracts (Kakizawa et al. 2004). The cloning and expression of the phytoplasma gene fragment in *E. coli* and purification of the protein were used as a means to overcome these limitations. Partial recombinant secA proteins were produced from six phytoplasma strains representing five 16S ribosomal groups, and the expressed, purified recombinant protein from Cape St. Paul wilt disease phytoplasma was used to immunize mice; eight monoclonal antibodies with varying degrees of specificity against recombinant proteins from different phytoplasma groups revealed in Western blotting two of them as specific for phytoplasmas. However ELISA gave negative results suggesting that either secA was not expressed at sufficiently high levels, or conformational changes of the reagents adversely affected detection and the *secA* gene is therefore not a suitable antibody target for routine detection (Hodgetts et al. 2014). Another antiserum raised against the SecA membrane protein of onion yellows phytoplasma detected eight phytoplasma strains from four distinct 16S ribosomal groups (16SrI, 16SrIII, 16SrXI, and 16SrV) in immunoblots; moreover immunohistochemical staining of thin sections prepared from infected plants detected phytoplasmas in the phloem tissues (Wei et al. 2004).

Partial sequences of the major immunodominant proteins of several phytoplasmas such as those associated with apple proliferation, Western X-disease, chlorantie-aster yellows, European stone fruit yellows, and onion yellows were used to produce antibodies to the respective phytoplasmas (Berg et al. 1999; Blomquist et al. 2001; Hong et al. 2001; Mergenthaler et al. 2001; Morton et al. 2003; Kakizawa et al. 2004). The expressed fusion proteins in *E. coli* obtained by cloning the immunodominant membrane protein (Imp) genes allowed also to obtain their *in planta* expression (Kakizawa et al. 2009). Following this procedure purified antibodies and conjugates were produced and applied for the lime witches' broom disease-associated phytoplasma detection in infected plants by DAS-ELISA and dot immunosorbent assay (DIBA) (Shahryari et al. 2013). Imp proteins of lime witches' broom and alfalfa witches' broom phytoplasmas from Iran were expressed as His-tagged recombinant proteins in *E. coli.* An antiserum raised against full-length recombinant Imp of lime witches' broom phytoplasmas reacted specifically in Western blots with membrane proteins from infected periwinkle and lime. Imp proteins of both strains were also recognized by an antiserum raised against an enriched preparation alfalfa witches' broom phytoplasmas (Siampour et al. 2013). A quantum dot (QD)-based nano-biosensor was developed for the sensitive detection of '*Ca.* P. aurantifolia' in infected trees using Imp as a target protein for the construction of a specific binding antibody conjugated to thioglycolic acid-modified cadmium-telluride quantum dots (CdTe-QDs). Dye (rhodamine) molecules were attached to the Imp, and then, the donor-acceptor complexes (QDs-Ab-Imp-Rhodamine) were formed based on the antigen-antibody interaction. The mutual affinity of the antigen and the antibody brought the CdTe-QDs and rhodamine close enough to allow fluorescence resonance energy transfer (FRET). The immunosensor constructed showed a high sensitivity and specificity so it could be used for the phytoplasma detection with consistent results (Rad et al. 2012).

Serological assays with the potential to be converted to lateral flow assays were recently developed from Imp of '*Ca.* P. oryzae' strains from napier grass stunt. Two specific monoclonal antibodies were designed, proved to be specific for the pathogen, and showed a high affinity to the target protein. The detection of native napier stunt Imp with picogram amounts of antibodies in microgram quantities of plant tissue extract demonstrates the power of these antibodies and the abundance of the protein (Wambua et al. 2017).

An antibody against the Amp of the Japanese hydrangea phyllody phytoplasma showed no cross-reactions with the antibody against the Amp protein from the closely related onion yellows phytoplasma, and the *in situ* detection of the Amp protein revealed that the phytoplasma was localized to the phloem tissues in the symptomatic flower. This study showed that this Amp protein is indeed expressed at detectable levels in the phytoplasma-infected hydrangea (Arashida et al. 2008).

7.7 Conclusion and Perspectives

The continuous effort to improve the phytoplasma diagnostic procedures aims at quicker, cheap and robust methods. Sensitivity is not an issue *per se*, as the current nested PCR protocols are extremely sensitive, but the achievement of high levels of sensitivity without the risk of false-positive results that can be associated with nested PCR, is highly desirable. The recent introduction of diagnostic assays based on quantitative PCR reduces the risk of amplicon contamination. Several other procedures have been proposed for the detection of phytoplasmas in infected plants and insects, including PCR-ELISA, PCR-dot blot, heteroduplex mobility assay (Wang and Hiruki 2000; Palmano and Firrao 2000), 16S–23S spacer length polymorphism, microarray, and NGS. Although these techniques may not have the characteristics of speed, sensitivity, and robustness required, they are nevertheless interesting for developing future assay methods with a higher multiplexing potential, thus improving the efficiency or ability to detect multiple phytoplasmas in a single step. It should be noted, however, that the major limitation to the development of high-throughput, robust diagnostic assays for phytoplasmas remains the difficulty in developing a rapid and cost/labour-effective preparation of suitable nucleic acids extracts. The most reliable diagnostic protocols, therefore, include the collection of samples as pools of subsamples taken from different parts of the individual plant to be tested. Due to the intrinsic characteristics of phytoplasma diseases, *i.e.* the low concentration and irregular distribution of the pathogens, it is unlikely that the field of diagnostics will see another boost such as that given by the introduction of the isothermal amplification techniques like LAMP and RPA, since there is still the need for further validation and confirmation with diverse plant/insect samples. The efforts for obtaining more efficient detection phytoplasma methods are still undergoing.

References

Ahrens U, Seemüller E (1992) Detection of DNA of plant pathogenic mycoplasma like organisms by a polymerase chain reaction that amplifies a sequence of the 16S rRNA gene. *Phytopathology* **82**, 828–832.

Aldaghi M, Massart S, Roussel S, Jijakli MH (2007) Development of a new probe for specific and sensitive detection of '*Candidatus* Phytoplasma mali' in inoculated apple trees. *Annals of Applied Biology* **151**, 251–258.

Angelini E, Bianchi GL, Filippin L, Morassutti C, BorgoM (2007) A new TaqMan method for the identification of phytoplasmas associated with grapevine yellows by real-time PCR assay. *Journal of Microbiological Methods* **68**, 613–622.

Anniballi F, Auricchio B, Delibato E, Antonacci M, De Medici D, Fenicia L (2012) Multiplex real-time PCR SYBR green for detection and typing of group III *Clostridium botulinum*. *Veterinarian Microbiology* **154**, 332–338.

Arashida R, Kakizawa S, Ishii Y, Hoshi A, Jung H-Y, Kagiwada S, Yamaji Y, Oshima K, Namba S (2008) Cloning and characterization of the antigenic membrane protein (Amp) gene and in situ detection of Amp from malformed flowers infected with Japanese hydrangea phyllody phytoplasma. *Phytopathology* **98**, 769–775.

Babini AR Fiumi E, Giunchedi L, Pignatta D, Poggi Pollini C, Reggiani N (2008) Investigations with real time PCR assay on the transmissibility of pear decline phytoplasma (PDP) with dormant buds. *Acta Horticulturae* **781**, 495–498.

Bahder BW, Helmick EE, De-Fen M, Harrison NA, Davis RE (2018) Digital PCR technology for detection of palm-infecting phytoplasmas belonging to group 16SrIV that occur in Florida. *Plant Disease* **102**, 1008–1014.

Baric S, Dallavia J (2004) A new approach to apple proliferation detection: a highly sensitive real-time PCR assay. *Journal of Microbiological Methods* **57**, 135–145.

Baric S, Berger J, Cainelli C, Kerschbamer C, Letschka T, Dalla Via J (2011) Seasonal colonization of apple trees by '*Candidatus* Phytoplasma mali' revealed by a new quantitative TaqMan real-time PCR approach. *European Journal of Plant Pathology* **129**, 455–467.

Bekele B, Hodgetts J, Tomlinson J, Boonham N, Nikolic P, Swarbrick P, Dickinson M (2011) Use of a real-time LAMP isothermal assay for detecting 16SrII and -XII phytoplasmas in fruit and weeds of the Ethiopian Rift Valley. *Plant Pathology* **60**, 345–355.

Bellardi MG, Vibio M, Bertaccini A (1992) Production of a polyclonal antiserum to CY-MLO using infected *Catharanthus roseus*. *Phytopathologia Mediterranea* **31**, 53–55.

Berg M, Davies DL, Clark MF, Vetten J, Maier G, Seemüller E (1999) Isolation of a gene encoding an immunodominant membrane protein gene in the apple proliferation phytoplasma and expression and characterization of the gene product. *Microbiology* **145**, 1937–1943.

Bertaccini A (2007) Phytoplasmas: diversity, taxonomy, and epidemiology. *Frontiers in Bioscience* **12**, 673–689.

Bertaccini A, Lee I-M (2018) Phytoplasmas: an update. In: Phytoplasmas: Plant Pathogenic Bacteria-I. Characterization and Epidemiology of Phytoplasma-Associated Diseases. Chapter 1. Ed Rao GP, Bertaccini A, Fiore N, Liefting LW. Springer, Singapore, 1–29 pp.

Bertaccini A, Davis RE, Hammond RW, Bellardi MG, Vibio M, Lee I-M (1992) Sensitive detection of mycoplasma like organisms in field-collected and *in vitro* propagated plants of *Brassica*, *Hydrangea* and *Chrysanthemum* by polymerase chain reaction. *Annals of Applied Biology* **121**, 593–599.

Bertaccini A, Duduk B, Paltrinieri S, Contaldo N (2014) Phytoplasmas and phytoplasma diseases: a severe threat to agriculture. *American Journal of Plant Sciences* **5**, 1763–1788.

Bhat AI, Siljo A, Deeshma KP (2013) Rapid detection of *Piper yellow mottle virus* and *Cucumber mosaic virus* infecting black pepper (*Piper nigrum*) by loop-mediated isothermal amplification (LAMP). *Journal of Virological Methods* **193**, 190–196.

Biabani R, Ghasemi S, Salehi M, Rahimian H (2009) Purification and serological study of sugarcane white leaf phytoplasma in Khuzestane province. *Plant Protection Journal* **113**, 43–45.

Biabani Khankahdani R, Ghasemi S (2011) Serological aspects of phytoplasma associated with Bermudagrass white leaf (BGWL) disease. International Conference on Asia Agriculture and Animal IPCBEE, Singapore, 106–110.

Bianco PA, Casati P, Marziliano N (2004) Detection of phytoplasmas associated with grapevine "flavescence dorée" disease using real-time PCR. *Journal of Plant Pathology* **86**, 257–261.

Blomquist CL, Barbara DJ, Davies DL, Clark MF, Kirkpatrick BC (2001) An immunodominant membrane protein gene from the western X-disease phytoplasma is distinct from those of other phytoplasmas. *Microbiology* **147**, 571–580.

Botti S, Bertaccini A (2003) Variability and functional role of chromosomal sequences in 16SrI-B subgroup phytoplasmas including aster yellows and related strains. *Journal of Applied Microbiology* **94**, 103–110.

Bosco D, Palermo S, Mason G, Tedeschi R, Marzachì C, Boccardo G (2002) DNA-based methods for the detection and the identification of phytoplasmas in insect vector extracts. *Molecular Biotechnology* 22, 9–18.

Boudon-Padieu E, Larrue J, Caudwell A (1989) ELISA and dot blot detection of "flavescence dorée" MLO in individual leafhopper vector during latency and inoculative state. *Current Microbiology* **19**, 357–364.

Boyle DS, Mcnerney R, Low HT, Leader BT, Perez-Osorio AC, Meyer JC (2014). Rapid detection of *Mycobacterium tuberculosis* by recombinase polymerase amplification. *Plos One* **9**, e103091.

Brzin J, Ermacora P, Osler R, Loi N, Ravnikar M, Petrovič N (2003) Detection of apple proliferation phytoplasma by ELISA and PCR in growing and dormant apple trees. *Journal of Plant Diseases and Protection* **110**, 476–483.

Bustin SA, Nolan T (2004) Chemistries. In: A–Z of quantitative PCR. Ed Bustin SA. International University Line, La Jolla, California, United States of America, 215–278 pp.

Cai H, Wei W, Davis RE, Chen H, Zhao Y (2008) Genetic diversity among phytoplasmas infecting *Opuntia* species: virtual RFLP analysis identifies new subgroups in the peanut witches' broom phytoplasma group. *International Journal of Systematic and Evolutionary Microbiology* **58**, 1448–1457.

Chang F, Chen CC, Lin CP (1995) Monoclonal antibody for the detection and identification of a phytoplasma associated with rice yellow dwarf. *European Journal of Plant Pathology* **101**, 511–518.

Chen KH, Guo JR, Wu XJ, Loi N, Carraro L, Guo HJ, Chen YD, Osler R, Pearson R, Chen TA (1993) Comparison of monoclonal antibodies, DNA probes, and PCR for detection of the grapevine yellows disease agent. *Phytopathology* **83**, 915–922.

Chen KH, Credi R, Loi N, Maixner M, Chen TA (1994) Identification and grouping of mycoplasma like organisms associated with grapevine yellows and clover phyllody diseases based on immunological and molecular analyses. *Applied and Environmental Microbiology* **60**, 1905–1913.

Christensen NM, Nicolaisen M, Hansen M, Schulz A (2004) Distribution of phytoplasmas in infected plants as revealed by real-time PCR and bioimaging. *Molecular Plant-Microbe Interaction* **17**, 1175–1184.

Chiykowski LN (1991) Vector-pathogen-host plant relationships of clover phyllody mycoplasma like organism and the vector leafhopper *Paraphlepsius irroratus*. *Canadian Journal of Plant Pathology* **13**, 11–18.

Choi YH, Tapias EC, Kim HK, Lefeber AW, Erkelens C, Verhoeven JT, Brzin J, Zel J, Verpoorte R (2004) Metabolic discrimination of *Catharanthus roseus* leaves infected by phytoplasma using 1H-NMR spectroscopy and multivariate data analysis. *Plant Physiology* **135**, 2398–2410.

Clark MF, Morton A, Buss SL (1989) Preparation of mycoplasma immunogens from plants and a comparison of polyclonal and monoclonal antibodies made against primula yellows MLO-associated antigens. *Annals of Applied Biology* **114**, 111–124.

Crosslin JM, Vandemark GJ, Munyaneza JE (2010) Development of a real-time, quantitative PCR for detection of the Columbia basin potato purple top phytoplasma in plants and beet leafhoppers. *Plant Disease* **90**, 663–667.

Contaldo N, Paltrinieri S, Makarova O, Bertaccini A, Nicolaisen M (2015) Q-bank phytoplasma: a DNA bar-coding tool for phytoplasma identification. *Methods in Molecular Biology* **1302**, 123–135.

Contaldo N, Paltrinieri S, Bellardi MG, Lesi F, Satta E, Bertaccini A (2018) Rapid screening for phytoplasma presence in flower crops using *tuf* gene barcode. *Acta Horticulturae* **1193**, 63–67.

Daire X, Clair D, Larrue J, Boudon-Padieu E (1997) Survey for grapevine yellows phytoplasmas in diverse European countries and Israel. *Vitis* **36**, 53–54.

Davis RE, Lee I-M (1993) Cluster-specific polymerase chain reaction amplification of 16S rDNA sequences for detection and identification of mycoplasma like organisms. *Phytopathology* **83**, 1008–1011.

Deng SJ, Hiruki C (1991) Amplification of 16S ribosomal-RNA genes from culturable and nonculturable mollicutes. *Journal of Microbiological Methods* **14**, 53–61.

Dickinson M (2015) Loop Mediated Isothermal amplification (LAMP) for detection of phytoplasmas in the field. In: Plant Pathology: Techniques and Protocols. Methods in Molecular Biology. Ed Lacomme C, Springer Science + Business Media, New York, United States of America, 99–111 pp.

Dufva M (2005) Fabrication of high quality microarrays. *Biomolecular Engeneering* **22**, 173–184.

Doyle JJ, Doyle JL (1990) Isolation of plant DNA from fresh tissue. *Focus* **12**, 13–15.

Errampalli D, Fletcher J, Claypool PL (1991) Incidence of yellows in carrot and lettuce and characterization of mycoplasma like organism isolates in Oklahoma. *Plant Disease* **75**, 579–584.

Errampalli D, Fletcher J (1993) Production of monospecific polyclonal antibodies against aster yellows mycoplasma-like organism associated antigen. *Phytopathology* **83**, 1279–1282.

Firrao G, Martini M, Ermacora P, Loi N, Torelli E, Foissac X, Carle P, Kirkpatrick BC, Liefting L, Schneider B, Marzachì C, Palmano S (2013) Genome wide sequence analysis grants unbiased definition of species boundaries in '*Candidatus* Phytoplasma'. *Systematic and Applied Microbiology* **36**, 539–548.

Franke-Whittle IH, Klammer SH, Mayrhofer S, Insam H (2006) Comparison of different labeling methods for the production of labeled target DNA for microarray hybridization. *Journal of Microbiological Methods* **65**, 117–126.

Fos A, Danet J-L, Zreik L, Garnier M, Bové J-M (1992) Use of a monoclonal antibody to detect the stolbur mycoplasmalike organism in plants and insects and to identity a vector in France. *Plant Disease* **76**, 1092–1096.

Galetto L, Bosco D, Marzachì C (2005) Universal and group-specific real-time PCR diagnosis of "flavescence dorée" (16SrV), "bois noir" (16SrXII) and apple proliferation (16SrX) phytoplasmas from field-collected plant hosts and insect vectors. *Annals of Applied Biology* **147**, 191–201.

Garcia-Chapa M, Batlle A, Rekab D, Rosquete MR, Firrao G (2004) PCR-mediated whole genome amplification of phytoplasmas. *Journal of Microbiological Methods* **56**, 231–242.

Gibbs KS, Padovan AC, Mogen BD (1995) Studies on sweet potato little-leaf phytoplasma detected in sweet potato and other plant species growing in northern Australia. *Phytopathology* **85**, 169–174.

Gibbs KS, Schneider B, Padovan AC (1998) Differential detection and genetic relatedness of phytoplasmas in papaya. *Plant Pathology* **47**, 325–332.

Gomez G, Conci L, Ducasse D, Nome S (1996) Purification of the phytoplasma associated with China-tree (*Melia azedarach* L.) decline and the production of a polyclonal antiserum for its detection. *Journal of Phytopathology* **144**, 473–477.

Goto M, Honda E, Ogura A, Nomoto A, Ken-Ichi Hanaki DVM (2009) Colorimetric detection of loop-mediated isothermal amplification reaction by using hydroxy naphthol blue. *Biotechniques* **46**, 167–172.

Green MJ, Thompson DA, MacKenzie DJ (1999) Easy and efficient DNA extraction from woody plants for the detection of phytoplasmas by polymerase chain reaction. *Plant Disease* **83**, 482–485.

Gundersen DE, Lee I-M (1996) Ultrasensitive detection of phytoplasmas by nested-PCR assays using two universal primer pairs. *Phytopathologia Mediterranea* **35**, 144–151.

Harrison NA, Womack M, Carpio ML (2002) Detection and characterization of a lethal yellowing (16SrIV) group phytoplasma in Canary Island date palms affected by lethal decline in Texas. *Plant Disease* **86**, 676–681.

Hebert PDN, Cywinska A, Ball SL, deWaard JR (2003) Biological identifications through DNA barcodes. *Proceedings of Biological Sciences* **270**, 313–321.

Heinrich M, Botti S, Caprara L, Arthofer W, Strommer S, Hanzer V, Katinger H, Laimer da Câmara Machado M, Bertaccini A (2001) Improved detection methods for fruit tree phytoplasmas. *Plant Molecular Biology Reporter* **19**, 169–179.

Herath P, Hoover G, Angelini E, Moorman GW (2010) Detection of elm yellows phytoplasma in elms and insects using real-time PCR. *Plant Disease* **94**, 1355–1360.

Hobbs HA, Reddy DVR, Reddy AS (1987) Detection of a mycoplasma-like organism in peanut plants with witches' broom using indirect enzyme-linked immunosorben assay (ELISA). *Plant Pathology* **36**, 164–167.

Hodgetts J, Boonham N, Mumford R, Dickinson M (2009) Panel of 23S rRNA gene based real-time PCR assays for improved universal and group-specific detection of phytoplasmas. *Applied Environmental Microbiology* **75**, 2945–2950.

Hodgetts J, Tomlinson J, Boonham N, González-Martín I, Nikolić P, Swarbrick P, Yankey EN, Dickinson M (2011) Development of rapid in-field loop-mediated isothermal amplification (LAMP) assays for phytoplasmas. *Bulletin of Insectology* **64**(Supplement), S41–S42.

Hodgetts J, Johnson G, Perkins K, Ostoja-Starzewska S, Boonham N, Mumford R, Dickinson M (2014) The development of monoclonal antibodies to the secA protein of cape St. Paul wilt disease phytoplasma and their evaluation as a diagnostic tool. *Molecular Biotechnology* **56**, 803–813.

Hong Y, Davies DL, Wezel RV, Ellerker BE, Morton A, Barbara D (2001) Expression of the immunodominant membrane protein of chlorantie-aster yellows phytoplasma in *Nicotiana benthamiana* from a potato virus X-based vector. *Acta Horticulturae* **550**, 409–415.

Hren M, Boben J, Rotter A, Kralj P, Gruden K, Ravnikar M (2007) Real-time PCR detection systems for "flavescence dorée" and "bois noir" phytoplasmas in grapevine: comparison with conventional PCR detection and application in diagnostics. *Plant Pathology* **56**, 785–796.

IRPCM 2004. '*Candidatus* Phytoplasma', a taxon for the wall-less, non-helical prokaryotes that colonise plant phloem and insects. *International Journal of Systematic and Evolutionary Microbiology* **54**, 1243–1255.

Jarausch W, Peccerella T, Schwind N, Jarausch B, Krczal G (2004) Establishment of a quantitative real-time PCR assay for the quantification of apple proliferation phytoplasmas in plants and insects. *Acta Horticulturae* **657**, 415–420.

Jawhari M, Abrahamian P, Abdel Sater A, Sobh H, Tawidian P, Abou-Jawdah Y (2015) Specific PCR and real-time PCR assays for detection and quantitation of '*Candidatus* Phytoplasma phoenicium'. *Molecular and Cellular Probes* **29**, 63–70.

Jiang YP, Chen TA (1987) Purification of mycoplasma-like organisms from lettuce with aster yellows disease. *Phytopathology* **77**, 949–953.

Jiang YP, Lei JD, Chen TA (1988) Purification of aster yellows agent from diseased lettuce using affinity chromatography. *Phytopathology* **78**, 828–831.

Kakizawa S, Oshima K, Kuboyama T, Nishigawa H, Jung H-Y, Sawayanagi T, Tsuchizaki T, Miyata S, Ugaki M, Namba S (2001) Cloning and expression analysis of phytoplasma protein translocation genes. *Molecular Plant-Microbe Interactions* **14**, 1043–1050.

Kakizawa S, Oshima K, Nishigawa H, Jung H-Y, Wei W, Suzuki S, Tanaka M, Miyata S, Ugaki M, Namba S (2004) Secretion of immunodominant membrane protein from onion yellows phytoplasma through the Sec protein-translocation system in *Escherichia coli*. *Microbiology* **150**, 135–142.

Kakizawa S, Oshima K, Ishii Y, Hoshi A, Maejima K, Jung H-Y, Yamaji Y, Namba S (2009) Cloning of immunodominant membrane protein genes of phytoplasmas and their *in planta* expression. *FEMS Microbiology Letters* **293**, 91–101.

Kakizawa S, Makino A, Ishii Y, Tamaki H, Kamagata Y (2014) Draft genome sequence of '*Candidatus* Phytoplasma asteris' strain OY-V, an unculturable plant-pathogenic bacterium. *Genome Announcements* **2**, e00944–14.

Khan AJ, Botti S, Al-Subhi AM, Gundersen-Rindal DE, Bertaccini A (2002) Molecular identification of a new phytoplasma associated with alfalfa witches' broom in Oman. *Phytopathology* **92**, 1038–1047.

Kenyon L, Henríquez NP, Harrison NA (1998) Diagnosis and detection of phytoplasma diseases of tropical crops. British Crop Protection Conference, Pests and Diseases, Brighton, England, United Kingdom, 779–787.

Kirkpatrick BC, Stenger DC, Morris TJ (1987) Cloning and detection of DNA from a nonculturable plant pathogenic mycoplasmalike organism. *Science* **238**, 197–200.

Kirkpatrick B, Smart C, Blomquist C, Guerra L, Harrison N, Ahrens U, Lorenz KH, Schneider B, Seemüller E (1994) Identification of MLO strain-specific primers obtained from 16/23S spacer sequences. *IOM Letters* **3**, 261–262.

Kirkpatrick BC, Harrison NA, Lee I-M (1995) Isolation of mycoplasma-like organism DNA from plant and insect hosts. In: Molecular and diagnostic procedures in Mycoplasmology. Eds Razin S,Tully JG. Volume I, Academic Press, San Diego, California, United States of America, 105–116 pp.

Kogovsek P, Hodgetts J, Hall J, Prezelj N, Nikolic P, Mehle N, Lenarcic R, Rotter A, Dickinson M, Boonham N, Dermastia M, Ravnikar M (2015) LAMP assay and rapid sample preparation method for on-site detection of "flavescence dorée" phytoplasma in grapevine. *Plant Pathology* **64**, 286–296.

Kollar A, Seemüller E, Bonnet F (1990) Isolation of the DNA of various plant pathogenic mycoplasma like organisms from infected plants. *Phytopathology* **80**, 233–237.

Lane D, Pace B, Olsen G, Stahl D, Sogin M, Pace N (1985) Rapid determination of 16S ribosomal RNA sequences for phylogenetic analyses. *Proceedings of the National Academy of Sciences – United States of America* **82**, 6955–6959.

Lee I-M, Davis RE (1992) Mycoplasmas which infect plants and insects. In: Mycoplasmas – Molecular Biology and Pathogenesis. Eds Maniloff J, McElhaney RN, Finch LR, Baseman JB. American Society for Microbiology, Washington DC, United States of America, 379–390 pp.

Lee I-M, Davis RE, Hiruki C (1991) Genetic interrelatedness among clover proliferation mycoplasma like organisms (MLOs) and other MLOs investigated by nucleic acid hybridization and restriction fragment length polymorphism analyses. *Applied and Environmental Microbiology* **57**, 3565–3569.

Lee I-M, Hammond RW, Davis RE Gundersen DE (1993a) Universal amplification and analysis of pathogen 16S rDNA for classification and identification of mycoplasma-like organisms. *Phytopathology* **83**, 834–842.

Lee I-M, Davis RE, Hsu H-T (1993b) Differentiation of strains in the aster yellows mycoplasma-like organisms strain cluster by serological assay with monoclonal antibodies. *Plant Disease* **77**, 815–817.

Lee I-M, Davis RE, Sinclair WA, DeWitt ND, Conti M (1993c) Genetic relatedness of mycoplasma like organisms detected in *Ulmus* spp. in the United States and Italy by means of DNA probes and polymerase chain reactions. *Phytopathology* **83**, 829–833.

Lee I-M, Gundersen DE, Hammond RW, Davis RE (1994) Use of mycoplasmalike organism (MLO) group-specific oligonucleotide primers for nested-PCR assays to detect mixed-MLO infections in a single host plant. *Phytopathology* **84**, 559–566.

Lee I-M, Bertaccini A, Vibio M, Gundersen DE (1995) Detection of multiple phytoplasmas in perennial fruit trees with decline symptoms in Italy. *Phytopathology* **85**, 728–735.

Lee I-M, Gundersen-Rindal DE, Davis RE, Bartoszyk IM (1998) Revised classification scheme of phytoplasmas based on RFLP analysis of 16S rRNA and ribosomal protein gene sequences. *International Journal of Systematic Bacteriology* **48**, 1153–1169.

Lee I-M, Davis RE, Gundersen-Rindal DE (2000) Phytoplasma: phytopathogenic mollicutes. *Annual Revue of Microbiology* **54**, 221–255.

Lee I-M, Martini M, Bottner KD, Dane RA, Black MC, Troxclair N (2003) Ecological implications from a molecular analysis of phytoplasmas involved in an aster yellows epidemic in various crops in Texas. *Phytopathology* **93**, 1368–1377.

Lee I-M, Zhao Y, Bottner KD (2006) *SecY* gene sequence analysis for finer differentiation of diverse strains in the aster yellows phytoplasma group. *Molecular and Cellular Probes* **20**, 87–91.

Lee I-M, Zhao Y, Davis RE, Wei W, Martini M (2007) Prospects of DNA-based systems for differentiation and classification of phytoplasmas. *Bulletin of Insectology* **60**, 239–244.

Lherminier J, Prensier G, Boudon-Padieu E, Caudwell A (1990) Immunolabeling of grapevine "flavescence dorée" MLO in salivary glands of *Euscelidius variegatus*: a light and electron microscopy study. *Journal of Histochemistry and Cytochemistry* **38**, 79–85.

Li R, Ling K-S (2014) Development of reverse transcription loop mediated isothermal amplification assay for rapid detection of an emerging Potyvirus: *Tomato necrotic stunt virus*. *Journal of Virological Methods* **200**, 35–40.

Li W, Hartung JS, Levy L (2007) Evaluation of DNA amplification methods for improved detection of '*Candidatus* Liberibacter species' associated with citrus "huanglongbing". *Plant Disease* **91**, 51–58.

Lin CP, Chen TA (1985) Monoclonal antibodies against the aster yellows agent. *Science* **227**, 1233–1235.

Lin CP, Chen TA (1986) Comparison of monoclonal antibodies and polyclonal antibodies in detection of aster yellows mycoplasma-like organism. *Phytopathology* **76**, 45–50.

Loi N, Ermacora P, Carraro L, Osler R, Chen TA (2002) Production of monoclonal antibodies against apple proliferation phytoplasma and their use in serological detection. *European Journal of Plant Pathology* **108**, 81–86.

Linck H, Krüger E, Reineke A (2017) A multiplexTaqMan qPCR assay for sensitive andrapid detection of phytoplasmas infecting *Rubus* species. *Plos One* **12**, e0177808.

Lobato IM, O'Sullivan CK (2018) Recombinase polymerase amplification: basics, applications and recent advances. *Trends in Analytical Chemistry* **98**, 19–35.

Londono MA, Harmon CL, Polston JE (2016) Evaluation of recombinase polymerase amplification for detection of *Begomoviruses* by plant diagnostic clinics. *Virology Journal* **13**, 48.

Lorenz KH, Schneider B, Ahrens U, Seemüller E (1995) Detection of the apple proliferation and pear decline phytoplasmas by PCR amplification of ribosomal and nonribosomal DNA. *Phytopathology* **85**, 771–776.

Makarova O, Contaldo N, Paltrinieri S, Kawube G, Bertaccini A, Nicolaisen M (2012) DNA barcoding for identification of '*Candidatus* Phytoplasmas' using a fragment of the elongation factor Tu gene. *Plos One* **7**, e52092.

Mandrioli M (2008) A cost-effective, simple and high-throughput method for DNA extraction from insects. *Insect Science* **17**, 465–470.

Manimekalai R, Soumya VP, Sathish Kumar R, Selvarajan R, Reddy K, Thomas GV, Sasikala M, Rajeev G, Baranwal VK (2010) Molecular detection of 16SrXI group phytoplasma associated with root (wilt) disease of coconut (*Cocos nucifera*) in India. *Plant Disease* **94**, 636.

Margaria P, Rosa C, Marzachì C, Turina M, Palmano S (2007) Detection of "flavescence dorée" phytoplasma in grapevine by reverse-transcription-PCR. *Plant Disease* **91**, 1495–501.

Margaria P, Turina M, Palmano S (2009) Detection of "flavescence dorée" and "bois noir" phytoplasmas, grapevine leafroll associated virus-1 and-3 and grapevine virus a from the same crude extract by reverse transcription – real time Taqman assays. *Plant Pathology* **58**, 838–845.

Martini M, Lee I-M, Bottner KD, Zhao Y, Botti S, Bertaccini A, Harrison NA, Carraro L, Marcone C, Khan AJ, Osler R (2007a) Ribosomal protein gene-based phylogeny for finer differentiation and classification of phytoplasmas. *International Journal of Systematic and Evolutionary Microbiology* **57**, 2037–2051.

Martini M, Loi N, Ermacora P, Carraro L, Pastore M (2007b) A real-time PCR method for detection and quantification of '*Candidatus* Phytoplasma prunorum' in its natural hosts. *Bulletin of Insectology* **60**(Supplement), S251–S252.

Marzachí C, Bosco D (2005) Relative quantification of chrysanthemum yellows (16SrI) phytoplasma in its plant and insect host using real-time polymerase chain reaction. *Molecular Biotechnology* **30**, 117–128.

Mehle N, Dreo T, Ravnikar M (2014) Quantitative analysis of "flavescence doreé" phytoplasma with droplet digital PCR. *Phytopathogenic Mollicutes* **4**, 9–15.

Mergenthaler E, Viczian O, Fodor M, Sule S (2001) Isolation and expression of an immunodominant membrane protein gene of the ESFY phytoplasma for antiserum production. *Acta Horticulturae* **550**, 355–360.

Minguzzi S, Terlizzi F, Lanzoni C, Poggi Pollini C, Ratti C (2016) A rapid protocol of crude RNA/DNA extraction for RT-qPCR detection and quantification of '*Candidatus* Phytoplasma prunorum'. *Plos One* **11**, e0146515.

Mirzai M, Heydarnejad J, Salehi M, Hosseinipour A, Massumi H, Shaabanian M (2009) Production of polyclonal antiserum against the causal agent of lime witches' broom. *Iran Journal of Plant Pathology* **45**, 155–159.

Monti M, Martini M, Tedeschi R (2013) EvaGreen real-time PCR protocol for specific '*Candidatus* Phytoplasma mali' detection and quantification in insects. *Molecular and Cellular Probes* **27**, 129–136.

Morton A, Davies DL, Blomquist CL, Barbara DJ (2003) Characterization of homologues of the apple proliferation immunodominant membrane protein gene from three related phytoplasmas. *Molecular Plant Pathology* **4**, 109–114.

Nagamine K, Hase T, Notomi T (2002) Accelerated reaction by loop-mediated isothermal amplification using loop primers. *Molecular and Cellular Probes* **16**, 223–229.

Namba S, Kato S, Iwanami S, Oyaizu H, Shiozawa H, Tsuchizaki T (1993) Detection and differentiation of plant-pathogenic mycoplasma like organisms using polymerase chain reaction. *Phytopathology* **83**, 786–791.

Nicolaisen M, Bertaccini A (2007) An oligonucleotide microarray-based assay for identification of phytoplasma 16S ribosomal groups. *Plant Pathology* **56**, 332–336.

Nicolaisen M, Nyskjold H, Bertaccini A (2013) Microarrays for universal detection and identification of phytoplasmas. In: Phytoplasma: Methods and Protocols, Methods in Molecular Biology. Eds Dickinson M, Hodgetts J. Springer Protocols, Humana Press, London, United Kingdom, 223–232 pp.

Nikolić P, Mehle N, Gruden K, Ravnikar M, Dermastia M (2010) A panel of real-time PCR assays for specific detection of three phytoplasmas from the apple proliferation group. *Molecular and Cellular Probes* **24**, 303–309.

Notomi T, Okayama H, Masubuchi H, Yonekawa T, Watanabe K, Amino N, Hase T (2000) Loop-mediated isothermal amplification of DNA. *Nucleic Acids Research* **28**, E63.

Obura E, Masiga D, Wachira F, Gurja B, Khan ZR (2011) Detection of phytoplasma by loop-mediated isothermal amplification of DNA (LAMP). *Journal of Microbiological Methods* **84**, 312–316.

Padovan AC, Gibb KS (2001) Epidemiology of phytoplasma diseases in papaya in Northern Australia. *Journal of Phytopathology* **149**, 649–658.

Padovan AC, Gibb KS, Bertaccini A, Vibio M, Bonfiglioli RE, Magarey PA, Sears BB (1995) Molecular detection of the Australian grapevine yellows phytoplasma and comparison with a grapevine yellows phytoplasma from Emilia-Romagna in Italy. *Australian Journal of Grape and Wine Research* **1**, 25–31.

Palmano S (2001) A comparison of different phytoplasma DNA extraction methods using competitive PCR. *Phytopathologia Mediterranea* **40**, 99–107.

Palmano S and Firrao G (2000) Diversity of phytoplasmas isolated from insects, determined by a DNA heteroduplex mobility assay and a length polymorphism of the 16S & 23S rDNA spacer region analysis. *Journal of Applied Microbiology* **89**, 744–747.

Pelletier C, Salar P, Gillet J, Cloquemin G,Very P, Foissac X, Malembic-Maher S (2009) Triplex real-time PCR assay for sensitive and simultaneous detection of grapevine phytoplasmas of the 16SrV and 16SrXII-A groups with an endogenous analytical control. *Vitis* **48**, 87–95.

Petrzik K, Sarkisova T, Čurnová L (2011) Universal primers for plasmid detection and method for their relative quantification in phytoplasma-infected plants. *Bulletin of Insectology* **64**(Supplement), S25–S26.

Piepenburg O, Williams CH, Stemple DL, Armes NA (2006). DNA detection using recombination proteins. *Plos Biology* **4**, e204.

Pignatta D, Poggi Pollini C, Giunchedi L, Ratti C, Reggiani N, Forno F, Mattedi L, Gobber M, Miorelli P, Ropelato E (2008) A real-time PCR assay for the detection of European stone fruit yellows phytoplasma (ESFYP) in plant propagation material. *Acta Horticulturae* **781**, 499–504.

Prabu GR, Kawar PG, Theertha Prasad D (2008) Differential filtration approach for isolation and enrichment of sugarcane grassy shoot phytoplasma. *Sugar Tech* **10**, 274–277.

Prince JP, Davis RE, Wolf TK, Lee I-M, Mogen BD, Dally EL, Bertaccini A, Credi R, Barba M (1993) Molecular detection of diverse mycoplasmalike organisms (MLOs) associated with grapevine yellows and their classification with aster yellows, X-disease and elm yellows MLOs. *Phytopathology* **83**, 1130–1137.

Rad F, Mohsenifar A, Tabatabaei M, Safarnejad MR, Shahryari F, Safarpour H,. Foroutan A, Mardi M, Davoudi D, Fotokian M (2012) Detection of '*Candidatus* Phytoplasma aurantifolia' with a quantum dots fret-based biosensor. *Journal of Plant Pathology* **94**, 525–534.

Rajan J, Clark MF (1985) Detection of apple proliferation and other MLOs by immunocapture PCR (IC-PCR). *Acta Horticulturae* **386**, 511–514.

Ramazzotti M, Bacci G (2018) 16S rRNA-based taxonomy profiling in the metagenomics era. In: Metagenomics Perspectives, Methods, and Applications. Ed Nagarajan M. Acadmic Press, Elsevier, The Netherlands, 103–119 pp.

Saeed EM, Roux J, Cousin M (1993) Studies of polyclonal antibodies for the detection of MLOs associated with faba bean (*Vicia faba* L.) using different ELISA methods and dot-blot. *Journal of Phytopathology* **37**, 33–43.

Salehi M, Izadpanah K, Siampour M, Esamilzadeh SA (2011) Polyclonal antibodies for the detection and identification of Fars alfalfa witches' broom phytoplasma. *Bulletin of Insectology* **64**(Supplement), S59–S60.

Sarindu N, Clark MF (1993) Antibody production and identity of MLOs associated with sugarcane white leaf disease and Bermudagrass white leaf disease from Thailand. *Plant Pathology* **42**, 396–402.

Satta E, Nanni IM, Contaldo N, Collina M, Poveda JB, Ramírez AS, Bertaccini A (2017) General phytoplasma detection by a q-PCR method using mycoplasma primers. *Molecular and Cellular Probes* **35**, 1–7.

Shahryari F, Safarnajad MR, Shams-Bakhsh M, Ataiee S (2011) Use of a recombinant protein for development of a DAS-ELISA serological kit for sensitive detection of witches' broom disease of lime. *Bulletin of Insectology* **64**(Supplement), S43–S44.

Shahryari F, Shams-Bakhsh M, Safarnejad MR, Safaie N, Ataei Kachoiee S (2013) Preparation of antibody Immunodominant Membrane Protein (IMP) of '*Candidatus* Phytoplasma aurantifolia'. *Iran Journal of Biotechnology* **11**, 14–21.

Schneider B, Seemüller E (1994) Presence of two set of ribosomal genes in phytopatogenic mollicutes. *Applied and Environmental Microbiology* **60**, 3409–3412.

Schneider B, Ahrens U, Kirkpatrick BC, Seemüller E (1993) Classification of plant pathogenic mycoplasma-like organisms using restriction-site analysis of PCR-amplified 16S rDNA. *Journal of General Microbiology* **139**, 519–527.

Schneider B, Seemüller E, Smart CD, Kirkpatrick BC (1995) Phylogenetic classification of plant pathogenic mycoplasma-like organisms or phytoplasmas. In: Molecular and Diagnostic Procedures in Mycoplasmology. Eds Razin S, Tully J. Academic Press, San Diego, California, United States of America, 369–380 pp.

Seddas A, Meignoz R, Daire X, Boudon-Padieu E, Caudwell A (1993) Purification of grapevine "flavescence dorée" MLO (mycoplasma-like organism) by immunoaffinity. *Current Microbiology* **27**, 229–236.

Shen WC, Lin CP (1993) Production of monoclonal antibodies against a mycoplasmalike organism associated with sweet potato witches' broom. *Phytopathology* **83**, 671–675.

Shen WC, Lin CP (1994) Application of immunofluorescent staining and tissue-blotting techniques for the detection of a mycoplasma-like organism associated with sweet potato witches' broom. *Plant Pathology Bulletin* **3**, 79–83.

Skrzeczkowski LJ, Howell WE, Eastwell KC (2001) Bacterial sequences interfering in detection of phytoplasma by PCR using primers derived from the ribosomal RNA operon. *Acta Horticulturae* **550**, 417–424.

Siampour M, Izadpanah K, Galetto L, Salehi M, Marzachí C (2013) Molecular characterization, phylogenetic comparison and serological relationship of the Imp protein of several '*Candidatus* Phytoplasma aurantifolia' strains. *Plant Pathology* **62**, 452–459.

Sinha RC (1979) Purification and serology of mycoplasmalike organisms antigens from aster yellows-diseased plants by two serological procedures. *Canadian Journal of Plant Pathology* **1**, 65–70.

Sinha RC, Benhamou N (1983) Detection of mycoplasma-like organisms antigens from aster yellows-diseased plants by two serological procedures. *Phytopathology* **73**, 1199–1202.

Smart CD, Schneider B, Blomquist CL, Guerra LJ, Harrison NA, Ahrens U, Lorenz KH, Seemüller E, Kirkpatrick BC (1996) Phytoplasma-specific PCR primers based on sequences of the 16S–23S rRNA spacer region. *Applied Environmental Microbiology* **62**, 2988–2993.

Sugawara K, Himeno M, Keima T, Kitazawa Y, Maejima K, Oshima K, Namba S (2012) Rapid and reliable detection of phytoplasma by loop mediated isothermal amplification targeting a housekeeping gene. *Journal of General Plant Pathology* **78**, 389–397.

Thomas S, Balasundaran M (2001) Purification of sandal spike phytoplasma for the production of polyclonal antibody. *Current Science Online* **80**, 1489–1494.

Tomlinson JA, Dickinson MJ, Boonham N (2010) Rapid detection of *Phytophthora ramorum* and *P. kernoviae* by two-minute DNA extraction followed by isothermal amplification and amplicon detection by generic lateral flow device. *Phytopathology* **100**, 143–149.

Torres E, Bertolini E, Cambra M, Monton C, Martin MP (2005) Real-time PCR for simultaneous and quantitative detection of quarantine phytoplasmas from apple proliferation (16SrX) group. *Molecular and Cellular Probes* **19**, 334–340.

Valasevich N, Schneider B (2017) Rapid detection of '*Candidatus* Phytoplasma mali' by recombinase polymerase amplification assays. *Journal of Phytopathology* **65**, 11–12.

Villamor DEV, Eastwell K (2019) Multilocus characterization, gene expression analysis of putative immunodominant protein coding regions, and development of recombinase polymerase amplification assay for detection of '*Candidatus* Phytoplasma pruni' in *Prunus avium*. *Phytopathology* **109**, 983–992.

Viswanathan R (1997) Detection of phytoplasmas associated with grassy shoot disease of sugarcane by ELISA techniques. *Zeitschrift PflanKrankheit PflanSchutz* **104**, 9–16.

Vu NT, Pardo JM, Alvarez E, Le HH, Wyckhuys K, Nguyen K-L, Le DT (2016). Establishment of a loop-mediated isothermal amplification (LAMP) assay for the detection of phytoplasma-associated cassava witches' broom disease. *Applied Biological Chemistry* **59**, 151–156.

Wambua L, Schneider B, Okwaro A, Wanga JO, Imali O, Wambua PN, Agutu L, Olds C, Jones CS, Masiga D, Midega C, Khan Z, Jores J, Fischer A (2017) Development of field-applicable tests for rapid and sensitive detection of '*Candidatus* Phytoplasma oryzae'. *Molecular and Cellular Probes* **35**, 44–56.

Wei W, Kakizawa S, Jung H-Y, Suzuki S, Tanaka M, Nishigawa H, Miyata S, Oshima K, Ugaki M, Hibi T, Namba S (2004) An antibody against the SecA membrane protein of one phytoplasma reacts with those of phylogenetically different phytoplasmas. *Phytopathology* **94**, 683–686.

Wei W, Davis RE, Lee I-M, Zhao Y (2007) Computer-simulated RFLP analysis of 16S rRNA genes: identification of ten new phytoplasma groups. *International Journal of Systematic and Evolutionary Microbiology* **57**, 1855–186.

Wei W, Lee I-M, Davis RE, Suo X, Zhao Y (2008) Automated RFLP pattern comparison and similarity coefficient calculation for rapid delineation of new and distinct phytoplasma 16Sr subgroup lineages. *International Journal Systematic Evolutionary Microbiology* **58**, 2368–2377.

Yvon M, Thébaud G, Alary R, Labonne G (2009) Specific detection and quantification of the phytopathogenic agent '*Candidatus* Phytoplasma prunorum'. *Molecular and Cellular Probes* **23**, 227–234.

Yan L, Zhou J, Zheng Y, Gamson AS, Roembke BT, Nakayama S, Sintim HO (2014) Isothermal amplified detection of DNA and RNA. *Molecular Biology Systems* **10**, 970–1003.

Yu Y, Yeh K, Lin C 1998. An antigenic protein gene of a phytoplasma associated with sweet potato witches' broom. *Microbiology* **144**, 1257–1262.

Wang K, Hiruki C (2000) Heteroduplex mobility assay detects DNA mutations for differentiation of closely related phytoplasma strains. *Journal of Microbiological Methods* **41**, 59–68.

Zhang Y, Uyemoto JK, Kirkpatrick BC (1998) A small-scale procedure for extracting nucleic acids from woody plants infected with various phytopathogens for PCR assay. *Journal of Virological Methods* **71**, 45–50.

Zhang L, Hurek T, Reinhold-Hurek B (2005) Position of the fluorescent label is a crucial factor determining signal intensity in microarray hybridizations. *Nucleic Acids Research* **33**, e166.

Zhang S, Ravelonandro M, Russell P, McOwen N, Briard P, Bohannon S, Vrient A (2014) Rapid diagnostic detection of plum pox virus in *Prunus* plants by isothermal AmplifyRP using reverse transcription-recombinase polymerase amplification. *Journal of Virological Methods* **207**, 114–120.

Zhao Y, Wei W, Lee I-M, Shao J, Suo X, Davis RE (2009) Construction of an interactive online phytoplasma classification tool, *i*PhyClassifier, and its application in analysis of the peach X-disease phytoplasma group (16SrIII). *International Journal of Systematic and Evolutionary Microbiology* **59**, 2582–2593.

Chapter 8
The Development and Deployment of Rapid In-Field Phytoplasma Diagnostics Exploiting Isothermal Amplification DNA Detection Systems

Mattew Dickinson and Jennifer Hodgetts

Abstract Detection and diagnosis of phytoplasmas is an important requirement, both for the identification of their hosts, alternate hosts and potential insect vector species and for the management of these devastating diseases of plants. Being able to detect pathogen presence at pre-symptomatic stages allows the removal of infected plants before they provide a significant reservoir of the pathogen for spread of disease to other plants or dissemination *via* propagation material, whilst detection in weeds from which the disease might spread into crops allows appropriate weed control to be undertaken. Similarly, detection in potential insect vectors allows for management practices to be considered that might reduce the spread of the diseases. As new technologies have developed over the years since phytoplasmas were first discovered, the methods for detection have evolved, as have their sensitivity, reliability and ease of use. In-field or point-of-care diagnostics, which allows tests to be conducted rapidly on site with minimal equipment and costs and which also allows easy and unambiguous interpretation of the results, is a goal that has driven the implementation of diagnostic techniques. This chapter will detail how, with the advent of isothermal DNA amplification methods, particularly loop-mediated isothermal amplification (LAMP) and recombinase polymerase amplification (RPA), this goal is now being attained.

Keywords DNA extraction · In-field diagnostics · Lateral flow devices · Loop-mediated isothermal amplification · Recombinase polymerase amplification

M. Dickinson (✉)
School of Biosciences, University of Nottingham, Loughborough, UK

J. Hodgetts
Elsoms Seeds Ltd, Spalding, UK

A. Bertaccini et al. (eds.), *Phytoplasmas: Plant Pathogenic Bacteria - III*,
https://doi.org/10.1007/978-981-13-9632-8_8

8.1 Introduction

Detection and diagnosis of the phytoplasmas associated with diseases on plants is important to aid with the development of appropriate disease management strategies. For routine, laboratory-based diagnostics, the polymerase chain reaction (PCR) is now the main method of choice, with many sets of primers having been developed both for universal detection of phytoplasmas and for ribosomal group specific assays (Gundersen and Lee 1996; Smart et al. 1996). These assays have typically involved nested PCR reactions and identification of amplification products through RFLP analyses in gel electrophoresis, although the various stages involved in these procedures make them both time-consuming and also prone to contamination if sufficient care is not taken, due to the repeated opening and closing of reaction tubes and potential production of spillages and aerosols.

Real-time PCR assays have greatly improved the reliability of these diagnostic methods, with the closed-tube assays reducing the risk of contamination problems, and there are now many validated, sensitive and reliable real time PCR assays published, both probe-based and SYBR-green-based, for universal (Christensen et al. 2004; Hodgetts et al. 2009) and ribosomal group-specific (Baric and Dalla-Via 2004; Galetto et al. 2005; Crosslin et al. 2006; Angelini et al. 2007; Hodgetts et al. 2009) detection of phytoplasmas. However, whilst the size and cost of real time PCR machines and reagents have come down over recent years, making this approach much more affordable not only for well-resourced large screening laboratories, but also for more basic laboratory settings, the equipment generally needs to be run off a main power supply. In addition, the *Taq* DNA polymerase enzyme used in PCR and real time PCR is prone to inhibition by compounds in crude plant and insect DNA extracts, such that the DNA samples generally have to be purified to ensure that they support the reaction. Suitable equipment and reagents such as vortex mixers and centrifuges are therefore normally required to extract this DNA from plant or insect samples, which makes the techniques unsuitable for in-field use. It is the ability to combine a rapid DNA extraction technique, that requires minimal equipment and handling stages, with a portable, battery-operated, closed-tube amplification product detection system, that makes isothermal amplification procedures the potential solution to rapid in-field diagnostics.

8.2 In-Field Antibody-Based Detection Systems

Before describing the DNA-based isothermal amplification detection systems, it is of value to also discuss potential alternative approaches for in-field detection of plant pathogens. One system that has been developed, and is widely used, is the lateral flow device (LFD) (Danks and Barker 2000), particularly for the detection of plant viruses, but also increasingly for bacteria (Braun-Kiewnick et al. 2011) and fungi (Thornton et al. 2004). These devices, which are based on the presence of antibodies for detection of specific proteins or antigens on the pathogens, are simple to use, requiring minimal equipment for extraction of samples from plants, give clear results that are easy to interpret and are well received by plant health inspectors and extension workers. A number of research groups have used enriched or partially purified phytoplasma preparations to raise polyclonal and monoclonal antibodies (Chapter 8). However, this approach has had limitations, such as low pathogen titre, low sensitivity and low specificity and cross-reactivity. More recently, there has been some progress made with attempting to grow phytoplasmas in culture (Contaldo et al. 2016, 2019), although this has only been demonstrated for a small number of strains and growth of colonies is difficult to reproduce in different laboratories.

The more systematic approach for raising antibodies against phytoplasmas for use in diagnostic systems has been based on using phytoplasma sequencing and genomics to identify suitable target genes or parts of genes that can be cloned into expression vectors to produce the designated peptide in sufficient quantities for immunisation into laboratory animals. Kakizawa et al. (2001) used this technique to produce secA antibodies from the onion yellows phytoplasma (16SrI-B) which detected phytoplasmas from four 16S ribosomal groups. In later work, Hodgetts et al. (2014) used a similar approach to obtain monoclonal antibodies against the secA of Cape St Paul wilt disease phytoplasma of coconut (16SrXXII). However, whilst these antibodies were shown to have good specificity for detection of the target protein, they were unable to detect it in extracts from infected plants. More promising targets have involved genes for the immunodominant membrane proteins (IDPs), for which polyclonal antibodies have been produced against a number of phytoplasmas such as apple proliferation, lime witches' broom, Western X disease, onion yellows and napier grass stunt. Some of these antibodies have also been developed into detection systems for use in enzyme-linked immunosorbent assay (ELISA) tests, but it is unclear how widespread their usage has been, and there are no examples of these antibodies having as yet been developed into LFD-based detection systems for use in the field. A relative lack of sensitivity compared to DNA-based methods is a typical limitation in the deployment of antibody-based detection systems.

8.3 In-Field DNA-Based Detection Systems

Over recent years, a number of isothermal nucleic acid amplification technologies have been described that have the potential to be developed into in-field detection systems for plant pathogens (Lau and Botella 2017; Donoso and Valenzuela 2018). Amongst these are loop-mediated isothermal amplification (LAMP) (Notomi et al. 2000), helicase-dependent amplification (HDA) (Vincent et al. 2004), rolling circle amplification (RCA) (Fire and Xu 1995), molecular inversion probes (MIPs) (Lau et al. 2014) and recombinase polymerase amplification (RPA) (Piepenburg et al. 2006). Of these, the most promising and the ones exploited to date for phytoplasmas are LAMP and RPA. Both methods have been patented and require the use of licensed reagents when applied to diagnostic testing, but not for the use in research and development. Furthermore much of the associated software, equipment and reagents which enable the suitability and applicability of the methods to in-field detection are available as commercial products (Table 8.1).

LAMP Originally developed by Notomi et al. (2000), LAMP is a method by which a specific target region of DNA can be amplified during incubation at a single tem-

Table 8.1 Patents, commercial kits and equipment associated with the use of LAMP and/or RPA

Type of patent	Used for	Details
Technology patent	LAMP	Eiken Chemical Company Ltd., Japan. First published by Notomi et al. (2000)
	RPA	ASM Scientific Ltd., United Kindgom (now TwistDx Ltd.). First published by Piepenburg et al. (2006)
Primer design	LAMP	Free software; Primer Explorer v4 or v5, http://primerexplorer.jp/elamp4.0.0/index.html Commercial programme; Premier Biosoft LAMP Designer, http://www.premierbiosoft.com/isothermal/lamp.html
Crude DNA extraction	LAMP	Nitrocellulose membrane-based LFD
		Cellulose-based dip-sticks, created in a lab from Whatman No.1 filter and Paraplast Plus wax
End-point detection strategies	LAMP or RPA	Hydroxynaphthol blue, multiple suppliers
		SYBR® Green I, multiple suppliers
		EvaGreen Dye, Biotium Inc., United States of America
		Fluorescent Detection reagent (FD), Eiken, Japan
		WarmStart® Colorimetric LAMP 2X Master Mix, New England Bio Labs Ltd., United Kingdom
Lateral flow device detection system	LAMP or RPA	PCRD nucleic acid detection lateral flow assays, Abingdon Health, United Kingdom
	RPA	TwistAMP®nfo kit, TwistDx Ltd., United Kingdom
		Milenia HybriDetect 1 lateral flow strips, TwistDx Ltd., United Kingdom
Real-time (fluorescent) detection reagents	LAMP	Isothermal master-mix ISO-001 and ISO-004, OptiGene Ltd., United Kingdom
		WarmStart® LAMP Kit (DNA & RNA), New England Biolabs, United States of America
	RPA	TwistAmp® exo, TwistDx Ltd., United Kingdom
Real-time in-field amplification/ detection	LAMP	Genie® platforms (e.g. II, III), OptiGene Ltd., United Kingdom
		Loop amp Real time Turbidimeter (LA-500), Eiken Chemical Company Ltd., Japan
	RPA	T8-ISO or T16-ISO Instrument, TwistDx Ltd., United Kingdom
Pathogen-/ host-specific assay kits	LAMP	*Guignardia citricarpa*, *Dothistroma septosporum*, *Clavibacter michiganensis* subsp. *sepedonicus*, *Chalara fraxinea* and *Botrytis cinerea* (pathogens) and plant COX gene (control assay), OptiGene Ltd., United Kingdom. Kits licensed for diagnostic use

perature. LAMP uses strand-displacing DNA polymerase (*Bst* DNA polymerase large fragment or equivalent) and, in its original format, uses a set of four primers that recognise six distinct sequences in the target DNA. Two of these primers, the F3 and B3, are relatively simple forward and reverse primers that delimit the target region of typically 200–300 bp in length, whilst the other two primers, the Forward Inner Primer (FIP) and Backward Inner Primer (BIP), contain sequences of the sense and antisense strands joined, so that when they bind to the target, they form stem loops. The single-stranded loop regions serve as primer binding sites, allowing amplification to proceed without thermal denaturation of the template, normally at an amplification temperature of around 65°C. The result is a cycling reaction that creates a mixture of stem loop DNAs, inverted repeats of the target and cauliflower-like structures with multiple loops, which result in a characteristic ladder-like appearance when the products are visualised by gel electrophoresis. It has been shown that in many cases, adding a further two primers, the Forward Loop (FL) and Backward Loop (BL), enhances the efficiency of the process (Nagamine et al. 2002), such that most published assays now use the six-primer system that targets eight regions within the target DNA sequence, and can produce detectable numbers of copies of the target sequence within 5–30 minutes.

It is this requirement for primers targeting six or eight target regions that is often perceived as making the design of the primers complicated and can put researchers off developing LAMP assays. However, there are free software packages or commercial programmes (Table 8.1) that can design primers from target sequences for testing and validating in LAMP, and in the authors' experience, these programmes have good success rates. However, it should be noted that with phytoplasmas, one of the complications can often be the high A+T content of the target sequences, which makes primer design of any kind difficult, particularly for certain 16S ribosomal groups, such as 16SrXI and 16SrXIV, which have particularly high A+T contents.

There are many advantages to LAMP as a diagnostic technique, once primers have been designed and validated, in that it can amplify with high efficiency and rapidity from low levels of target sequence, with a great specificity because of the number of primers. It also requires minimal reagents and equipment, potentially just a primer mix, the DNA polymerase and reaction buffer and a means of heating the samples to 65°C, such that heated blocks, water baths, flasks of hot water or even disposable pocket warmers (Hatano et al. 2010) have been used in the field. In addition, the polymerase has also been shown to be less sensitive to inhibitors in crude DNA extracts than the *Taq* DNA polymerases used in PCR (Sugawara et al. 2012). As a result, LAMP assays have been developed for a broad range of plant pathogens, and numerous papers have been published detailing assays for different pathogens. Indeed a number of commercial kits are now available for specific pathogens (Table 8.1) that provide all the necessary materials and reagents for the rapid detection of these pathogens from plant samples. In addition, and of importance for providing the essential controls for tests applied to plant samples, an assay has been designed for the cytochrome oxidase *cox* gene from plants that appear to amplify from most plant species (Tomlinson et al. 2010a) and is also available as a commercial kit. This means that samples can be tested with the pathogen-specific primers and plant-specific primers in LAMP reactions run in parallel, so that any reactions that are negative for the pathogen but positive with the COX primers can be confirmed as true negatives for the pathogen and not false negatives resulting from poor quality DNA/presence of inhibitors in the extract/failed reactions.

RPA An alternative approach that also has the potential to be used in-field with similar approaches and equipment for detection of the amplification products as LAMP is the RPA method, first described by Piepenburg et al. (2006), in which the reaction begins with the T4 UvsX recombinase protein, assisted by the T4 UvsY loading factor binding to the primers. This complex then searches the homologous sequences to the primers in the target DNA and then invades the double-stranded DNA to form D-loop structures. In each of the D-loop structures, formed by the forward and reverse primers, the primer hybridises to the one side of the D-loop to form double-stranded DNA, and a DNA polymerase (*Bsu* or *Sau*) then initiates strand synthesis from this primer site. The recombinase disassociates from the complex and becomes immediately available to initiate another strand displacement with more primer. Incorporation of both forward and reverse primers in the reaction ensures that polymerisation occurs in both directions simultaneously, resulting in exponential amplification of duplex DNA delineated by the primer binding sites. This amplification occurs rapidly, within the same timeframes as LAMP of 5–30 minutes, and optimal temperatures for RPA are between 37 and 42°C. Whilst this technique has only recently started to be used for plant pathogens, in those cases where it has been used, such as for root-infecting fungi of turfgrass (Karakkat et al. 2018), it has been shown to be comparable to LAMP in both speed and accuracy.

8.4 Amplification Product Detection Systems

Both LAMP and RPA are excellent isothermal amplification techniques that can be developed with relative ease for phytoplasma detection with appropriate primer design. However, what is also needed to make these approaches applicable for in-field detection is a method for detecting and visualising the amplification product with minimal equipment that can be used easily in remote locations to deliver results in a way that can be easily interpreted.

In laboratory settings, gel electrophoresis can be used to detect amplification products; however, there are inherent problems with this approach if care is not taken, since the amplification procedures produce such large quantities of product that it is very easy to produce aerosols and contaminate equipment when opening tubes after amplification, which can lead to problems with subsequent experiments in the same laboratory and the generation of false-positive results. Therefore, as with PCR, the drive with these isothermal amplification techniques has been towards closed-tube detection methods in which amplification products are detected without the need to reopen the tubes once the reactions have been prepared, and a number of closed-tube detection approaches, both indirect and direct, have been developed. Amongst the indirect approaches that have been developed for LAMP are those based on detecting the pyrophosphate that accumulates during the LAMP reaction, such as through visualisation of turbidity or changes of pH that result from this build-up of pyrophosphate (Mori et al. 2001). Dyes that can be incorporated into the reactions when they are set up (Table 8.1), so that the assays can be run in a closed-tube format, are particularly popular for providing binary (*i.e.* plus/minus) results from a LAMP reaction. For example, hydroxynaphthol blue (HNB) is a dye that can be added into the LAMP reaction mix when the reaction is set up, and the mix can then be incubated at 65°C for a set time (usually 30–60 minutes) and then observed by eye for the subtle blue colour change in positive amplifications (Tomlinson et al. 2010b). The colour change is caused through the HNB detecting reductions in the levels of alkaline earth metals such as Mg^{2+} and Ca^{2+}, which are converted by the pyrophosphate produced in the LAMP reaction into insoluble magnesium pyrophosphate. However, the colour change can be quite difficult to discern, so alternative dyes are now on the market, such as the Fluorescent Detection Reagent (Table 8.1) or calcein, which can also be added at the start of the reaction, and then show a light orange to yellowish-green colour change in visible light or clear fluorescence if observed under UV light (Sugawara et al. 2012), following the incubation. Similarly a commercial master mix is available (Table 8.1) that uses pH-sensitive phenol red as the end-point indicator incorporated into the polymerase master mix, so that all the users have to add is their primer mix and the sample DNA. This system is designed to detect the pH changes that occur through the production of protons from the DNA polymerase activity, producing a change in the colour of the solution from pink to yellow.

Turbidity meters, including portable battery-operated meters, have also been used successfully to detect LAMP products through the pyrophosphate build-up (Table 8.1) (Mori et al. 2001). However, the drawback of all these indirect binary detection methods is the lack of validation that it is the correct product causing the pyrophosphate build-up/pH change and not an artefact of aberrant priming, so whilst they have advantages in being easy and relatively cheap to use in field settings, the results have to be treated with some level of caution, and additional tests may need to be carried out subsequently in a laboratory to confirm their validity.

An alternative method that has been used for both LAMP and RPA is an approach to provide results that can be visualised by lateral flow devices, whereby one line appears for negative and two lines for positive, in order to provide a system that plant health inspectors are familiar with (Tomlinson et al. 2010a; Mekuria et al. 2014). In the case of LAMP, the approach has been to incorporate biotin labels on the 5′ ends of the Backward Loop (BL) primers for the pathogen-specific and plant COX gene assays and then to incorporate a digoxigenin (DIG) label at the 5′ end of the pathogen-specific Forward Loop (FL) primer and a fluorescein isothiocyanate (FITC) label at the 5′ end of the equivalent COX assay FL primer. Following the LAMP reactions, which can be undertaken on any device in the field that heats the reactions to 65°C for 30–60 minutes, the tubes are opened, and the product is diluted into an appropriate amount of LFD buffer and then applied to the LFD devices that have latex-coloured beads coated with a reagent that specifically binds to the target molecule (FITC or DIG) in the release pad. The samples then pass through the device by lateral flow and the test lines contain reagents that also bind to the target-latex complex (either DIG or FITC). Therefore, in the DIG-detecting LFD device that is used for detecting the pathogen, coloured latex beads will accumulate at the test line if the pathogen is present, as the reagent in the test line binds to the DIG and the latex binds to the biotin, and nothing will accumulate at this line if pathogen is absent and there is no LAMP product containing both the DIG and biotin (*i.e.* only unincorporated primers). The COX assay, using the FITC-detecting device, works in the same way, with the test line only showing if the FITC and biotin have been combined into an amplification product, *i.e.* as a test to validate the DNA extract from the infected plant supported LAMP, and did not contain inhibitors of the LAMP reaction. Then in both assays, to confirm that the LFD device is working, the second control line imprinted on the LFD device contains a reagent that binds directly to the coated latex, so that the control line will show up as long as the lateral flow of solution through the device has worked correctly. Thus, a positive result for the pathogen on the DIG-LFD will show up as two lines and a negative as one line, and on the FITC-LFD, a positive for amplification of plant DNA will show up as two lines and a negative for amplification of plant DNA as a single line.

The lateral flow system for RPA assays works in a similar way but is a probe-based system, whereby the reaction contains the RPA forward and reverse primers that delimit the sequence to be amplified plus an internal probe. The procedure is to label the reverse primer with biotin at the 5′ end and the probe with FAM (6-carboxyfluerescein) at the 5′ end. The probe also contains a blocker sequence at the 3′ end to prevent extension from it and tetrahydrofuran (TFA) incorporated in the middle, which will be cleaved by the *Escherichia coli* endonuclease IV (*Nfo*) in the reaction mix. Cutting at the TFA site releases the blocker, allowing incorporation of the probe into the amplicon and extension from it, and because the opposing (reverse) primer is biotin labelled, the result is a dual-FAM-biotin-labelled amplicon that can then be detected via a lateral flow device. All of the necessary reagents for the assays can be purchased as a kit (Table 8.1) so that the only additional reagents required are the labelled probes and primers for the specific assay. As with the LAMP system, a separate set of primers and probes can also be designed for amplification of plant DNA to confirm that samples do not contain inhibitors of the RPA reaction. The advantage of these LFD-based systems over incorporation of dyes into reaction mixtures is that the labelling of the probes/primers ensures that the reaction products being detected are from specific amplification of the target sequences and therefore not artefacts. However, the procedures do involve opening of reaction tubes after amplification to be able to apply product to the LFD strips, so have additional manual handling steps and the associated contamination risk.

The third approach that has therefore been developed for detection of LAMP and RPA products in the field, and which combines the advantages of validation steps to ensure that the products are not artefacts, and a closed-tube system with minimal manual handling, is to use real-time detection methodologies, analogous to the approaches used for real time PCR. The methods used for LAMP and RPA are different, using either intercalating dyes or labelled probes, respectively. The LAMP-based approach is effectively the same as the SYBR Green real time PCR approach, where an intercalating fluorescent dye is added into the reaction mix that uses conventional LAMP primers with no added labels. The dye then effectively incorporates into the newly synthesised double-stranded DNA during the amplification, and fluorescence is then detected by real time LAMP or real time PCR machines over time, such that the data can be visualised and recorded as the Tp, time to positive (time of product formation). Following amplification, the machines can then be set to heat the samples to melt the DNA and allow reannealing to generate anneal curves, and this is used as validation, since the annealing temperature for the correct LAMP products will occur at a specific temperature, whilst if any amplification of artefacts has occurred, these will give different annealing temperatures. Real time, battery-operated LAMP product detection machines are commercially available and can be used for in-field detection (Table 8.1). These machines use a rechargeable battery, which has sufficient power to be able to complete multiple runs in the field before the battery needs recharging. In addition, it is possible to buy single-tube master mixes that contain polymerase, buffer and fluorescent dye and are stable at ambient temperature, such that all that needs to be added in the field is the primer mix for the target and the sample DNA, and the whole analysis can then be undertaken as a closed-tube assay to provide results in under 30 minutes. Potential disadvantages of these assays are the costs of the battery-operated machines and the fact that they can only undertake 8–16 reactions at a time. However, it is also possible to undertake these real time LAMP assays on any real time PCR machine, which potentially allows more samples to be analysed at any one time.

In the case of real time assays for RPA, the approach taken is different and more similar to TaqMan®-type real time PCR assays in that labelled probes are used. Generally, the probes are labelled with a fluorescent label such as FAM adjacent to the tetrahydrofuran (THF) and a quencher, located in the middle of the probe sequence, and then with the blocker sequence at the 3′ end of the probe. The *E. coli* endonuclease IV (*Nfo*) in the RPA reaction mix will cleave the probe at the THF to release the fluorescent label from the quencher and blocker, and allow the fluorescent label-containing part of the probe to bind to the target extending to the reverse primer, and this can then be detected as fluorescence using the same types of machines as for real time LAMP and real time PCR. In this case no anneal steps are required in the detection assays, since the specificity of the product is determined by the probe itself, so the overall assay time is quicker than for LAMP. A reagent mix containing all the necessary enzymes and buffer for the RPA reaction in a lyophilised stable form is available (Table 8.1), so that again, all that needs to be added in the field is the primer mix and the DNA sample. In terms of overall costs per assay, the RPA probe, coupled with the costs of the enzyme mixes for the LAMP versus RPA assays, makes the RPA assays slightly more expensive (at the time of writing). However, it should also be noted that the RPA system has the advantage that multiplexing is possible within reactions by using probes with different fluorescence wavelengths for different assays within the same reaction. For example, it would be possible to include an assay for the pathogen labelled with FAM combined with primers and a probe for the plant gene labelled with a different fluorophore such as VIC in a single reaction mix and undertake the plant and pathogen tests simultaneously, whereas in the LAMP system, it would be necessary to run the pathogen assays in one set of tubes and the plant assays in a separate set of tubes in parallel.

To date, the detection systems described above have been the most widely used for detection of LAMP and RPA products in plant pathogen diagnostics. However, a number of biosensor/electrochemical sensor-type detection systems are also being designed and developed for LAMP and RPA (Lau and Botella 2017), which may become applicable in the future, although whether these will be cost-effective for plant pathogen detection systems in resource-poor settings is not yet clear. Indeed, one of the concerns that has already been expressed regarding real time assays is the cost of the hardware for detection of the amplification products (Pérez-López et al. 2017), which is why the incorporation of dyes/LFD-type detection systems might be preferred in some settings.

8.5 In-Field DNA Extraction

One of the key requirements for in-field detection of plant pathogens is a simple, quick and reliable method for extracting DNA from the samples of sufficient quality to be amplifiable. This is not straightforward, since there can be a number of chemicals in plant extracts that inhibit DNA polymerases. However, one of the key advantages of LAMP and RPA are that the polymerases in these systems tend to be less prone to inhibitors than the *Taq* DNA polymerases used in PCR reactions (Sugawara et al. 2012), such that much cruder DNA preparations/extracts can be used in these isothermal methods.

One approach for rapid DNA extraction has been to use the reagents that have been developed for LFD devices (Tomlinson et al. 2010a, b). In this approach, the plant material is added into extraction bottles containing a buffer and metal ball bearings, and then the bottles are shaken for 2 minutes to macerate the plant tissue. Some of the lysate solution is then placed onto the release pad of an LFD nitrocellulose membrane, and then, after lateral flow, a small piece (approximately 1 mm x 1 mm) is excised from the membrane and placed directly into the LAMP reaction mix. One advantage of this approach is that the DNA on the membranes is relatively stable, so these can be returned to a laboratory for any further studies that might be required. It is possible to use 1–2 μl of the lysate solution directly in the LAMP reaction without the need for the LFD device (M. Dickinson, unpublished).

Other similar methods that have been developed include simple cellulose-based dip sticks that allow sample processing in as little as 30 seconds (Zou et al. 2017). In this method, the tissues are macerated in a similar way by shaking in a tube containing extraction buffer and ball bearings for a few seconds. The dip stick is then inserted into the lysate and then into a tube containing a wash buffer, before being inserted into the amplification mix; in this case sufficient DNA washes into the reaction mix off the dip stick, which is then removed, unlike in the LFD approach, where the piece of the membrane remains in the reaction mix. This dip stick method has been shown to work for a broad range of difficult plant tissues and can also be used for extracting RNA for virus assays.

A simpler DNA extraction approach that has been piloted in a broad range of phytoplasma assays for different plant tissues including cassava, sugarcane and coconut trunk borings has also been developed (M. Dickinson, unpublished). In this approach, around 10–20 mg of plant material is added directly into an alkaline polyethylene glycol (PEG) solution (Chomczynski and Rymaszewski 2006) in a tube. The material is then macerated for 20–30 seconds with a disposable micropestle, and then 1–2 μl of the solution is placed directly into the LAMP reaction mixture. Alternatively, plant material can be shaken in a 5–7 ml tube with ball bearings. This approach uses minimal equipment and has been successfully piloted in remote locations to provide fast accurate results. One point to note is that the alkaline nature of the DNA extract added in to the reaction mix does alter the normal annealing temperature of the amplification product in real time LAMP by 1–2°C. Diluting the extracts 1: 10 in molecular biology grade water before use removes this effect on the annealing temperature. Whilst this DNA extraction has not yet been tested in RPA, it is possible that it could provide a fast and effective in-field DNA extraction approach that can be incorporated with rapid isothermal amplification and product detection systems.

8.6 LAMP and RPA Assays for Phytoplasmas

The first use of LAMP for detection of phytoplasmas was reported in 2010 (Tomlinson et al. 2010b), when six-primer set assays were published for the 16SrI group ('*Ca.* P. asteris') and for the 16SrXXII group ('*Ca.* P. palmicola') with primers designed based on the 16S-23S rRNA (Table 8.2). These primers were shown to be specific for the groups from which they were designed and to have levels of detection comparable to nested PCR. In these assays the detection of product was demonstrated through gel electrophoresis and also through the incorporation of hydroxynaphthol blue (HNB) dye into the reagent mixes, which showed a subtle blue colour change in positive reactions. Methods were also presented to show the potential for rapid in-field DNA extraction, in this case using lateral flow device membranes as a means of purifying the crude DNA extracts and then excising a small section of the LFD membrane to place directly into the LAMP master mix as DNA template.

The first use of real time LAMP assays for phytoplasmas was reported in 2011 for 16SrII and 16SrXII group phytoplasmas (Bekele et al. 2011). These assays were also developed using six-primer sets from 16S-23S rRNA sequences, tested using real time LAMP machines and optimised to be specific for the phytoplasma groups that they were designed to. In addition, these assays used the control LAMP assay for the detection of plant genes, based on the COX sequence as described by

Table 8.2 Primers designed for phytoplasma LAMP assays

Phytoplasma ribosomal group target	Sequences 5′-3′	Target gene	References
Phytoplasma universal	**F3** GAA GTC TGC AAC TCG ACT TC **B3** CCT TAG AAA GGA GGT GAT CC **FIP** ACG GGC GGT GTG TAC AAA CCG GAA TCG CTA GTA ATC GCG AAT C **BIP** GTC TAA GGT AGG GTC GAT GCA CCT TCC GGT AGG GAT AC **FL** CGA GAA CGT ATT CAC CGC GAC **BL** GGG GTT AAG TCG TAA CAA G	rRNA	Dickinson 2015
16SrI	**F3** TAA TAT TAA GGG CCT ATA GAT CAG TTG G **B3** CAC GGA TCT TCA CTT ATT TAC AGC TT **FIP** TAT TTT GCC TAT TTG TGG TTA TGG TGT TAT AGA GTA CAC ACC TAA TAA TTG TGA GG **BIP** GAA GGT TAA AAA ATC AAA GGA ACT AAG GGT TAA TTG CGT CCT TCA TCG G **FL** CTAAATGTGTAACTTGAACCACCGA **BL** ACAGTGGATGCCTTGGCACT	rRNA	Tomlinson et al. 2010b
16SrI	**F3** AGGTACCCGAAAAACCTCACC **B3** TCCCCACCTTCCTCCAATT **FIP** TGC ACC ACC TGT GCA ACT GAT AAG GTC TTG ACA TGC TT CTG C **BIP** TGG GTT AAG TCC CGC AAC GAG CTT GCT AAA GTC CCC ACC AT	rRNA	Vu et al. 2016
16SrI	**F3** GTTGCCAAAAAACTTTTAGCT **B3** ACTTCTAATTCTGTTTCAAAACC **FIP** CTA CCT GAT GAA ACA GCA GCC AAT CTA AAA AAG TAGACGCCCA **BIP** AAT TGG TAA AAT CAT TGC CCA AGC GGA TTC ATC AAC ATT AAT AAC TCC A	*groEL*	Sugawara et al. 2012
16SrI	**F3** ACTCAATATGCAGACAGACTT **B3** CGTAAAAACACCTGTTTTGTCT **FIP** AGG GCG AGT GGT GAA AGT TTT ATT GGC AAA AAT CAA GGT GC **BIP** TTA GTA ATA GCG CCC GAA CAC GGT TAA CTG CTT GTT GGT ATT CG **FL** TCAGAAACTGGAAAAGAGACGA **BL** TAGCCTTACAACTCACCAAACC	*leuS*	Al-Jaf 2016
16SrII	**F3** CCGAATGGGGCAACCTAC **B3** CTCGTGTCTCGCCGTACTT **FIP** TTC CTA CAG TTA CTT AGA TAT TTC AGT TAT AAG TAG TTA TAG TAT CAA TTG TTA GG **BIP** CTA GTT ATC AAA TTT TAA TTA ACT CGA TTT TAA TCT ACC TCT ACA GGA TTT TCA C **FL** CAC TGC GTT CCT TTC AAT C **BL** AAT AGT TGG AAA ACT ATA TCC TAG A	rRNA	Bekele et al. 2011

(continued)

Table 8.2 (continued)

Phytoplasma ribosomal group target	Sequences 5′-3′	Target gene	References
16SrII	**F3** TGG TTT GCC TAC AGA ACG **B3** CAT TTC CTT AAT AGA CTC TGG C **FIP** CCG AAG TCG CTA ATT CTC TAC TAC AGG TCA ACA TCC TTC TTA TG **BIP** GCT GTT TTG CAA GAT GTT GAG GGA TGA CCA CCA CGT TCT G **FL** CCA ATC AAT ACC TTT TCC AAG C **BL** ACA GTA TTA GCG AAT GAA GAG G	*leuS*	Al-Jaf 2016
16SrIII	**F3** GGA TAC GCA GTG AAC TGA A **B3** GAT TTC TCG TGT CTC GCC **FIP** GTA TCA GGC TCC TCC GGA TTG TAA CTG CAG GAA AAG AAA GT **BIP** AAC CAA CTT GAA GTT TTT GGG AAA ACG ATA CCC TAG ATT TTA ACT TAC C **FL** GTC RCT ACT RCC AGA ATC GTT ATT **BL** AAC ACC AAA GAA GGT GAT AGT CC	rRNA	Dickinson 2015
16SrIII	**F3** CCA GAT CGT TGG AAT GAA GT **B3** ACA GCG ATA ATT TTA CGA TTC G **FIP** AAG GAA CGG TCA TAA TGG CTC CTG CTA TTA ACT TCG CTA ATC AA **BIP** ACC TGA AGA AAC CCA AAA TTC TCT TCT TCA TAA TCA AGA CCG TT **FL** AAT CAG CGA TCC AAA TAG GGA T **BL** AGA AAT TAA GGT GGA AGA CGG T	*leuS*	Al-Jaf 2016
16SrV	**F3** CGT GTC GTG AGA TGT TAG GTT AAG **B3** CGC GAT TAC TAG CGA TTC CAG **FIP** TAT CCC CAC CTT CCT CCA ATG TTT AAT TCT AAA ACG AAC GCA ACC CC **BIP** TCA AAT CAT CAT GCC CCT TAT GAT CTG GCA GAC TTC AAT CCG TAC TGA GAC TA **FL** ACC ATT ACG TGC TGG CAA CTA G **BL** GCT ACA AAC GTG ATA CAA TGG CTA	rRNA	Kogovšek et al. 2015
16SrV	**F3** CCC AAT ATC TTT GGC GAG AA **B3** TTG GAT ATC TGA TTA TGT CCT T **FIP** ACG GTC TTA ATG TTG AAG AAG CAG CCA ATC ACG CAT TTG A **BIP** AGC GAA ACA AAA ATC TCT TTC ACC TTC TTA TGC TAC AGG AGC **FL** TTA GGT CAT GTT CAT TAT GCT T **BL** TCA CAA AAA GGA ACC AAC ATA A	*leuS*	Al-Jaf 2016

(continued)

Table 8.2 (continued)

Phytoplasma ribosomal group target	Sequences 5′-3′	Target gene	References
16SrVI	**F3** GCT TCT TCA AAA GTA AAT CCG T **B3** GGA AGT TTT TAC GAC AAA ACC A **FIP** CGG TGT TGT TAT GTC TGT TCC TGC ATC ATT AGT GGA TTC GGA **BIP** AGG ATG AAT AGC ATA ACT ACC AAC AGA TTT TGT CGA AGG TGT **FL** TGA AAG AGA TTT TGC TTT TGC T **BL** ACT CCT GTT TGA TCT TTG TTC A	*leuS*	Al-Jaf 2016
16SrX	**F3** CCT GCC TCT TAG ACG AGG AT **B3** CAA TGT GGC CGT TCA ACC T **FIP** AGC ATA CCC TTG CGG GTC TTT TTT TTA CAG TTG GAA ACG ACT GCT A **BIP** AAG AGA TGG GCT TGC GGC ACT TTT CTC AGT CCA GCT ACA CAT CA	rRNA	de Jonghe et al. 2017
16SrX	**F3** AAC TGG CAA CAA TCC TAG AA **B3** TCCAATCCAATTACGTTGAATT **FIP** ACA GTT CCT AAT CCT TCA CAA ATT AGC TAC TTC TGA TAA CGA TT **BIP** CCG AAA GAG GCG ATT TTC CGA AGC CTT TCT GCA TAA TCA GT **FL** TCT TGA GAA TCC ATT GTG TCC A **BL** ACC TTA AAA CAG TGG GTT TTG A	*leuS*	Al-Jaf 2016
16SrXI/ XIV	**F3** AAG AAG GAG GGC CTA TAG CTC AGT **B3** ATA TCG CTG TTA ATT ACG TC **FIP** AGA AAG ATG ACC TTT TTC AGT TGG TGT GGT TAG AGC ACA CGC CTG ATA AG **BIP** CAA AGT AAA ATA ATA AAA TCA AAG GAC ATC GGC TCT TAG TGC CAA G **FL** ATG GAC TTG AAC CAT CGA CC **BL** AAG GGC GTA CAG TGG ATG C	rRNA	Dickinson 2015
16SrXI/ XIV	**F3** ACT AAG TCT TTC AAA ACT GAA GA **B3** CTC GGG GTT TGT ACA CAC **FIP** CGT AAC AAG GTA TCC CTA CCG GAA AAT TAT TCC TTA GAA AGG AGG TG **BIP** ACT TAA CCC CAA TCA TCG ATC CTA GTT GGT AAT ACT CAA AAA CGG	rRNA	Nguyen et al. unpublished
16SrXI/ XIV	**F3** CGG AAT TCC ATG TGT AGC G **B3** CCT TTG AGC CAT GAC TTC A **FIP** TAT ATG GAG GAA CAC TTT TAA TGA CCC AGA AAT GAC TGC GAC T **BIP** GCA CCC CTC GTT TGT CCT TTT TCG GCA TTT GCT ACT CAT GAT	rRNA	Obura et al. 2011

(continued)

Table 8.2 (continued)

Phytoplasma ribosomal group target	Sequences 5′-3′	Target gene	References
16SrXI/ XIV	**F3** GCG GTG GAT CAT GTT GTT **B3** CAT CAT GAC ATG CTG GCA **FIP** TCC CTG ATA ACC TCC ACT ATA TTG CAT TCG AAG ATA CAC GAA AAA CC **BIP** TAC AGG TGG TGC ATG TTG TCG TAA TGA CAA GGG TTG CGC **FL** GCA GAG TAT GTC AAG ACC TGG TAA **BL** CAG CTC GTG TCG TGA GAT GTT A	rRNA	Nair et al. 2016
16SrXII	**F3** GGA TGC CTT GGC ACT AAG **B3** GTG TCT ACG CCG TAC TTA TC **FIP** AAC GGG TTG TCC CAT TCG GCC GAT GAA GGA CGC AAT T **BIP** TTC TGG TAG TAG TGA CGA GCG ATT ACA GGA CTG TCA CCT TCT **FL** AAT CCA CGG ATC TCC ACT TAT **BL** CGG AAG AGC CTG ATG CTA TT	rRNA	Bekele et al. 2011
16SrXII	**F3** TTC CAC CAA ATC TTT GAG CT **B3** ACA ATA GCT ACC AAT ATG GCA G **FIP** AAT TAA GAG GAC GTG CCG GTC GTT CGT CTT CGC TGG AAA **BIP** TCC TAA AAC CGC CAA ACC TCC GAA GAG GAA CTG ATA TTC GCT T **FL** GAG ATC CTG GTT ATT CTC GCT T **BL** TTC AAC AAC GCC TTC ACC T	*secA*	Kogovšek et al. 2017
16SrXIII	**F3** AAA GGA TTT GAA ACC GAG TTA **B3** CCC TTA ATT TAT TAG CAA CCA A **FIP** ATT CAA CCG CCA TAG ATT CTC TGA AGT AGT TCA AGG TTT ACA AT **BIP** AAC CAT TAG TTT TAG TGA CCG ATT CAA CTG ATT CAG CTA CAA TT **FL** AAT ACG GAG AAG CAT ATC CTT **BL** AGA AGG TGT TGT TAA AGA ATC T	*groEL*	Pérez-López et al. 2017
16SrXXII	**F3** TAG AGG AAG GGC CTA TAG CTC AGT **B3** GTA TCG CCG TTA ATT GCG TC **FIP** TGA ATA AGA GGA ATA TGG TAT GGG TGT GGT AGA GCA CAC GCT TGA TAA G **BIP** TCT CTA ATG ACA CAC CAA TGA AGG ACA TCG GCT CTT AGT GCC AAG **FL** GGA CTT GAA ACC ATT GAC CG **BL** AAG GGC GTA CAG TGG ATG C	rRNA	Tomlinson et al. 2010b

(continued)

Table 8.2 (continued)

Phytoplasma ribosomal group target	Sequences 5′-3′	Target gene	References
'*Candidatus* Phytoplasma novoguineense'	**F3** CGC CAC ATT AGT TAG TTG GTA **B3** TTC ATC GAA TAG CGT CAA GG **FIP** GTT TGG GCC GTG TCT CAG TGC CTA CCA AGA CGA TGA TG **BIP** TAC GGG AGG CAG CAG TAG GAG TAC TTC ATC GTT CAC GC **FL** GTG GCT GTT CAA CCT CTCA **BL** AAC TCT GAC CGA GCA ACG	rRNA	Lu et al. 2016
Cox universal	**F3** TAA TAT TAA GGG CCT ATA GCT CAG TTG G **B3** CAC GGA TCT TCA CTT ATT TAC AGC TT **FIP** ATT TTG CCT ATT TGT GGT TAT GGT GTT ATA GAG CAC ACG CCT AAT AAG CGT GAG G **BIP** GAA GGT TAA AAA ATC AAA GGA ACT AAG GGT TAA TTG CGT CCT TCA TCG G **FL** CTA AAT GGA CTT GAA CCA CCG A **FB** ACA GTG GAT GCC TTG GCA CT	*Cox*	Tomlinson et al. 2010a

Tomlinson et al. (2010a) as a means of confirming that DNA extracts supported LAMP and that negative results with the phytoplasma assays were the result of the phytoplasma absence and not due to poor-quality DNA.

Real time LAMP assays have subsequently been developed for on-site detection of the "flavescence dorée" phytoplasma in grapevine (Kogovšek et al. 2015) for in-field testing or for testing in the laboratory, and the performance of LAMP was directly compared with real time PCR methods. Primers were designed and tested against both the 16S rRNA and the 23S rRNA genes, and whilst the 16S rRNA assay was more sensitive, it lacked the specificity of the 23S rRNA assay, which was the one subsequently developed for further use. The assay was estimated to detect as few as 9–27 copies of the phytoplasma genome per assay and gave good semi-quantitative data, whereby the Tp, which varied between 11 and 24 minutes, showed inverse correlation with the concentration of the phytoplasma DNA (*i.e.* the highest concentration gives the fastest amplification times) and correlated well with the C_q scores obtained by testing the same samples in real time PCR. Importantly, a rapid DNA extraction, in which berries, leaf veins or flowers were homogenised directly in the field in either ELISA or Na-acetate buffer, and then the homogenate was diluted in water and added directly into the LAMP mix, was tested and shown to provide good results, although with approximately tenfold less sensitivity than purified DNA samples. Subsequently, a set of primers were designed based on the *secA* gene for the "bois noir" phytoplasma of grapevine (Kogovšek et al. 2017). These primers were shown to be specific to the phytoplasma; show similar sensitivity to real time PCR assays, detecting as few as three to nine copies and work reliably on crude homogenate extracts.

Real time assays have also been designed for some of the 16SrX phytoplasmas (de Jonghe et al. 2017), where it was shown that a four-primer set was sufficient for amplification from the 16SrX-A '*Ca.* P. mali', 16SrX-B '*Ca.* P. prunorum' and 16SrX-C '*Ca.* P. pyri' targets within 30 minutes. Assays were undertaken both on portable real time machines suitable for in-field use and on real time PCR machines. However, it was noted that the detection limit was around tenfold lower with this LAMP assay than with the probe-based real time PCR assay that it was compared against. For the 16SrXI napier grass stunt phytoplasma of East Africa, a set of four primers designed from the 16S rRNA were published by Obura et al. (2011), and whilst these primers were shown not to detect the 16SrI, 16SrII, 16SrIII and 16SrXXII phytoplasmas tested in their validation, they did detect the 16SrVI, 16SrX, 16SrXII and 16SrXIV phytoplasmas tested along with the napier grass stunt target. In these experiments the results of amplification were visualised by gel electrophoresis, and the primers were subsequently used to screen napier grass accessions to help select those with diverse susceptibility to the disease (Wamalwa et al. 2017). However, these primers do not appear to work for real time LAMP assays (M. Dickinson, unpublished), and an alternative six-primer set from the 16S-23S rRNA of 16SrXI group phytoplasmas that are more specific and only detect 16SrXI and 16SrXIV phytoplasmas was recently developed (Dickinson 2015), coupled with a four-primer set that are only able to detect the 16SrXI sugarcane white leaf/sugarcane grassy shoot and 16SrXIV Bermuda grass white leaf phytoplasmas, and do not detect the 16SrXI napier grass stunt phytoplasma (BQ Nguyen et al. unpublished). The reasons for this apparently odd specificity may be because recent work (Abeysinghe et al. 2016) using multilocus sequence typing (MLST) has suggested that the 16SrXI sugarcane phytoplasmas are more closely related to the 16SrXIV Bermuda grass white leaf type than they are to the supposedly 16SrXI napier grass stunt. Primers based on the 16S rRNA have also been designed and published for the closely related coconut root wilt and areca nut yellow leaf diseases of southern India, which are also in the 16SrXI/16SrXIV groups. In these assays, the amplification products were identified through gel electrophoresis and incorporation of hydroxynaphthol blue and in real-time assays, with amplification occurring within 10–15 minutes. However, no data on testing these primers on other 16S ribosomal groups was reported, so the specificity of the assay is unclear, although it was able to detect the sugarcane grassy shoot phytoplasma, which is closely related to the coconut root wilt/areca nut yellow leaf-associated phytoplasmas (Nair et al. 2016). LAMP assays have also been reported for the cassava witches' broom disease phytoplasma in Southeast Asia, using a four-primer set based on the 16S rRNA gene (Vu et al. 2016). Primers were designed for the cassava actin gene as a control, and amplification products were detected by gel electrophoresis in laboratory and also using hydroxynaphthol blue or the presence of a visual precipitate in field-based studies. The pilot studies showed good potential for on-site screening, although it was not clear how much validation had been undertaken on the phytoplasma primers to confirm which specific 16Sr group phytoplasmas they were able to detect and what their sensitivity was.

LAMP assays have also been used successfully for the detection of phytoplasmas in insect samples, again using the real time systems. The 23S rRNA gene "flavescence dorée" assay (Kogovšek et al. 2015) was successfully used on laboratory DNA extracts of field-collected specimens of the vector species *Orientus ishidae* and *Scaphoideus titanus*. Primers designed from the 16S rRNA gene of the phytoplasma associated with the Bogia coconut syndrome in Papua New Guinea ('*Ca*. P. novoguineense') and the surrounding region were used on DNA extracted from insect heads and also from sucrose feeding solution on which the potential vectors had been fed (Lu et al. 2016). In the case of the insect heads, the DNA was extracted by kits, whilst from the sucrose feeding solution, the extraction involved an incubation step in an alkaline SDS solution followed by ethanol precipitation and centrifugation. This produced DNA that was suitable for both LAMP and PCR and identified a number of potential insect vectors. It would have been interesting to test whether heating and ethanol precipitation were essential to produce DNA of suitable quality for LAMP or whether a rapid extraction of head samples in alkaline PEG buffer or mixing sucrose solution with alkaline PEG buffer followed by the use of the solution directly in the LAMP reaction would have worked. Certainly, it has been shown that LAMP can detect the presence of phytoplasma DNA from samples of whole insects extracted directly into alkaline PEG buffer, although care must be taken to ensure that only 10–20 mg of insect are added per 500 μl of alkaline PEG buffer, since addition of too much insect material can result in inhibition (M. Dickinson, unpublished). Indeed, the main unresolved issue with using these assays on insects compared to on plants has been the lack of suitable control primers equivalent to the COX gene primers to use for confirming that samples negative for phytoplasma DNA are negative for the lack of phytoplasma and not due to inhibitors in the extracts. A generic invertebrate control assay has been developed and validated for use alongside a yellow-legged Asian hornet (*Vespa velutina nigrithorax*) species-specific assay (Stainton et al. 2018) which amplifies the D10 region of the 28S ribosomal RNA gene. This assay was tested on *Vespa* sp. (hornets), *Vespula* sp. and *Dolichovespula* sp. (wasps), *Urocerus* sp. (sawflies) and *Apis* sp. (bees) and successfully amplified all species tested. However this has not been tested with phytoplasma vector species, nor is there are any other published universal insect primers for LAMP to present knowledge. Without a generic insect assay, assays have to be designed for the specific insect species that are being tested to be able to undertake these essential control experiments, or alternatively, sub-samples from the insect extracts can be spiked with phytoplasma DNA that is known to amplify to determine whether they contain inhibitors. A further series of six-primer sets for 16SrI, 16SrII, 16SrIII, 16SrXII and 16SrXXII phytoplasmas have been published based on the 16S-23S rRNA (Dickinson 2015). The primers for 16SrI have subsequently been used in Costa Rica in real time LAMP to detect phytoplasma presence in two new native host plant species, *Genipa americana* and *Ageratina anisochroma* (Villalobos et al. 2018).

Whilst the rRNA gene has been used for the majority of LAMP assays published to date for phytoplasmas, equally efficient primers have been designed from other phytoplasma genes. For example, Sugawara et al. (2012) designed primers based on the '*Ca*. P. asteris' *groEL* gene and compared the speed of reaction and detection limits of these to primers designed from the 16S rRNA and against PCR assays. In their analysis, they found a set of primers designed against the 16S rRNA; the RRN8 primers were the most sensitive, with 100-fold more sensitivity than PCR, whilst the best *groEL* primer set, GL2, were 10-fold less sensitive than the RRN8 primers but still 10-fold more sensitive than PCR. However, whilst the RRN8 primers took between 10 and 13 minutes to detect the product in the real time system from the infected plant templates tested, the GL2 primer set was able to detect products from the same samples in 4–10 minutes. In these assays, both sets of primers had similar specificity, being able to detect all the '*Ca.* P. asteris'-infected samples tested and none of the other phytoplasmas tested from other ribosomal groups.

In a similar study, Pérez-López et al. (2017) used primers based on the single copy *cpn60* (= *groEL*) gene to develop and compare real time PCR and real time LAMP for the 16SrXIII group phytoplasmas in Mexico. The LAMP assay was shown to be more reliable than PCR and comparable with the real time PCR, with an advantage of amplification speed, since the assays detected the product in as little as 9 minutes. The real time LAMP assay was also shown to provide a good relationship between amplification time and concentration of phytoplasma, detecting as few as 100 copies of the phytoplasma, such that it could be used to a certain extent for providing an estimate of phytoplasma levels, moreover the importance of the validation provided by the anneal analysis was also noted. However, the authors also noted that the requirement and costs for dedicated portable real time LAMP instruments for field analysis may be a limitation to take up in resource-limited settings, even though the costs of consumables, including the requirement for less sophisticated materials for DNA extraction, would potentially make the consumables costs for individual assays less than for other PCR-based assays. The authors also used incorporation of calcein and observation under UV light as an alternative field-based detection system and noted that this has value as a rapid, cheaper binomial (positive/negative) detection approach, but as noted above, this does lack the anneal curve validation provided by the real time based assays. In similar experiments to target alternative genes to the rRNA genes for phytoplasma assays, the *leuS* gene, encoding leucyl tRNA synthetase, was used to design six-primer sets for a range of phytoplasmas from different ribosomal groups (16SrI, 16SrII, 16SrIII, 16SrV, 16SrVI and 16SrX) that have excellent specificity to the groups from which they were designed and comparable levels of speed and sensitivity to assays designed based on the 16S-23S rRNA genes, indicating that in principal, any gene could be a successful target for the design of sensitive LAMP primers (Al-Jaf 2016).

A six-primer set for universal phytoplasma amplification have also been designed based on the 16S rRNA (Dickinson 2015) that have been shown to work on all phytoplasma groups tested, although with slightly slower amplification times than the group-specific assays (typically 15–30 minutes). It was postulated that these universal primers might be able to give 16S ribosomal group identification within the same assay through differences in the anneal temperatures for different ribosomal groups. However, because the region that is being amplified is only around 200 bp, it was found that there was not enough difference in the G+C content between the F3 and B3 primers for the different ribosomal groups to give significant differences in annealing temperatures that could be reliably used for group identification. In addition, and as noted in the DNA extraction section, differences in pH of the added DNA solution can also affect the annealing temperatures.

More recently, RPA assays have been developed for the detection the 16SrX '*Ca.* P. mali' (Valasevich and Schneider 2017) that uses primers for the *imp* gene. Two approaches were used for detection of the product, a real time approach, using a fluorophore probe that was used in a real time PCR detection machine, but could also potentially be used in field-based fluorescence detection machines, and a probe for direct detection in the lateral flow-based system developed for RPA assays. Both assays were as sensitive as real time PCR assays detecting down to 10 copies and showed excellent specificity, with real time amplification times as quick as 2.38 minutes for 10^6 copies and 10.19 minutes for 10 copies at temperatures between 39 and 44°C. The assays were also tested on crude homogenates of plant tissue, obtained using CTAB buffer in disposable nylon mesh bags and hand homogenisation; however, the homogenate was cleared by centrifugation before use in the assays, which potentially makes this less appropriate as a method for in-field assays in remote locations and adds to the number of steps required for DNA preparation. In a comparable RPA assay developed for the napier grass stunt phytoplasma (Wambua et al. 2017), also using the *imp* gene as target, similar sensitivity and specificity was shown, and the assay was also shown to work on dilutions in water of crude sap extracts and crushed insect preparations, confirming the applicability of RPA assays for rapid in-field detection of phytoplasmas. An equivalent RPA assay, also based on the *imp* gene, has been developed for '*Ca.* P. pruni', also using crude sap extracts and showing comparable sensitivity to PCR (Villamor and Eastwell 2019).

8.7 Conclusions

The desire for detection away from a centralised laboratory has numerous drivers. Rapid, specific detection of pathogens allows appropriate plant health action to be taken, helping to limit the spread of disease, to prevent propagation from infected plants and to allow appropriate control measures to be implemented. These benefits can be leveraged in both high- and low-resource settings, as alternatives to, or to complement, laboratory testing. Therefore the development of in-field diagnostic tools has remained a focus of phytoplasma research over the decades. Whilst immunological methods remain the simplest to perform and are cost-effective, the development of suitable antibodies presents a major up-front cost, and it is often very difficult to achieve the desired specificity. Therefore DNA-based approaches have become the method of choice.

Whilst a range of methods are available, at the present time, both LAMP and RPA have emerged as frontrunners primarily due to the optimal performance characteristics, principally in terms of rapidity and sensitivity, alongside their amenability for in-field testing with minimal equipment requirements. In the literature, there are conflicting reports on the relative performance of LAMP versus RPA and which is the more sensitive and reliable technique; for example, Karakkat et al. (2018) found that RPA had lower levels of false positives than LAMP for turfgrass fungal disease detection, whilst Howson et al. (2017) found that RPA was tenfold less sensitive and less amenable to simple nucleic acid extraction techniques than LAMP in their studies on the foot-and-mouth disease virus. As noted earlier, assays for phytoplasmas have been developed with both techniques, and the comparative levels of sensitivity are probably more a reflection on the particular primer sets that have been developed than on inherent differences in the relative performance of the two techniques, since it has been found that even within a technique such as LAMP, there can be significant differences in performance depending on which primers are designed and from what gene.

In terms of field detection, both techniques have been shown to work for phytoplasmas with crude DNA preparations, and methods are available for cheap in-field detection of amplification products, such as through the incorporation of dyes that change colour, or for more expensive and sophisticated detection of amplification products through portable real time machines. The real time methods are undoubtedly more reliable, since they have in-built product validation, but the costs of the equipment may be prohibitive for resource-poor settings. The consumable reagents for doing the isothermal amplifications themselves are relatively inexpensive, and costs are comparable with nested PCR, particularly when the costs of extracting the higher-quality DNA required for PCR are built in to the PCR costings. In comparison to each other, LAMP is less expensive than RPA per reaction, but with the potential for multiplexing different assays into a single reaction, RPA could potentially require fewer assays to be undertaken, since the plant control assay and different phytoplasma group assays could be combined into a single test, whilst each would have to be done separately with LAMP.

There are no limitations that prevent the development of new LAMP or RPA assays, although to date, LAMP has the advantage that more assays are currently published and available, including assays for most phytoplasma ribosomal groups and a universal phytoplasma assay, whilst for RPA there are currently fewer assays available. With careful primer design and optimisation to ensure specificity, assays to target particular ribosomal groups can be developed for either technique. Thorough validation of newly developed assays is an essential component to enable the performance characteristics of an assay to be established and therefore to understand the diagnostic value of a test result. Exploiting rapid, simple, sample preparation processes in lieu of complex laboratory-based DNA extractions is key to the implementation of DNA-based in-field detection methods. A range of methods have been developed and tested with phytoplasma-infected plants. The combination of all of these factors now means that phytoplasma diagnostics away from the laboratory is not only possible but practical. Furthermore, the methods used even provide advantages over laboratory tests in terms of the time taken to produce a result. With careful application from DNA extraction, through assay design, optimisation and deployment, in-field DNA-based detection of phytoplasmas is now not only possible but a suitable alternative to laboratory testing.

References

Abeysinghe S, Abeysinghe PD, Kanatiwela-de Silva C, Udagama P, Warawichanee K, Aljafar N, Kawicha P, Dickinson M (2016) Refinement of the taxonomic structure of 16SrXI and 16SrXIV phytoplasmas of gramineous plants using multilocus sequence typing. *Plant Disease* **100**, 2001-2010.

Al-Jaf B (2016) Development of improved methods for phytoplasma diagnostics. PhD Thesis, University of Nottingham, UK.

Angelini E, Bianchi GL, Filippin L, Morassutti C, Borgo M (2007) A new TaqMan method for the identification of phytoplasmas associated with grapevine yellows by real-time PCR assay. *Journal of Microbiological Methods* **68**, 613-622.

Baric S, Dalla-Via J (2004) A new approach to apple proliferation detection: a highly sensitive real-time PCR assay. *Journal of Microbiological Methods* **57**, 135-145.

Bekele B, Hodgetts J, Tomlinson J, Boonham N, Nikolic P, Swarbrick P, Dickinson M (2011) Use of a real-time LAMP isothermal assay for detecting 16SrII and XII phytoplasmas in fruit and weeds of the Ethiopian Rift Valley. *Plant Pathology* **60**, 345-355.

Braun-Kiewnick A, Altenbach D, Oberhansli T, Bitterlin W, Duffy B (2011) A rapid lateral-flow immunoassay for phytosanitary detection of *Erwinia amylovora* and on-site fire blight diagnosis. *Journal of Microbiological Methods* **87**, 1-9.

Chomczynski P, Rymaszewski M (2006) Alkaline polyethylene glycol-based method for direct PCR from bacteria, eukaryotic tissue samples, and whole blood. *BioTechniques* **40**, 454-458.

Christensen NM, Nicolaisen M, Hansen M, Schulz A (2004) Distribution of phytoplasmas in infected plants as revealed by real-time PCR and bioimaging. *Molecular Plant-Microbe Interactions* **17**, 1175-1184.

Contaldo N, Satta E, Zambon Y, Paltrinieri S, Bertaccini A (2016) Development and evaluation of different complex media for phytoplasma isolation and growth. *Journal of Microbiological Methods* **127**, 105-110.

Contaldo N, D'Amico G, Paltrinieri S, Diallo HA, Bertaccini A, Arocha Rosete Y (2019) Molecular and biological characterization of phytoplasmas from coconut palms affected by the lethal yellowing disease in Africa. *Microbiological Research* **223–225**: 51-57.

Crosslin JM, Vandemark GJ, Munyaneza JE (2006) Development of a real-time, quantitative PCR for detection of the Columbia basin potato purple top phytoplasma in plants and beet leafhoppers. *Plant Disease* **90**, 663-667.

Danks C, Barker I (2000) On-site detection of plant pathogens using lateral-flow devices. *Bulletin OEPP/EPPO Bulletin* **30**, 421-426.

de Jonghe K, De Roo I, Maes M (2017) Fast and sensitive on-site isothermal assay (LAMP) for diagnosis and detection of three fruit tree phytoplasmas. *European Journal of Plant Pathology* **147**, 749-759.

Dickinson M (2015) Loop Mediated Isothermal amplification (LAMP) for detection of phytoplasmas in the field. In: Plant Pathology: Techniques and Protocols. Methods in Molecular Biology, ed Lacomme C, Springer Science + Business Media, New York, United States of America, 99-111 pp.

Donoso A, Valenzuela S (2018) In-field molecular diagnosis of plant pathogens: recent trends and future perspectives. *Plant Pathology* **67**, 1451-1461.

Fire A, Xu SQ (1995) Rolling replication of short DNA circles. *Proceedings of the National Academy of Science United States of America* **92**, 4641-4645.

Galetto L, Bosco B, Marzachì C (2005) Universal and group-specific real-time PCR diagnosis of "flavescence dorée" (16SrV), "bois noir" (16SrXII) and apple proliferation (16SrX) phytoplasmas from field-collected plant hosts and insect vectors. *Annals of Applied Biology* **147**, 191-201.

Gundersen DE, Lee I-M (1996) Ultrasensitive detection of phytoplasmas by nested-PCR assays using two universal primer pairs. *Phytopathologia Mediterranea* **35**, 144-151.

Hatano B, Maki T, Obara T, Fukumoto H, Hagisawa K, Matsushita Y, Okutani A, Bazartseren B, Inoue S, Sata T, Katano H (2010) LAMP using a disposable pocket warmer for anthrax detection, a highly mobile and reliable method for anti-bioterrorism. *Japanese Journal of Infectious Disease* **63**, 36-40.

Hodgetts J, Boonham N, Mumford R, Dickinson M (2009) Panel of 23S rRNA gene-based real-time PCR assays for improved universal and group-specific detection of phytoplasmas. *Applied and Environment Microbiology* **75**, 2945-2950.

Hodgetts J, Johnson G, Perkins K, Ostoja-Starzewska S, Boonham N, Mumford R, Dickinson MJ (2014) The development of monoclonal antibodies to the SecA protein of Cape St Paul wilt disease phytoplasma and their evaluation as a diagnostic tool. *Molecular Biotechnology* **56**, 803-813.

Howson ELA, Kurosaki Y, Yasuda J, Takahashi M, Goto H, Gray AR, Mioulet V, King DP and Fowler VL (2017) Defining the relative performance of isothermal assays that can be used for rapid and sensitive detection of foot-and-mouth virus. *Journal of Virological Methods* **249**, 102-110.

Kakizawa S, Oshima K, Kuboyama T, Nishigawa H, Jung H-Y, Sawayanagi T, Tsuchizaki T, Miyata S, Ugaki M, Namba S (2001) Cloning and expression analysis of phytoplasma protein translocation genes. *Molecular Plant Microbe Interactions* **14**, 1043-1050.

Karakkat BB, Hockemeyer K, Franchett M, Olson M, Mullenberg C, Koch PL (2018) Detection of root-infecting fungi on cool-season turfgrass using loop-mediated isothermal amplification and recombinase polymerase amplification. *Journal of Microbiological Methods* **151**, 90-98.

Kogovšek P, Hodgetts J, Hall J, Prezelj N, Nikolic P, Mehle N, Lenarcic R, Rotter A, Dickinson M, Boonham N, Dermastia M, Ravnikar M (2015) LAMP assay and rapid sample preparation method for on-site detection of "flavescence dorée" phytoplasma in grapevine. *Plant Pathology* **64**, 286-296.

Kogovšek P, Mehle N, Pugelj A, Jakomin T, Schroers H-J, Ravnikar M, Dermastia M (2017) Rapid loop-mediated isothermal amplification assays for grapevine yellows phytoplasmas on crude leaf-vein homogenate has the same performance as qPCR. *European Journal of Plant Pathology* **148**, 75-84.

Lau HY, Botella JR (2017) Advanced DNA-based point-of-care diagnostic methods for plant diseases detection. *Frontiers in Plant Science* **8**, 2016.

Lau HY, Palanisamy R, Trau M, Botella JR (2014) Molecular inversion probe: a new tool for highly specific detection of plant pathogens. *Plos One* **9**, e111182.

Lu H, Wilson BAL, Ash GJ, Woruba SB, Fletcher MJ, You M, Yang G, Gurr GM (2016) Determining putative vectors of the Bogia coconut syndrome phytoplasma using loop-mediated isothermal amplification of single-insect feeding media. *Scientific Reports* **6**, 35801.

Mekuria TA, Zhang SL, Eastwell KC (2014) Rapid and sensitive detection of little cherry virus 22 using isothermal reverse transcription-recombinase polymerase amplification. *Virological Methods* **205**, 24-30.

Mori Y, Nagamine K, Tomita N, Notomi T (2001) Detection of loop-mediated isothermal amplification reaction by turbidity derived from magnesium pyrophosphate formation. *Biochemical and Biophysical Research Communications* **289**, 150-154.

Nagamine K, Hase T, Notomi T (2002) Accelerated reaction by loop-mediated isothermal amplification using loop primers. *Molecular and Cellular Probes* **16**, 223-229.

Nair S, Manimekalai R, Raj PG, Hegde V (2016) Loop mediated isothermal amplification (LAMP) assay for detection of coconut root wilt disease and arecanut yellow leaf disease phytoplasma. *World Journal of Microbiology and Biotechnology* **32**, 108.

Notomi T, Okayama H, Masubuchi H, Yonekawa K, Amino N, Hase T (2000) Loop-mediated isothermal amplification of DNA. *Nucleic Acids Research* **28**, e63.

Obura E, Masiga D, Wachira F, Gurja B, Khan ZR (2011) Detection of phytoplasmas by loop-mediated isothermal amplification of DNA (LAMP). *Journal of Microbiological Methods* **84**, 312-316.

Pérez-López E, Rodriguez-Martinez D, Olivier CY, Luna-Rodriguez M, Dumonceaux TJ (2017) Molecular diagnostic assays based on *cpn60* UT sequences reveal the geographic distribution of subgroup 16SrXIII-(A/I)I phytoplasma in Mexico. *Scientific Reports* **7**, 950.

Piepenburg O, Williams CH, Stemple DL, Armes NA (2006) DNA detection using recombination proteins. *Plos Biology* **4**, e204.

Smart CD, Schneider B, Blomquist CL, Guerra LJ, Harrison NA, Ahrens U, Lorenz K-H, Seemuller E, Kirkpatrick B (1996) Phytoplasma-specific PCR primers based on sequence of the 16S-23S rRNA spacer region. *Applied and Environmental Microbiology* **62**, 2988-2993.

Stainton K, Hall J, Budge GE, Boonham N, Hodgetts J (2018) Rapid molecular methods for in-field and laboratory identification of the yellow-legged Asian hornet (*Vespa velutina nigrithorax*). *Journal of Applied Entomology* **142**, 610-616.

Sugawara K, Himeno M, Keima T, Kitazawa Y, Maejima K, Oshima K, Namba S (2012) Rapid and reliable detection of phytoplasmas by loop-mediated isothermal amplification targeting a housekeeping gene. *Journal of General Plant Pathology* **78**, 389-397.

Thornton CR, Groenhof AC, Forrest R, Lamotte R (2004) A one-step, immunochromatographic lateral flow device specific to *Rhizoctonia solani* and certain related species, and its use to detect and quantify *R. solani* in soil. *Phytopathology* **94**, 280-288.

Tomlinson JA, Dickinson M, Boonham N (2010a) Rapid method for detection of *Phytophthora ramorum* and *P. kernoviae* by a two-minute DNA extraction method followed by isothermal amplification, and amplicon detection by generic lateral flow device. *Phytopathology* **100**, 143-149.

Tomlinson JA, Boonham N, Dickinson M (2010b) Development and evaluation of a one-hour DNA extraction and loop-mediated isothermal amplification assay for rapid detection of phytoplasmas. *Plant Pathology* **59**, 465-471.

Valasevich N, Schneider B (2017) Rapid detection of '*Candidatus* Phytoplasma mali' by recombinase polymerase amplification assays. *Journal of Phytopathology* **165**, 762-770.

Villalobos W, Montero-Astua M, Coto T, Sandoval I, Moreira L (2018) *Genipa Americana* and *Ageratina anisochroma*, two new hosts of '*Candidatus* Phytoplasma asteris' in Costa Rica. *Australasian Plant Disease Notes* **18**, 31.

Villamor DEV, Eastwell K (2019) Multilocus characterization, gene expression analysis of putative immunodominant protein coding regions, and development of recombinase polymerase amplification assay for detection of '*Candidatus* Phytoplasma pruni' in *Prunus avium*. *Phytopathology* **109**, 983-992.

Vincent M, Xu Y, Kong HM (2004) Helicase-dependant isothermal DNA amplification. *EMBO Reports* **5**, 795-800.

Vu NT, Pardo JM, Alvarez E, Le HH, Wyckhuys K, Nguyen K-L, Le DT (2016). Establishment of a loop-mediated isothermal amplification (LAMP) assay for the detection of phytoplasma-associated cassava witches' broom disease. *Applied Biological Chemistry* **59**, 151-156.

Wamalwa NIE, Midega CAO, Ajanga S, Omukunda NE, Muyekho FN, Asudi GO, Mulaa M, Khan ZR (2017) Screening napier grass accessions for resistance to napier grass stunt disease using the loop-mediated isothermal amplification of DNA. *Crop Protection* **98**, 61-69.

Wambua L, Schneider B, Okwaro A, Wanga JO, Imali O, Wambua PN, Agutu L, Olds C, Jones CS, Masiga D, Midega C, Khan Z, Jones J, Fischer A (2017) Development of field-applicable tests for rapid and sensitive detection of '*Candidatus* Phytoplasma oryzae'. *Molecular and Cellular Probes* **35**, 44-56.

Zou Y, Mason MG, Wang Y, Wee E, Turni C, Blakall P, Trau M, Botella JR (2017) Nucleic acid purification from plants, animals and microbes in under 30 seconds. *Plos Biology* **15**, e2003916.

Chapter 9
Multilocus Genetic Characterization of Phytoplasmas

Marta Martini, Fabio Quaglino, and Assunta Bertaccini

Abstract Classification of phytoplasmas into 16S ribosomal groups and subgroups and '*Candidatus* Phytoplasma' species designation have been primarily based on the conserved 16S rRNA gene. However, distinctions among closely related '*Ca.* Phytoplasma' species and strains based on 16S rRNA gene alone have limitations imposed by the high degree of rRNA nucleotide sequence conservation across diverse phytoplasma lineages and by the presence in a phytoplasma genome of two, sometimes sequence heterogeneous, copies of this gene. Thus, in recent years, moderately conserved genes have been used as additional genetic markers with the aim to enhance the resolving power in delineating distinct phytoplasma strains among members of some 16S ribosomal subgroups. The present chapter is divided in two parts: the first part describes the non-ribosomal single-copy genes less conserved (housekeeping genes) such as ribosomal protein (*rp*), *secY*, *secA*, *rpoB*, *tuf*, and *groEL* genes, which have been extensively used for differentiation across the majority of phytoplasmas; the second part describes the differentiation of phytoplasmas in the diverse ribosomal groups using multiple genes including housekeeping genes and variable genes encoding surface proteins.

Keywords Molecular differentiation · Genetic markers · Non-ribosomal genes · Variable genes · RFLP analysis · Sequencing

---

M. Martini (✉)
Department of Agricultural, Food, Environmental and Animal Sciences, University of Udine, Udine, Italy

F. Quaglino
Department of Agricultural and Environmental Sciences – Production, Landscape, Agroenergy, University of Milan, Milan, Italy

A. Bertaccini
Department of Agricultural and Food Sciences, *Alma Mater Studiorum* – University of Bologna, Bologna, Italy

A. Bertaccini et al. (eds.), *Phytoplasmas: Plant Pathogenic Bacteria - III*,
https://doi.org/10.1007/978-981-13-9632-8_9

9.1 Introduction

Phytoplasmas classification into 16S ribosomal groups and subgroups (Lee et al. 1998) and '*Candidatus* Phytoplasma' species designation are largely based on analysis of the conserved 16S rRNA gene sequences, which allowed so far the identification of 34 ribosomal groups and more than 200 subgroups, and the description of 43 '*Candidatus* Phytoplasma' species (Bertaccini and Lee 2018). At present the species designation is primarily based on an arbitrary threshold of 2.5% dissimilarity of 16S rDNA sequences among phytoplasmas (IRPCM 2004); however, distinctions among closely related '*Ca.* Phytoplasma' species and strains based on 16S rRNA gene alone have limitations imposed by the high degree of rRNA nucleotide sequence conservation across diverse phytoplasma lineages and by the presence in the phytoplasma genome of two, sometimes sequence heterogeneous, copies of the 16S rRNA gene (Davis et al. 2013). Moreover, many ecologically or biologically distinct phytoplasma strains with a sequence similarity higher than 97.5%, some of which may warrant designation as a new taxon, may be excluded by the 2.5% threshold (Lee et al. 2010; Duduk and Bertaccini 2011). It is not uncommon indeed that closely related phytoplasma strains, on the basis of the 16S rRNA gene sequences, have unique ecological niches encompassing both plant host range and insect vectors. Therefore, additional unique biological or geographic characteristics, as well as other molecular criteria, need to be included for speciation as it was done for some '*Candidatus* Phytoplasma' species in the 16SrX (Seemüller and Schneider 2004) and in the 16SrV (Win et al. 2013) groups.

Similarly to most DNA-based bacterial classifications, which use multilocus sequence typing (MLST) to provide more detailed differentiation with as many as 10 genes examined, in the recent years in phytoplasma genetic characterization studies, interest has focused on conserved and less conserved non-ribosomal single-copy genes for finer differentiation of closely related strains (Martini et al. 2002; Lee et al. 2010). Some of the additional markers used to improve phytoplasmas differentiation are the rp (ribosomal protein) operon, *tuf*, *secY*, *secA*, *groEL* (or *cpn60*), and *rpoB* genes (Marcone et al. 2000; Botti and Bertaccini 2003; Lee et al. 2006, 2010; Martini et al. 2002, 2007; Hodgetts et al. 2008; Valiunas et al. 2013; Mitrovic et al. 2011, 2015). Just a few of them, like the *tuf*, *rp*, and *secY* genes, have been extensively characterized through the majority of phytoplasma groups and include a quite comprehensive database, primarily due to the difficulty in designing universal primers that can be used to amplify these genes across all 16S ribosomal groups (Martini et al. 2007; Hodgetts et al. 2008; Lee et al. 2010; Makarova et al. 2012).

For epidemiological studies and quarantine purposes, the 16S rDNA is rarely used due to the lack of variability; in recent years, besides less conserved non-ribosomal genes, interest has also focused on the use of variable genes encoding surface proteins such as *vmp*1 (Cimerman et al. 2009; Fialová et al. 2009), *imp* (Danet et al. 2011), *amp* (Kakizawa et al. 2006), *stamp* (Fabre et al. 2011), and *hfl*B (Schneider and Seemüller 2009) which appears more discriminant at the strain level.

The analyses of these variable genes are especially recommended in epidemiological studies at local and/or regional scale.

The present chapter is divided in two parts: the first part describes the non-ribosomal single-copy genes less conserved (housekeeping genes) which have been extensively used for differentiation across the majority of phytoplasmas; the second part describes the differentiation of phytoplasmas in the diverse ribosomal groups using multiple genes including housekeeping genes and variable genes encoding surface proteins.

9.2 Genes Other Than 16S rRNA for Phytoplasma Differentiation

Ribosomal Protein Genes These genes are part of the large rp operon, which contains at least 21 genes in phytoplasma genomes (Hodgetts and Dickinson 2010); they represent the first non-ribosomal marker used for phytoplasma strain differentiation. The first phytoplasma *rp* genes to be cloned were those encoding the proteins rpL2 and rpS19 from a phytoplasma infecting *Oenothera* spp. by using a heterologous probe derived from the *Mycoplasma capricolum* (Lim and Sears 1991). Subsequently, the genes encoding proteins rpL22 and rpS3 were cloned with the same approach, and primers rpF1 (5′-GGA CAT AAG TTA GGT GAA TTT-3′) and rpR1 (5′-ACG ATA TTT AGTTCT TTT TGG-3′) were initially designed to amplify a 1,245–1,389 bp region encompassing the *rplV* (L22), *rpsC* (S3), and part of the *rplP* (L16) gene from 16SrI, 16SrIII, 16SrIV, 16SrV, 16SrVII, 16SrVIII, 16SrIX, and 16SrXIII phytoplasmas (Lim and Sears 1992; Martini et al. 2007). Successively, additional primers, referred to as semi-universal, have been designed to amplify ribosomal genes from phytoplasmas belonging to several phylogenetic groups (Martini et al. 2007). In particular, two degenerate forward primers were designed: rpL2F3 (5′-WCC TTG GGG YAA AAA AGC TC-3′) and rpF1C (5′-ATG GTD GGD CAY AAR TTA GG-3′) to be used with the reverse primer rp(I)R1A, previously described by Lee et al. (2003). Primer pairs rpF1C/rp(I)R1A and rpL2F3/rp(I)R1A have been shown to amplify products that were, respectively, 1,212–1,386 (amplifying the same genes) and 1,600 bp long (encompassing the 3′ end of the *rplB* gene; the *rpsS*, *rplV*, *rpsC*, and *rplP* genes; and the 5′ end of the *rpmC* gene) from groups 16SrI, 16SrIII, 16SrIV, 16SrV, 16SrVI, 16SrVII, 16SrIX, 16SrX, 16SrXII, 16SrXIII, and 16SrXVIII, whereas members of phytoplasma groups 16SrVIII and 16SrXI didn't yield amplicons of the expected size. Moreover, only primer pair rpF1C/rp(I)R1A amplified rp genes from strains of group 16SrII. In addition, primers for these genes have been designed that are group specific.

A phylogenetic tree for 87 phytoplasma strains belonging to the above-mentioned 16S ribosomal groups was constructed based on the analysis of *rplV* and *rpsC* genes. This *rp* gene-based phylogenetic tree, which was congruent with that inferred from the 16S rRNA gene, yielded more clearly defined phylogenetic interrelationships among phytoplasma strains and delineated more distinct phytoplasma subclades and distinct lineages than those resolved by the 16S rRNA gene-based tree. For instance, '*Ca.* P. mali', '*Ca.* P. pyri', and '*Ca.* P. prunorum', which share 98.9–99.1% 16S rDNA sequence similarity, shared 94.3–94.6% *rp* gene sequence similarity and were readily delineated by analysis of the *rp* gene sequences. The average nucleic acid sequence similarity of the *rp* genes among members of two ribosomal phytoplasma groups ranged from 50.4% (among members of groups 16SrXII-A and 16SrII) to 83.5% (among members of groups 16SrV and 16SrVI), compared with 85.1% (groups 16SrIX and 16SrX) to 96.9% (groups 16SrVI and 16SrVII) similarity for the 16S rRNA gene sequences. The greater sequence variation makes the *rp* genes a better molecular tool for differentiation of genetic closely related but distinct phytoplasma strains within a given 16S ribosomal group or '*Candidatus* Phytoplasma' species. They have been used as molecular markers especially for finer differentiation of phytoplasma strains in the groups 16SrI (Lee et al. 2004a, 2006), 16SrIII (Davis et al. 2013), 16SrV (Lee et al. 2004b, Arnaud et al. 2007), and 16SrIX (Lee et al. 2012). In these studies, the PCR products amplified by using these rp operon primers have been used in RFLP analysis in combination with 16S rRNA gene RFLP analysis to assign phytoplasmas to 16Sr-rp subgroups (Lee et al. 1998, 2004a, 2004b, 2012). The results indicated that the analysis of the *rp* gene sequences not only delineated subgroups that are consistent with 16S ribosomal subgroups but also identified, within some subgroups, additional distinct strains that could not be resolved by the 16S rRNA gene sequence analyses. For example, 10 and 12 RFLP subgroups were differentiated on the basis of ribosomal protein gene sequences in 16SrI and 16SrV phytoplasma strains, respectively (Lee et al. 2004a, 2004b). Most of the additional strains identified have distinct biological and ecological properties; *e.g.* subgroup 16SrV-C can be further differentiated into several rp subgroups (Lee et al. 2004b, Martini et al. 2012).

***SecY* Gene** This gene, similarly to *secA* and *secE* genes, encodes for an essential component of the Sec protein translocation system (Kakizawa et al. 2001). The *secY* gene was initially identified in phytoplasmas in a randomly cloned 16SrV "flavescence dorée" DNA fragment that had originally been used as a hybridization probe (Daire et al. 1997). Based on the sequencing of this fragment, three primers FD9f2, FD9f3, and FD9r2 in addition to primer set FD9f/r (Daire et al. 1997) were designed to amplify a 1,150 bp fragment in nested PCR with primer set FD9f3/FD9r2 (Angelini et al. 2001). Primers FD9f/r were also used and allowed to differentiate the elm yellows group phytoplasmas into 13 secY-V-based RFLP subgroups (Lee et al. 2004b). These FD9 primers were then adapted and used in nested PCR to amplify a 1,174 bp fragment from over 40 16SrV phytoplasma strains infecting grapevine and alders in Europe (Arnaud et al. 2007). A phylogenetic tree was constructed from the sequences obtained, and the resultant analyses supported the existence of three distinct "flavescence dorée" clusters (FD1, FD2, and FD3), similarly to the phylogenetic trees obtained based on the *map* gene and the uvrB-degV region. It is noteworthy that the same phytoplasma strains had already been differentiated into two ribosomal subgroups (16SrV-C and 16SrV-D) only based on one SNP in the 16S rDNA gene (Martini et al. 1999). The *secY* gene has also been used for differentiation of the AY group phytoplasmas (Lee et al. 2006). In this study, primers were designed based on the published AY and OY genome sequences (Oshima et al. 2004; Bai et al. 2006) to amplify a 1.4 kb near-full-length *secY* gene. Twenty representative 16SrI strains from ten 16SrI subgroups were used, and the sequencing and RFLP analysis of the resultant PCR products grouped the phytoplasmas into 10 SecY subgroups. Overall the SecY subgroups delineated by RFLP analyses of the *secY* gene sequences from phytoplasma strains in groups 16SrI and 16SrV generally coincided with those delineated with the *rp* gene sequences (Lee et al. 2004a, 2004b, 2006). This gene has also been used as molecular marker for finer differentiation of phytoplasma strains of other groups such as 16SrIII (Davis et al. 2013), 16SrIX (Lee et al. 2012), 16SrX (Danet et al. 2011), and 16SrXII (Fialová et al. 2009).

Recently Lee et al. (2010) constructed a comprehensive phylogenetic tree for 83 phytoplasma strains belonging to twelve 16S ribosomal groups based on the analysis of *secY* gene using semi-universal primers. This *secY* gene-based phylogenetic tree, which was congruent with that inferred from the 16S rRNA gene, yielded more clearly defined phylogenetic interrelationships among phytoplasma strains, and delineated more distinct phytoplasma subclades and lineages than those resolved by the 16S rRNA gene-based tree. The semi-universal degenerate primer pair L15F1/MapR1 was initially designed and used to amplify the partial spc ribosomal protein operon [which contains the *rpl15* gene, the adenylate kinase gene (*adk*), a protein translocase gene (*secY*), and the map gene] from groups 16SrI–VIII, 16SrX, 16SrXII, 16SrXIII, and 16SrXVIII. Additional primers, including 16S ribosomal group-specific primers, were designed within the L15F1/MapR1 amplicon for the amplification of the partial *spc* operon and of DNA fragments containing the complete *secY* gene and partial flanking genes of selected ribosomal groups (Lee et al. 2010). The *secY* gene sequence variability is similar to that of the *rp* genes. The *secY* gene nucleic acid sequence similarity between members of two given 16S ribosomal phytoplasma groups ranged from 53.5 to 77.9 %, compared with 85.1–96.9 % for 16S rRNA gene sequences. However, the resolving power of *secY* is slightly better than *rp* gene sequences, and it was found to be more efficient than other gene markers especially for resolving closely related strains within the same 16S rRNA gene RFLP group (Hodgetts et al. 2008; Lee et al. 2006; Martini et al. 2007).

***SecA* Gene** Another protein translocase subunit encoding gene, *secA*, was employed for phytoplasma differentiation (Hodgetts et al. 2008). SecA is a highly conserved protein due to its involvement in a basic cellular process consuming ATP for surface and secreted protein translocation outside the cell. The *secA* gene was originally sequenced from OY (Kakizawa et al. 2001), and three degenerate primers were designed that can be used in a semi-nested PCR assay to amplify a portion of the gene sequence (480 bp): SecAFor1 (5′-GAR ATG AAA ACT GGR GAA GG-3′), SecAFor2 (5′-GAY GAR GSW AGA ACK CCT-3′), and SecARev3 (5′-GTT TTR GCA GTT CCT GTC ATN CC-3′). Degenerate primers were designed in the conserved regions of the protein by aligning the OY *secA* gene sequence with that of AY witches' broom and the equivalent gene from coconut lethal yellowing classified in the 16SrIV group. When the primer pair SecAFor1/SecARev3 was used in a semi-nested PCR followed by primer pair SecAFor2/SecARev3, products of expected size were generated from 34 phytoplasma DNA samples tested from various phytoplasma strains representing 12 16S ribosomal groups. However, these nested primers can also generate other PCR products of different sizes making the detection more difficult; thus, new primers SecAFor5 and SecARev2 were redesigned to produce in nested PCR a clear and unambiguous product (Dickinson and Hodgetts 2013). Sequencing of these PCR products has showed that the region for the 16SrI group phytoplasmas is two amino acids longer than that from all the phytoplasmas enclosed in the other described ribosomal groups.

Phylogenetic analysis based on the *secA* gene sequences has supported the results obtained for other genes and provided improved resolution of ribosomal groups and subgroups, when compared with the 16S rRNA gene. The subgroups of 16SrI are clearly defined, and 16SrII clearly splits into a cluster that contains 16SrII-B and 16SrII-C strains and a cluster that contains the 16SrII-D strains. These results support the *in silico* results of Wei et al. (2007), which were supporting the classification of the 16SrII-B strains as '*Ca.* P. aurantifolia' (Zreik et al. 1995) whilst classifying the 16SrII-D strains as '*Ca.* P. australasiae' (White et al. 1998). The results using the *secA* gene have shown a clear distinction between strains within the coconut lethal yellowing-type disease group (Hodgetts et al. 2008). Based on the *secA* sequence analysis, Bila et al. (2015) separated phytoplasmas affecting coconut palms into at least three distinct groups, previously described by Hodgetts et al. (2008), reflecting the strains' geographic origins: an American group typified by LY (classified as 16SrIV-A, 16SrIV-C, 16SrIV-D, and 16SrIV-E), an East African group typified by Tanzanian LD (classified as 16SrIV-B), and a West African group typified by lethal decline (LD) from Nigeria and Cape St. Paul Wilt Disease (CSPWD) from Ghana. These latter phytoplasmas are now classified as 16SrXXII-A, 16SrXXII-B, and 16SrXXII-C that are all related to '*Ca.* P. palmicola' (Harrison et al. 2014; Kra et al. 2017). The sequence similarity ranged from 69.7 to 84.4% between two given 16S ribosomal groups. The resolving power of the *secA* gene as a parameter for phytoplasma differentiation is similar to those of *rp* and *secY* genes. The amplification of a 480 bp-long fragment of the *secA* gene can be considered as the first attempt to use a shorter non ribosomal marker for universal phytoplasma identification (Hodgetts et al. 2008).

***RpoB* Gene** The use of the single-copy DNA-dependent RNA polymerase (DpRp) b-subunit gene (*rpoB* gene) has also been explored for phytoplasma classification and phylogenetic analysis, since this gene has been successfully used for other prokaryotes (Kim et al. 2003; Valiunas et al. 2013). A clover phyllody (CPh) phytoplasma has been cloned and sequenced containing a complete *rpoB* gene, ribosomal protein genes, and a partial *rpoC* gene. Based on the alignment of the CPh phytoplasma *rpoB* gene sequence with those of other phytoplasmas, conserved regions were identified and primers were designed that can be used in direct or nested PCR assays to amplify a segment of the *rpoB* gene from several diverse phytoplasma strains. The following primers were designed: rpoBF1 5′-TGC CCA ATT TAA TTG AAA TTC-3′, rpoBF2 5′-GAT TGG TTT TTA AAA CAC GG-3′, rpoBF4 5′-TTT CTC AAA TTG TAC GTT CC-3′, rpoBR3 5′-TTA CCT AAA TGA TCG ATA TCA TC-3′, rpoBR2 5′-ATT GGT TTT TTA ACA ATT CTC C-3′, and rpoBR1 5′-AAG ACC AAT TCG AAA TTG G-3′. Partial *rpoB* gene sequences from phytoplasma strains classified in group 16SrI ('*Ca.* P. asteris'-related strains) were amplified using the primer pair rpoBF2/rpoBR1. Partial *rpoB* gene sequences of phytoplasma strains from groups 16SrIII, 16SrX, and 16SrXII were amplified using primer pair rpoBF1/rpoBR3, whereas the amplification of a *rpoB* fragment from '*Ca.* P. fragariae' (strain StrawY, member of subgroup 16SrXII-E) was accomplished by using primer pair rpoBF4/rpoBR2. Primer pair rpoBF1/rpoBR1 was used to prime the first reaction in nested PCRs, followed by one of the previously mentioned primer pairs. Successively, the primer pair rpoBF1/rpoBR3 was used by Pérez-López and Dumonceaux (2016) to amplify the *rpoB* gene sequence from the subgroup 16SrXIII-(A/I)I phytoplasmas associated with strawberry green petal disease and Mexican periwinkle virescence. In the study by Valiunas et al. (2013), the *rpoB* gene sequences from phytoplasmas classified in groups 16SrI, 16SrII, 16SrIII, 16SrX, and 16SrXII were compared for sequence similarity and by phylogenetic analyses. Alignment of 1.3 kb fragments of the *rpoB* gene revealed 61.6 to 79.4% sequence identity among 16Sr groups and 95.1 to 98.8% sequence identity among strains of a given 16Sr subgroup. The greater sequence variability, compared with that of the 16S rRNA gene, indicates that the *rpoB* gene provides a more informative molecular tool for the differentiation of closely related phytoplasma strains. Comparative study of several molecular markers revealed that *rpoB* gene nucleotide sequence identities among different phytoplasma lineages approached the discriminating levels observed for *rpl22*, *rps3*, *secY*, and *secA* genes (Valiunas et al. 2013). Phylogenetic trees based on 16S rRNA and *rpoB* gene sequences had similar topologies, and branch lengths in the *rpoB* tree facilitated distinctions among closely related phytoplasmas. Virtual RFLP analysis of *rpoB* gene sequences also improved distinctions among closely related lineages. Overall, the *rpoB* gene resulted a useful additional molecular marker for phytoplasma differentiation and epidemiology.

***Tuf* Gene** This gene, encoding the elongation factor, EF-Tu, a key protein involved in the process of translation, is relatively well conserved and found as a single-copy gene in the full phytoplasma genomes or draft genomes sequenced. This gene has often been used in phylogenetic studies for other bacteria. In 1997, Schneider et al. designed two primers fTuf1 (5′-CAC ATT GAC CAC GGT AAA AC-3′) and rTuf1 (5′-CCA CCT TCA CGA ATA GAG AAC-3′) that can be used for amplifications of the *tuf* gene sequences from most phytoplasma ribosomal groups. These primers amplified products of the expected size (1,000 bp) from phytoplasmas of 16SrI, 16SrIII, and 16SrXII groups, but failed to amplify from phytoplasmas of 16SrII and 16SrX groups (Schneider et al. 1997), and were found to fail the amplification of the *tuf* gene from phytoplasmas of 16SrIV and 16SrXXII groups. The nucleotide sequence similarities among the phytoplasmas of 16SrI, 16SrIII, and 16SrXII groups ranged from 87.8 to 97.0%. Additional primers have been designed on a conserved region within the *tuf* gene to amplify a fragment of the *tuf* gene, and the potential of this fragment as a DNA barcode was examined (Makarova et al. 2012, 2013; Contaldo et al. 2015). For amplification of the *tuf* barcode, two pairs of primer cocktails (Tuf340/Tuf890 for direct PCR and Tuf400/Tuf835 for nested PCR), which allowed minor sequence variations among individual groups of phytoplasmas, were used and amplified a 420–444 bp fragment of the *tuf* gene from all phytoplasmas strains tested (16SrI–VII, 16SrIX–XII, 16SrXV, and 16SrXX). Comparison of phylogenetic trees confirmed that the *tuf* tree is highly congruent with the 16S rRNA tree and had higher inter- and intra-group sequence divergence. The use of the *tuf* barcode allowed the separation of the main ribosomal groups and most of their subgroups, suggesting that this barcode performs as well or better than the 1.2 kbp fragment of the 16S rRNA gene and thus provides an easy procedure for phytoplasma identification. Phytoplasma *tuf* barcodes were deposited in the publicly available EPPO-Q-bank database that can be used by plant health services and researchers for online phytoplasma identification. The main use of the *tuf* gene was to establish differentiation within the 16SrI (Marcone et al. 2000) and the 16SrXII groups (Langer and Maixner 2004; Streten and Gibb 2005). In a study on phytoplasmas in the 16SrI group, the specific primers fTufAy (5′-GCT AAA AGT AGA GCT TAT GA-3′) and rTufAy (5′-CGT TGT CAC CTG GCA TTA CC-3′) (Schneider et al. 1997), which amplify a 940 bp product, were used on 70 phytoplasma strains in comparison with 16S rRNA gene primers (Marcone et al. 2000). RFLP analyses of the 16S rRNA gene PCR products divided the strains into ten 16S ribosomal subgroups, whilst the *tuf* gene RFLP profiles only resulted seven subgroups. Thus, the resolving efficacy for separation of distinct phytoplasmas in group 16SrI was lower than that of the 16S rRNA gene (Schneider et al. 1997; Marcone et al. 2000). However, in some cases, the *tuf* gene was found to be useful in the differentiation of various ecological strains or strain variants within 16S rRNA subgroups (Langer and Maixner 2004; Contaldo et al. 2011; Makarova et al. 2012). For example, several strain variants were recognized within 16XII-A and 16XII-B subgroups, based on analysis of *tuf* gene sequences (Langer and Maixner 2004; Streten and Gibb 2005; Andersen et al. 2006; Pacifico et al. 2007; Riolo et al. 2007; Iriti et al. 2008).

***GroEL* Gene** This gene, also known as *cpn60* or *hsp60* gene, is a conserved gene used for prokaryotic strain characterization (Vermette et al. 2010; Desai et al. 2009). It is proposed that it may act as an adhesin-invasin in mycoplasma and speculated that its function in *Mollicutes* could be that of a virulence factor (Clark and Tillier 2010). The *cpn60* universal target (*cpn60* UT) (Goh et al. 1996) is a fragment of approximately 550 bp that has been suggested as a molecular barcode for the domain *Bacteria* (Links et al. 2012). Similarly to some species of *Mollicutes* which are missing the *cpn*60 gene within their genomes (Clark and Tillier 2010), genes encoding Cpn60 have been found in all complete and draft phytoplasma genomes reported to date with the exception of the draft genomes of phytoplasma strains from the 16SrIII group suggesting that this group may lack this gene (Dumonceaux et al. 2014).

Several publically available sequences of the 3.6 kb fragments of '*Ca.* P. asteris'-related strains contain *groES*, *groEL*, *amp*, and *nadE* genes; the full-genome sequences of two '*Ca.* P. asteris'-related strains and two full-genome sequences of '*Ca.* P. mali' and '*Ca.* P. australiense' allowed design primers that specifically amplify DNA fragments inside the phytoplasma *groEL* gene of '*Ca.* P. asteris'-related strains. Both primers were designed inside the *groEL*, and expected length amplicons (about 1.4 kb) of partial *groEL* gene were obtained from aster yellows phytoplasmas. No amplification was obtained with 16SrXII-A and 16SrXII-D, 16SrV-C and 16SrV-D, 16SrII-D, and 16SrX-B phytoplasmas; therefore, these primers show some degree of specificity to 16SrI phytoplasma group (Mitrovic et al. 2011). Successively, two degenerate primer sets were designed and used to amplify the *cpn60* (=*groEL*) gene from a wide variety of '*Ca.* Phytoplasma' species that are highly divergent in sequences (61–98% identity). One set of primers H279p/H280p (5′-GAT III GCA GGI GAT GGA ACM ACI AC-3′/5′-TGR TTI TCI CCA AAA CCA GGI GCA TT-3′) was based on 18 full-length cpn60 sequences primarily representing the 16SrI group. The second set of PCR primers D0317/D0318 (5′-GAT III KCI GGI GAY GGI ACI AC-3′/5′-TGR TKI TCI CCA AAA CYW GGI GCW TC-3′) was designed based on the full-length *cpn60* reported in the peanut witches' broom (PnWB) phytoplasma genome sequence (Dumonceaux et al. 2014).

The *cpn60* gene sequences of 43 phytoplasma strains were determined and used to generate a phylogenetic tree that showed a topology congruent with the one reported for the 16S rRNA-encoding genes (Dumonceaux et al. 2014). In the study by Pérez-López et al. (2016), 96 *cpn60* gene sequences of phytoplasmas from the groups 16SrI, 16SrII, 16SrV, 16SrVII, 16SrIX, 16SrX, 16SrXII, 16SrXIII, and 16SrXIV were used to propose a phytoplasma classification method based on RFLP analysis of *cpn60* gene sequences with a set of seven restriction endonucleases. This scheme allows a finer differentiation of phytoplasma strains, and it was suggested as a supplementary tool to the existing classification scheme based on the 16S rRNA gene (Lee et al. 1998).

9.3 Differentiation of Phytoplasmas in Single Ribosomal Groups Using Multiple Genes

Group 16SrI Aster yellows (AY) group (16SrI) phytoplasmas belong to the species '*Candidatus* Phytoplasma asteris', are associated with over 100 economically important diseases worldwide and represent one of the most diverse and widespread phytoplasma group (Lee et al. 2004a). Up to now, almost 30 ribosomal subgroups have been described by actual and/or virtual RFLP analyses within 16SrI group (Bertaccini and Lee 2018).

RplV-rpsC, tuf, secY, and nusA Genes Phylogenetic analysis based on ribosomal protein (*rp*) gene sequences (*rplV-rpsC*) revealed substantial genetic variations among members of the aster yellows phytoplasma group, implying that the phytoplasmas in this group are more genetically diverse than indicated by the analysis based on 16S rRNA gene sequences. *Rp*-gene based phylogeny delineates 10 lineages or *rp* gene subgroups within the 16SrI group. The *rp* gene subgroups are considered consistent with their unique ecological niches and biological properties (Lee et al. 2004a, 2006). Moreover, also the *secY* gene exhibited greater sequence variation than 16S rRNA gene among members of this phytoplasma ribosomal group. Sequence homologies ranged from 94.7 to 98.8% based on *secY* gene sequences compared to 98.5 to 99.5% based on 16S rDNA sequences between two 16SrI subgroups. A phylogenetic tree, derived by analysis of the *secY* gene sequences and *in silico* translated SecY amino acid sequences, delineated ten distinct lineages. The *SecY* gene proved to be an efficient molecular tool for differentiation of phytoplasma strains. For example, strains of subgroups 16SrI-B and 16SrI-D, which share 98% 16S rDNA sequence homology with members of subgroup 16SrI-B, clustered with 16SrI-B strains by the 16S rDNA-based phylogeny, but the *secY* gene-based phylogeny clearly indicated that they represent two distinct lineages, reflecting their unique biological features. Phylogeny inferred by the *secY* gene sequence analysis was nearly congruent with that inferred by *rp* gene analysis (Lee et al. 2006). Phylogenetic analyses based on *tuf*, *rplV-rpsC*, and *secY* genes reinforced the notion that most subgroups identified by RFLP analysis of 16S rDNA sequences represent distinct phylogenetic lineages (Marcone et al. 2000; Lee et al. 2004a, 2006). '*Ca.* P. asteris' has been proposed to represent the almost entire 16SrI group phytoplasma. In light of the substantial genetic variability and differences in biological properties among these phytoplasmas, it should be possible to assign more '*Candidatus* Phytoplasma' species in this group according to the guidelines (IRPCM 2004). Both *secY* and *rp* genes are useful phylogenetic parameters that can be employed for accurate identification of closely related, but distinct 16SrI phytoplasma strains. Thus, each distinct lineage can be better defined by employing multiple phylogenetic parameters and single-nucleotide polymorphism (SNP) analyses (Bagadia et al. 2013). Furthermore, results from PCR, RFLP, and phylogenetic analyses also indicated that *nusA* gene may be useful in detection and differentiation of phytoplasma lineages within group 16SrI (Shao et al. 2006).

***GroEL* and *amp* Genes** *The groEL* gene was evaluated for its effectiveness towards detecting variability in 27 '*Ca.* P. asteris'-related strains assigned to different 16SrI subgroups and originated from different hosts and geographical areas. The RFLP analyses of the amplified fragments confirmed differentiation among 16SrI-A, 16SrI-B, 16SrI-C, 16SrI-F, and 16SrI-P subgroups and showed further differentiation in strains assigned to 16SrI-A, 16SrI-B, and 16SrI-C subgroups. However, analyses of *groEL* gene failed to discriminate strains in subgroups 16SrI-L and 16SrI-M (described on the basis of 16S rDNA interoperon sequence heterogeneity) from strains in subgroup 16SrI-B. On the contrary, the 16SrI unclassified strain ca2006/5 from carrot (showing interoperon sequence heterogeneity) (Duduk et al. 2009) was differentiable on both *rp* and *groEL* genes from the strains in subgroup

16SrI-B. Phylogenetic analyses carried out on *groEL* gene are in agreement with 16S ribosomal, *rp*, and *secY* gene-based phylogenies and confirmed the differentiation obtained by RFLP analyses on *groEL* gene amplicons. The average nucleic acid similarity of 16S rDNA of examined aster yellows strains was 99.5% ranging from 98.7 to 99.5% among the 16SrI subgroups and from 99.1 to 100% inside the subgroups. The lowest similarities of 16S rDNA (98.7–99.3%) were observed between PopD phytoplasma (16SrI-P) and the strains from the other 16SrI subgroups. On the other hand, the average nucleic acid similarity of *groEL* gene of examined strains was 98.1% ranging from 93.8 to 98.1% and from 98 to 100% among and inside the 16SrI subgroups, respectively. Again, the lowest similarities of *groEL* gene (93.8–94.8%) were detected between PopD phytoplasma (16SrI-P) and the strains from the other 16SrI subgroups. The average similarity of *groEL* gene predicted amino acid sequences was 98.8% ranging from 96.0 to 97.9% and from 97.7 to 100% among and inside the 16SrI subgroups, respectively, whilst for PopD phytoplasma, it was lower ranging from 94.7 to 96.2%. The seven lineages delineated on *groEL* gene are consistent with a *groEL* gene RFLP subgrouping, except that the split of 16SrI-C in subgroups groELI-VI and groELI-VII is not supported. Phylogenetic analyses carried out on the whole *groEL* gene (1,610 bp) confirmed the finer differentiation by RFLP analyses on the specifically amplified and sequenced 1,397 bp fragments (Mitrovic et al. 2011). The strawberry green petals phytoplasma (16SrI-C) was characterized also in the *groEL* gene and affiliated to GroELI-VII group (Contaldo et al. 2012). Cyclamen phyllody and virescence phytoplasma (16SrI-B) was further characterized on *groEL*, *secY*, *rp*, *tuf*, and *amp* genes, resulting identical to strains belonging to 16SrI-B subgroups in all the studied genes except *amp* gene, and it was classified as rpI-B, SecYI-B, and GroELI-III (Satta et al. 2013). Finally, collective RFLP characterization of the *groEL*, *amp*, and *rp* genes, together with sequence data, distinguished the aster yellows strain detected in Colombian oil palm showing a severe decline from other aster yellows phytoplasmas, in particular, from an aster yellows strain infecting corn in the same country (Fig. 9.1) (Alvarez et al. 2014).

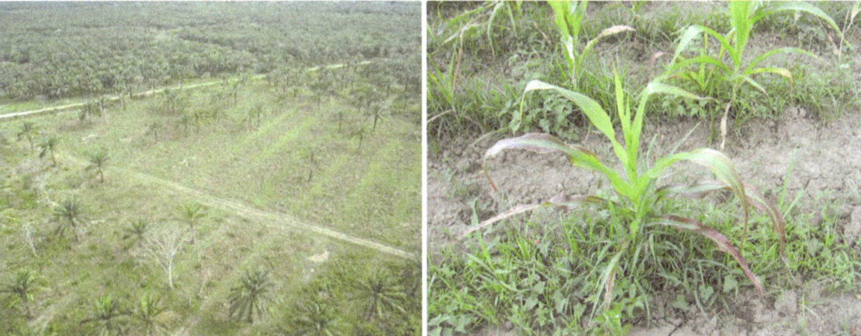

Fig. 9.1 On the left oil palm cultivation in Colombia in which the plants were destroyed by the presence of the aster yellows phytoplasmas (Alvarez et al. 2014) and on the right corn young plants in Colombia infected by an aster yellows strain that is differentiable from the one detected in the oil palms for its *rp* gene RFLP profile (A. Bertaccini et al. unpublished)

Group 16SrII Phytoplasmas in the ribosomal group 16SrII (peanut witches' broom) were mainly detected in crops and fruit trees in all the continents except Europe where their presence can be considered linked to a few wild species or occasional in flower crops (Tolu et al. 2006; Davino et al. 2007). Following the 16S rDNA classification, up to 21 ribosomal subgroups were identified (16SrII-A to 16SrII-U) by the restriction fragment length polymorphism analyses on the R16F2n/R16F2 amplicons (Bertaccini and Lee 2018). However, a number of variants were reported as separated subgroups infecting cactus plants in China (16SrII-G to -L) (Cai et al. 2008), whilst this species is usually infected by phytoplasmas classified in the 16SrII-C subgroup (Granata et al. 2006), allowing the speculation about very high variability of these phytoplasmas in some areas of the world. The 16SrII group encloses two '*Candidatus* Phytoplasma' species: '*Ca.* P. aurantifolia' associated with lime witches' broom in Oman (GenBank accession number U15442; 16SrII-B) (Zreik et al. 1995) and '*Ca.* P. australasia' associated with papaya mosaic disease in New Zealand (GenBank accession number Y10096; 16SrII-D) (White et al. 1998). Several studies (Lee et al. 2006; Martini et al. 2002, 2007) indicate that phylogenetic analyses based on moderately variable genes are able to increase the resolving power for closely related but biologically distinct phytoplasma strains. In the case of this ribosomal group, fine differentiations were achieved with *rp*, *secY*, *tuf*, and *secA* genes.

Rp genes from selected strains of the 16SrII group were amplified using the primer pairs rpF1C/rp(I)R1A and by rpL2F2/rp(I)R1A however, in many cases the amplification resulted in multiple bands. The amplicons from members of phytoplasma group 16SrII ranged from 1,284 to 1,309 bp, and their sequences allow to resolve four rp subclades; moreover, their average sequence similarities resulted ranging from 86 to 93.1% on the *rp* genes, whilst on the 16S rRNA gene, the range was from 97.4 to 98.8% (Martini et al. 2007). Overall the use of this gene is not fully exploited for the differentiation of members of this ribosomal group. The sequence similarities detected in the *secY* gene for members of the 16SrII group ranged from 87.7 to 100% compared to the 97.4 to 99.6% for the 16S rRNA gene. Four members of subgroup 16SrII-A were differentiated into three secY(II) genotypes, with two strains having the same genotype, whilst four members of subgroup 16SrII-C were differentiated into four secY(II) genotypes (Lee et al. 2010). Results from phylogenetic and virtual restriction fragment length polymorphism analyses of the 16S rRNA gene sequence indicated that a cauliflower-infecting phytoplasma in Yunnan (a southwest province of China), enclosed in 16SrII-A subgroup, was differentiated from a broad host range of phytoplasmas found only in East Asia by multilocus genotyping. In particular, genes coding ribosomal proteins S19, L22, and S3 (*rpsS-rplV-rpsC*) of three phytoplasma strains compared with those of the other 16SrII-A phytoplasma strains showed that this caulifower phytoplasma possessed a unique SNP within the recognition site of the restriction enzyme *HpyCH4*V that distinguishes this strain from all other known members in the subgroup 16SrII-A. Moreover, in the sequences of the *secY* gene, distinctive SNPs were found, which could distinguish by RFLP analyses these cauliflower phytoplasma strains from the reported strain of the subgroup 16SrII-A (Cai et al. 2016).

***Tuf* and *secA* Genes** Both genes were used for general phytoplasma differentiation purposes in particular for the barcoding of quarantine phytoplasmas in Europe (Makarova et al. 2012). Taken together, the results suggest that although the *tuf* gene barcode demonstrates a variable level of ribosomal subgroup resolution, its overall performance is comparable with the one of the full 16S ribosomal sequence or better. A high variation was found in the *tuf* gene dataset of '*Ca*. P. aurantifolia' group 16SrII (4.9%), suggesting the presence of phytoplasma subgroups which were not identified based on 16S rRNA gene. In particular 16SrII phytoplasmas could be split into the subgroups reported previously (Lee et al. 1998). Phytoplasma primers from *secA* gene produce sequences that in a phylogenetic tree indicate a clear split within the 16SrII group that is consistent with the current classification of '*Ca*. P. aurantifolia' and related strains (subgroup 16SrII-B) and '*Ca*. P. australasiae' (16SrII-D and 16SrII-A subgroups) (Hodgetts et al. 2008; Cai et al. 2016).

***RpoB* Gene** A grouping of phytoplasmas belonging to subgroup 16SrII-A by computer-simulated RFLP analysis of partial *rpoB* gene PCR products was implemented (Valiunas et al. 2013). Since the phytoplasma *rpoB* gene is characterized by higher divergence than that exhibited by the 16S rRNA gene, it should permit finer resolution of related phytoplasmas that may possess biologically distinct properties; therefore, further exploitation of its usefulness in the differentiation of other 16SrII subgroups for epidemiological studies should be done.

Multiple Genes Analyses Four witches' broom diseases from Hainan Province in China such as *Arachis hypogaea* (peanut), *Crotalaria pallida*, *Tephrosia purpurea*, and *Cleome viscosa* were identified as associated 16SrII-A phytoplasmas. Sequences of the *rplV-rpsC*, *rpoB*, *gyrB*, *dnaK*, *dnaJ*, *recA*, and *secY* from two strains for each of the four phytoplasmas were amplified, and the seven concatenated gene regions indicated that these phytoplasmas cluster most closely with one another and are also closely related to a 16SrII-A strain from Taiwan (Li et al. 2014). A multilocus characterization was achieved on *tuf*, *secY*, *dnaK*, and *dppA* genes amplified from several horticultural infected species from Egypt, and it indicates no diversity among the studied phytoplasma strains all enclosed in the 16SrII-D subgroup (El-Sisi et al. 2018). The gene sequences including *secA*, *tuf*, *imp*, and *SAP11* were employed for *Crotalaria aegyptiaca* witches' broom phytoplasma characterization from Oman (subgroup 16SrII-W). Sequences from *C. aegyptiaca* and *Orosius* sp. leafhoppers samples were 100% identical to each other. The lack of amplification with specific primers for the *SAP11* gene of this phytoplasma indirectly confirms its uniqueness since they are amplifying strains from subgroups 16SrII-A, 16SrII-B, and 16SrII-D (Al-Subhi et al. 2017).

Group 16SrIII Phytoplasmas enclosed in this ribosomal group are mainly detected in both American and European continents with scattered detection of some strains in Japan, Africa, and Asia. However, they were never reported in crops from Asia, Australia, or New Zealand. Ribosomal group 16SrIII encloses 26 subgroups (16SrIII-A until 16SrIII-Z) and is infecting both herbaceous and woody species. It

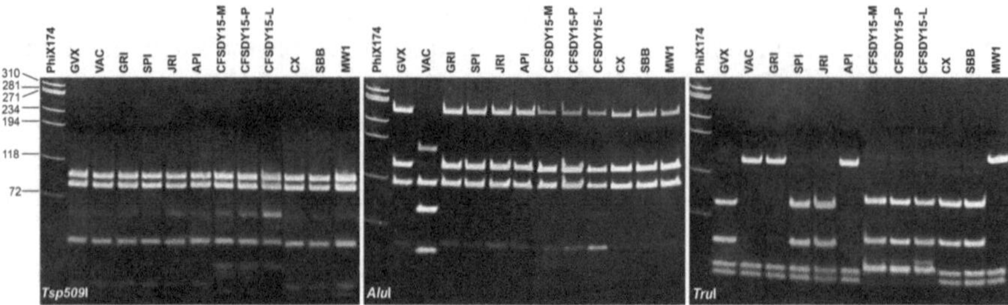

Fig. 9.2 Polyacrylamide gels showing RFLP patterns of phytoplasma DNAs from cassava and phytoplasma strains of the subgroup 16SrIII-L amplified in nested PCR with the rpIIIF1/rpIIIR1 primer pair. The restriction endonucleases used are listed at the bottom of each gel. Phytoplasma acronyms: CFSDY15-M, CFSDY15-P, and CFSDY15-L, Colombian cassava frogskin infected samples; GVX, 16SrIII-A (green Valley X disease from United States of America); VAC, 16SrIII-F (vaccinium witches' broom from Germany); GRI, 16SrIII-D (goldenrod yellows from United States of America); SPI, 16SrIII-E (spirea stunt from United States of America); JRI, 16SrIII-H (poinsettia branch-inducing from United States of America); API, 16SrIII-B (phytoplasma from *Euscelidius variegatus* from Italy); CX, 16SrIII-A (peach X-disease from Canada); SBB, 16SrIII-F (*Solanum marginatum* big bud from Ecuador); and MW1, 16SrIII-F (milkweed yellows from United States of America). PhiX174, marker DNA *Hae*III digested

only accommodates the '*Ca*. P. pruni' (16SrIII-A) that was associated with devastating diseases in stone fruits and cherry in the United States of America and Canada (Fiore et al. 2018).

The *rp* and *secY* Genes One of the first finer differentiations for phytoplasmas enclosed in group 16SrIII was achieved by nested PCR using the group III rp-specific primers (Gundersen et al. 1996). The RFLP analyses of this amplicon with *Alu*I restriction enzyme showed no differences between the cassava strain CFSDY15 and most of the 16SrIII strains; however, its *Tru1*I and *Tsp509*I profiles were unique, confirming that the phytoplasma infecting cassava in Colombia could be assigned to the new ribosomal protein subgroup, rpIII-H (Fig. 9.2) (Alvarez et al. 2009).

Among the members of the 16SrIII ribosomal group the sequence similarities ranged from 93.6 to 99.8% for the *secY* gene and 99.0 to 99.8% for the 16S rRNA gene. Up to 11 secY subgroups were therefore differentiated, but for a couple of them, the resolution of this gene was lower than the one of the 16S ribosomal gene (Lee et al. 2010). The differentiation of phytoplasmas in 16SrIII group was achieved on both *rp* and *secY* genes in order to help solving the tricky presence of 16S ribosomal interoperon heterogeneity. The amplified *rp* genomic regions of the studied strains were identical in the nucleotide sequence, except for a single-base difference located in *rplV* region encoding the L22 ribosomal protein (Davis et al. 2013). Direct and nested PCRs with primer pair secYF1(III)/secYR1(III) yielded 1.7 kb DNA fragments that resulted to contain a partial (3′-end) *rpl15* ribosomal protein (rp) gene, a complete *secY* gene, and a partial (5′-end) *map* (methionine aminopeptidase) gene. Like various 16SrIII phytoplasma strains and unlike some other phytoplasmas (Lee et al. 2010), the X-disease phytoplasma strains studied lacked an *adk* (adenylate kinase) gene between the *secY* and *map* genes. A 9-base insertion/deletion (indel) in the *secY* genomic locus distinguished two groups in the 16SrIII phytoplasma strain cluster, one containing strains having a *secY* gene length of 1,263 bp and the other containing strains having *secY* length of 1,272 bp. Single-nucleotide polymorphisms (SNPs) further distinguished the X-disease strains from group 16SrIII strains that are associated with other diseases, and are classified in other than subgroup 16SrIII-A. The topologies of the *secY* and *rp* gene trees suggested at least two divergent branches, with strain clover yellow edge (16SrIII-B) possibly representing a third. Phylogenetic analyses of SecY proteins and single-nucleotide polymorphism analyses of *secY* and *rp* genes further distinguished two grapevine sequevars in North America grapevine yellows (NAGY) phytoplasmas from one another and from '*Ca.* P. pruni'. The NAGYIIIα and NAGYIIIβ sequevars also differed from the '*Ca.* P. pruni' in regions of the folded SecY protein that are predicted to be near or exposed at the outer surface of the phytoplasma membrane (Davis et al. 2015). The group 16SrIII is widespread in South America where RFLP analysis with *Alu*I, *Dra*I, and *Tru1*I distinguished three new rp profiles within subgroup 16SrIII-B, one for subgroup 16SrIII-J, and one shared with strains of the subgroups 16SrIII-W and 16SrIII-X. The combined profiles of the 11 strains analysed were different from previously reported 16SrIII phytoplasmas (Galdeano et al. 2013). Such diversity was previously observed in subgroup 16SrIII-B by Gundersen et al. (1996) between the strains clover yellow edge (16SrIII-B) and milkweed yellows (16SrIII-F). Phytoplasmas belonging to subgroups 16SrIII-J and 16SrIII-U from South America resulted to be closely related to others that were detected in Mexico and characterized based on 16S rRNA and/or *rp* genes (Pérez-López et al. 2017).

***RpoB* Gene** Partial *rpoB* gene sequences were amplified, cloned, and sequenced from 17 phytoplasma strains in subgroups 16SrIII-A, 16SrIII-E, 16SrIII-F, and 16SrIII-H, and a comparative study revealed that these sequences approached the discriminating levels observed for the 16S rRNA, *rpl22*, *rps3*, *secY*, and *secA* genes. The size of the complete gene and of the deduced amino acid sequence of the RpoB protein differ not only between phytoplasmas and other wall-less and walled bacteria but also among different phytoplasmas and may provide another tool for differentiating among closely related phytoplasmas (Valiunas et al. 2013).

Group 16SrIV Phytoplasmas in group 16SrIV were only detected in lethal yellowing disease of palms in several areas of the world and in particular in the Caribbean areas in America, and in a few areas in Africa. Very recently in Papua New Guinea, a '*Ca.* P. novoguineense' was reported and, although not classified as ribosomal group, it results phylogenetically congruent with phytoplasma enclosed in the 16SrIV group (96.08% of identity in the 16S ribosomal gene). The *secY* and *rp* gene sequences were analysed, and the comparative analysis of the rp operon sequences shows that they are identical among the phytoplasma strains from different regions of the area (Miyazaki et al. 2018). No other official '*Candidatus*

Phytoplasma' species are published for this ribosomal group, whilst several subgroups were designed (16SrIV-A to 16SrIV-E) with diverse geographical distribution areas enclosing west Africa and central and north America (Bertaccini and Lee 2018).

***GroEl* Gene** A SybrGreen system based on the *groEL* gene was developed and accurately detected the LY phytoplasmas (16SrIV-A) in both qPCR using TaqMan probes and conventional PCR. The primers groELR1 (5′-CTT TAG GAC CAA AAG GTA CT-3′)/F1 (5′-GAA GAA CAA CAA CCA CTA TC-3′) and groELR2 (5′-CGA TAA TGC TGG AGA TGG GAC TAC T-3′)/F2 (5′-GAA CTA CAG CGG CTC CTG TTG TAA T-3′) (Myrie et al. 2011) also allowed their differentiation that resulted congruent with the one based on 16S ribosomal gene and also was able to distinguish among strains in the same subgroup (Paredes-Tomas et al. 2019).

Group 16SrV This ribosomal group is divided into nine subgroups (16SrV-A to 16SrV-I) (Fránová et al. 2016) including phytoplasma strains that are detected mainly in the Eurasian areas of the world and mainly on woody or shrub species. Phytoplasmas from this group are able to infect forest trees such as elm and alder (Marcone et al. 2018), and profitable cultivations such as grapevine and small fruits in Europe and jujube and stone fruits in Asia (Angelini et al. 2018; Fiore et al. 2018). The group encloses four '*Candidatus* species': '*Ca*. P. ulmi' (Lee et al. 2004b), '*Ca*. P. ziziphi' (Jung et al. 2003), '*Ca*. P. rubi' (Malembic-Maher et al. 2011), and '*Ca*. P. balanitae' (Win et al. 2013). However, since phytoplasma strains associated with the most important quarantine disease infecting grapevine in Europe fall into two distinct 16SrV subgroups, 16SrV-C and 16SrV-D (Martini et al. 1999), there is no space to support these phytoplasmas as a new '*Candidatus* species', and they must still be named as "flavescence dorée" (FD) until new rules or possibly their consistent cultivation on artificial media for collection deposition of the pathogen will be achieved (Contaldo et al. 2016). For this reason, research on additional markers has been specifically developed to distinguish FD strains for almost 20 years (Angelini et al. 2001; Martini et al. 2002; Arnaud et al. 2007; Quaglino et al. 2010).

Multilocus Sequences for Strain Differentiation in '*Ca*. P. ulmi' and '*Ca*. P. rubi' Genetic diversity of *Rubus* stunt and elm yellows phytoplasmas in Europe was determined also by the variability of *map* and *uvrB-degV* for 15 phytoplasmas either from *Rubus* with stunting or from elm and dog rose with yellowing. For *Rubus* and dog rose phytoplasmas, no sequence variability in the *map* locus was found, and only one SNP was detected in *uvrB-degV* gene sequences. All these phytoplasmas had a monophyletic origin, as they clustered in a single group supported by bootstrap values of 98% and 99%. They were all characterized by a specific ATT insertion at position 344 in the *uvrB-degV* intergenic sequence. Nucleotide sequence similarity between this strain cluster and other phytoplasmas ranged between 96 and 98% for the *map* gene and between 95 and 97% for *uvrB-degV* genes. The '*Ca*. P. ulmi' strains are enclosed in an homogenous strain cluster according to *map* sequences, which differed by eight SNPs over 674 bp. The variability of *uvrB-degV* genes was lower and only reached three SNPs over 1,025 bp. Sequence similarities with the other phytoplasma tested were in the range of 96–97% for the *map* gene and about 95% for *uvrB-degV* genes. Parsimony analyses of both genes indicated a single monophyletic origin as all '*Ca*. P. ulmi' strains clustered on one branch supported by 99% bootstrap value (Arnaud et al. 2007). Elm yellows phytoplasmas originating from localities in northeast, east, and southwest Serbia were characterized by RFLP analysis and DNA sequencing of *rp*, *secY*, and *map* genes. In total, five genotypes were identified based on collective sequencing. Based on their high degree of genetic variability, the Serbian strains were assigned to four different subtypes of '*Ca*. P. ulmi' (EY-S1, EY-S2, EY-S3, and EY-S4) (Jović et al. 2011). Diversity of '*Ca*. P. ulmi' in Croatia was determined by sequencing *rp*, *secY*, and *secY-map* genes, in 62 phytoplasma strains. Phylogenetic analysis indicated that these Croatian strains share a common origin and are closely related to '*Ca*. P. ulmi' from southeastern Europe. However, comparative sequence analysis revealed new genotypes based on the sequenced genes pointing a significantly higher genetic diversity than previously reported (Katanić et al. 2016).

Differentiation on *rp*, *tuf*, and *secY* Genes in '*Ca*. P. ziziphi' Phytoplasmas were detected in *Sophora japonica* and *Robinia pseudoacacia* with witches' broom in China and identified by sequence of partial gene 16S rRNA, *rp*, and *secY*. They resulted closely related to subgroup 16SrV-B, rpV-C, and secYV-C of the jujube witches' broom (JWB) phytoplasma (Ren et al. 2014). The *rp* and *tuf* genes of *Bischofia polycarpa* witches' broom phytoplasma in China were used to confirm its close relatedness to the subgroup 16SrV-B (Lai et al. 2014). Sweet cherry virescence disease in China was consistently associated with infection by a phytoplasma belonging to subgroup 16SrV-B. Further analysis of *rp* and *secY* genes revealed that it was essentially indistinguishable from the phytoplasmas associated with jujube witches' broom (JWB) and other diseases of many other plant species in China (Wang et al. 2018).

Multilocus Differentiation of Phytoplasmas in subgroups 16SrV-C and 16SrV-D The sequences of *map* and *uvrB-degV* genes, along with the sequence of the *secY* gene, were determined among a collection of "flavescence dorée" (FD) and FD-related phytoplasmas infecting grapevine, alder, elm, blackberry, and Spanish broom in Europe. Sequence comparisons and phylogenetic analyses consistently indicated the existence of three phytoplasma strain clusters. Strain cluster FD1 displayed low variability, and strain cluster FD2 (16SrV-D) displayed no variability, whereas the more-variable strain cluster was the FD3 (16SrV-C). German Palatinate grapevine yellows phytoplasmas (PGY) appeared variable and were often related to some of the alder phytoplasmas (AldY), and phylogenetic analyses concluded that these strains are members of the same phylogenetic subclade (Arnaud et al. 2007). Molecular characterization and phylogenetic analyses of the *map* gene sequences of FD and related strains in Switzerland revealed the prevalence of the FD2 phytoplasmas in grapevines (97%) and in *Orientus ishidae* pools (72%). Such map type was found also in hazel and in *Tamnotettix dilutior*, but not in *Scaphoideus titanus*. Moreover, map types FD1 and FD3 were identified in several host plants and phytoplasma insect vectors (Casati et al. 2017). The analysis on *tuf*, *secY*, and *rp* genes was carried out also for 22 strains identified as 16SrV-C and for 11 strains identified as 16SrV-D collected in grapevines in Italy. RFLP analyses on *tuf* gene showed the presence of two profiles for phytoplasma strains belonging to 16SrV-C and 16SrV-D subgroups, whilst higher molecular variability was detected on *secY* gene. The RFLP analyses carried out with *Tru1*I and *Tsp509*I restriction enzymes on the amplicons obtained with the primer pairs rp(V)F1A/rpR1 showed a higher variability when compared with the parallel analyses carried out on rp(V)F1A/rp(V)R1A amplicons. In the first case, five profiles in 16SrV-C strains and three profiles in 16SrV-D strains were obtained in comparison to a unique profile detected in the strains from both subgroups when amplified with the latter primer pairs. The FD variability was mainly found on the *rp* gene in agreement with previously reported data in Emilia-Romagna and Tuscany (Botti and Bertaccini 2006, 2007), France (Martini et al. 2002), and Serbia (Paltrinieri et al. 2012). The presence of a number of FD variants could be related to an endemic phase of the disease, when the pathogen is subjected to a low selective pressure (Zambon et al. 2018). The rp-based phylogeny enabled identification of four phytoplasma strains among the AldY strains from Macedonia. Three strains clustered within the rpV-E subgroup, whilst one belonged to rpV-L subgroup. Phylogenetic analysis of the map sequence showed the presence of five phytoplasma strains belonging to FD1 and FD2 clusters in Macedonia (Atanasova et al. 2014).

Groups 16SrVI, 16SrVII, and 16SrVIII Phytoplasmas in these ribosomal groups are mainly reported from America (16SrVI and 16SrVII) and from Taiwan '*Ca.* P. luffae' (16SrVIII-A). In the first two ribosomal groups there are three '*Candidatus* species': '*Ca.* P. trifolii' (16SrVI-A), '*Ca.* P. sudamericanum' (16SrVI-I), and '*Ca* P. fraxini' (16SrVII-A) (Bertaccini and Lee 2018). Based on actual and virtual RFLP analysis of the 1.2 kb 16S rRNA gene, eight 16SrVI subgroups (16SrVI-A to 16SrVI-F, 16SrVI-H, and 16SrVI-I) have been described (Davis et al. 2012; Bertaccini and Lee 2018): clover proliferation (CP) (16SrVI-A), strawberry multiplier disease (16SrVI-B), Illinois elm yellows (16SrVI-C), Indian brinjal little leaf (16SrVI-D), Italian *Centaurea solstitialis* virescence (16SrVI-E), Sudanese periwinkle phyllody (16SrVI-F), Indian portulaca little leaf (16SrVI-H), and Brazilian passionfruit (16SrVI-I).

The *rp* and *secY* Genes Phylogenetic analyses of the moderately conserved *rplV-rpsC* and *secY* genes corroborated the distinction among 16SrVI subgroups and allowed the identification of distinct genetic lineages within subgroup 16SrVI-A. These lineages cannot be readily differentiated based on analysis of 16S rRNA gene sequences alone. To evaluate the efficacy of the *rp* and *secY* genes for finer differentiation of phytoplasma strains, virtual RFLP analyses was performed using *rp* and *secY* gene sequences allowing unambiguous identification of 16SrVI subgroups (Martini et al. 2007; Lee et al. 2010). In particular, based on virtual RFLP patterns of *rp* gene sequences with four restriction enzymes (*Alu*I, *Dra*I, *Taq*I, and *Tsp*509I), representative strains CP (subgroup 16SrVI-A), PWB (potato witches' broom, 16SrVI-A), VR (vinca virescence, 16SrVI-A), EYIL (16SrVI-C), LUM (lucerne virescence), BLL-In (16SrVI-D), and CPS (16SrVI-F) in group 16SrVI were differentiated into six distinct rp(VI) subgroups. Strain VR, classified along with strains CP and PWB into the same subgroup 16SrVI-A, has been clearly separated based on *rp* gene sequences in a distinct rp(VI) subgroup (Martini et al. 2007). Based on computer-simulated RFLP patterns of the *secY* gene sequences, ten members of subgroup 16SrVI-A were differentiated into six secY(VI) genotypes represented by the following individual strains or strain clusters: CP and AKpot1 (Potato purple top-AK); AKpot2, AKpot4, and AKpot5; DBPh2 and DBPh3 (Dry bean phyllody); PWB; LUM; BLL-In; and VR (Lee et al. 2010). When *rp* and *secY* gene sequences were analysed among members of the CP phytoplasma group, the average sequence similarities ranged from 97.8 to 98.8% and from 94.4 to 99.8%, respectively, compared with 98.4 to 99.2% for the 16S rRNA gene sequence similarities. Moreover, molecular characterization on *rp* gene was carried out for phytoplasma strains belonging to 16SrVI and 16SrVII ribosomal groups from grapevine. The four grapevine cultivar Glera samples collected from three vineyards showed restriction profiles identical to each other's and to the '*Ca*. P. trifolii', 16SrVI-A. The phytoplasma strains identified as belonging to 16SrVII group and amplified on the *rp* gene enclosing strain from both grapevine end insect vectors showed also identical RFLP patterns to each other's and to the '*Ca*. P. fraxini', 16SrVII-A (Zambon et al. 2018).

Group 16SrIX Phytoplasmas in the ribosomal group 16SrIX (pigeon pea witches' broom) are associated with diseases affecting crop and wild plants in different geographic areas worldwide. Based on actual and virtual RFLP analysis of the 16S rDNA fragment amplified using primer pair R16F2n/R16R2, ten 16SrIX subgroups (16SrIX-A to 16SrIX-J) have been described (Bertaccini and Lee 2018; Salehi et al. 2018). Phytoplasmas of subgroup 16SrIX-B and their genetic variants (formerly classified in subgroups 16SrIX-D, 16SrIX-F, and 16SrIX-G), associated with almond witches' broom disease in Lebanon and Iran and with apricot yellows in Iran, were all classified within the species '*Ca*. P. phoenicium' (Verdin et al. 2003). This grouping was based on the rule that indicates that strains can be classified as '*Ca*. P. phoenicium' if they share with the reference strain of the species (strain A4, GenBank accession number AF515636): (i) 16S rDNA sequence identity >97.5%, (ii) the unique species-specific signature sequence 5′-CCT TTT TCG GAA GGT ATG-3′, and (iii) biological features (plant and insect host range) and molecular features (similarity in other genes). Within group 16SrIX, only strains of subgroup 16SrIX-B and their variants share all these features with the reference strain and can be defined '*Ca*. P. phoenicium'. Strains classified in other 16SrIX subgroups share with the '*Ca*. P. phoenicium' reference strain A4 only a 16S rDNA sequence identity >97.5% (i); they do not have an identical signature sequence (ii) or coomon biological/molecular features (iii). For this reason, those other strains can only be defined as related to '*Ca*. P. phoenicium'.

The *rp* and *secY* Genes Comparative phylogenetic analyses of the conserved 16S rRNA gene and the moderately conserved *rplV-rpsC* and *secY* genes corroborated the distinction among 16SrIX subgroups and allowed the identification of distinct lineages within subgroups 16SrIX-C and 16SrIX-E. These lineages cannot be readily differentiated based on analysis of 16S rRNA gene sequences alone. In particular, virtual RFLP analyses using *rp* and *secY* gene sequences allowed unambiguous identification of 16SrIX subgroups and lineages (Lee et al. 2012). Sequence and phylogenetic analysis of 16S rDNA indicated that the 16SrIX phytoplasmas were closely related to the juniper witches' broom (JunWB) phytoplasma detected in the United States of America, representing a 16SrIX-E variant. Ribosomal protein (*rp*) and *secY* gene-based phylogenies revealed that the blueberry stunt (BBS-NJ) and JunWB 16SrIX-E strains represented two distinct lineages. Single-nucleotide polymorphism (SNP) analyses of *rp* and *secY* gene sequences further revealed that BBS-NJ 16SrIX-E strains had 15 *consensus* rp SNPs and 28 *consensus* secY SNPs that separated them from the JunWB strains (Bagadia et al. 2013).

The *tuf*, *groEL*, and *inmp* Genes Alignment of 16S rDNA nucleotide sequences of '*Ca.* P. phoenicium' (subgroup 16SrIX-B and variants) strains, identified in almond, peach, and apricot in Lebanon and Iran, revealed the presence of 19 single-nucleotide polymorphisms (SNPs) which combination allowed the recognition of nine and six lineages in Lebanon and Iran, respectively (Salehi et al. 2018). This result reinforces recent evidences figuring out the usefulness of 16S rDNA to resolve the genetic complexity within phytoplasma populations. Furthermore, the identification of distinct lineages in Lebanon and in Iran suggests that, as reported in previous studies about phytoplasma strain populations (Cai et al. 2008; Quaglino et al. 2009, 2017; Cheng et al. 2015), climatic and geographical features in the ecosystems may be significant, directly or indirectly, in determining the strain composition of the phytoplasma populations in the different regions. Based on draft genome sequence of '*Ca.* P. phoenicium' strain SA213 from *Prunus dulcis* affected by almond witches' broom (AlmWB) disease (Quaglino et al. 2015), genes *tuf*, *groEL*, and *inmp* were selected for investigating the genetic diversity among strain populations.

Primer pairs allowed the amplification of these genes exclusively from 16SrIX-B phytoplasma strains, and not from the closely related 16SrIX-C strains. Multiple gene typing analyses revealed a substantial genetic homogeneity within the analysed phytoplasma populations, based on housekeeping gene sequences, and allowed the identification of distinct AlmWB-associated phytoplasma strains from diverse host plants based on *inmp* (integral membrane protein) gene sequences. This evidence, along with prior reports of multiple insect vectors of AlmWB phytoplasma, suggests that AlmWB could be associated with phytoplasma strains derived from the adaptation of an original strain to diverse host plant species (Lee et al. 2012; Quaglino et al. 2015). Due to the role of phytoplasma membrane proteins in determining the vectoring activity of insects and the interaction with plant hosts, it is notable that, within an extremely homogeneous population of AlmWB phytoplasma strains, the sole differences were detected on integral membrane protein gene sequences. Alignments of nucleotide (358 nt) and amino acid (119 aa) sequences of the *inmp* gene evidenced the presence of 21 nucleotide substitutions all non-synonymous. Codon-based test of positive selection (dN/dS) carried out using different methods rejected the null hypothesis of strict neutrality (dN = dS) in favour of the alternative hypothesis (dN > dS), indicating the presence of a positive selection. Intriguingly, these preliminary data seem to indicate that AlmWB phytoplasma strains identified in peach plants are distinct from strains infecting almond and nectarine based on molecular markers within the *inmp* nucleotide sequences (Fig. 9.3). Recently, insect species of the families Cicadellidae and Cixiidae have been reported

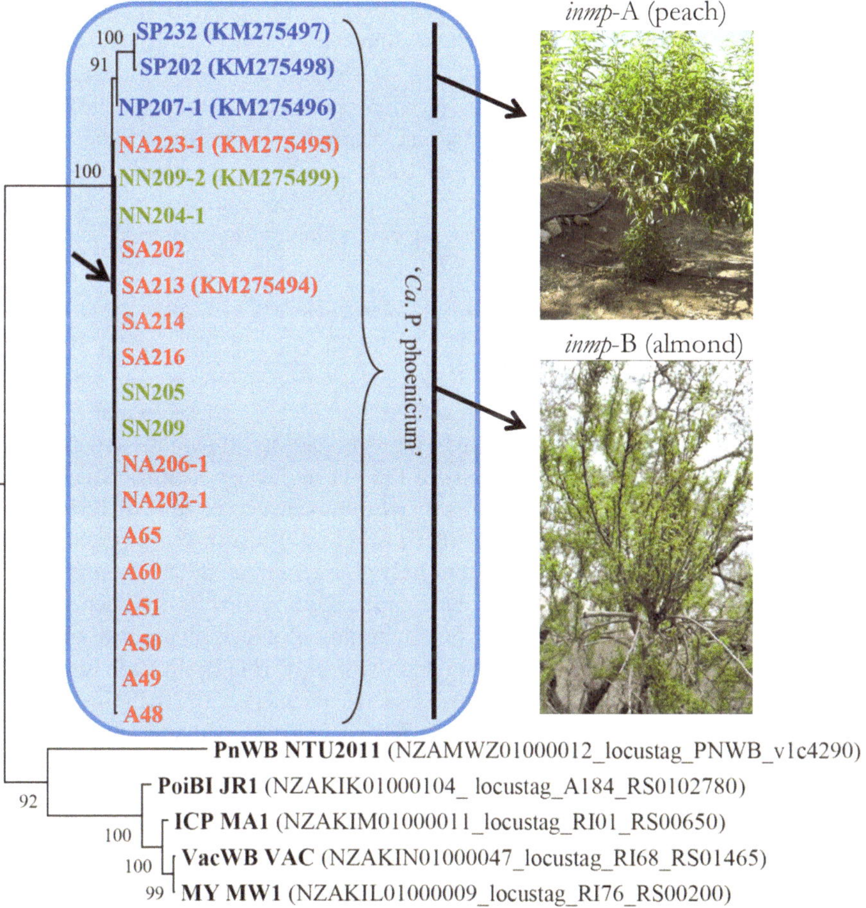

Fig. 9.3 Branch of an unrooted phylogenetic trees inferred from '*Ca.* P. phoenicium' strain nucleotide sequences of gene *inmp*. Minimum evolution analysis was carried out using the neighbour-joining method and bootstrap replicated 1,000 times. '*Ca.* P. phoenicium' strains identified in almond, peach, and nectarine are written in bold red, blue, and green colour, respectively. Names of other phytoplasmas included in the phylogenetic analysis are written on the tree image. GenBank accession number of each sequence is given in parenthesis

to be capable of transmitting AlmWB phytoplasmas. Considering this evidence, it is reasonable to hypothesize the possible implication of *inmp* gene diversity on multiple vector-specific epidemiological cycles of AlmWB phytoplasmas in the diverse plant hosts (Quaglino et al. 2015).

Group 16SrX Phytoplasmas in the ribosomal group 16SrX (apple proliferation) are associated with diseases affecting mainly fruit crops in Europe (Fiore et al. 2018). Based on actual and virtual RFLP analysis of the 16S rDNA fragment amplified using primer pair R16F2n/R16R2, four 16SrX subgroups have been described: 16SrX-A, including phytoplasmas associated with apple proliferation; 16SrX-B, including phytoplasmas associated with European stone fruit yellows; 16SrX-C, including phytoplasmas associated with pear decline; and 16SrX-D, including phytoplasmas associated with spartium witches' broom. According to IRPCM guidelines (2004), phytoplasmas sharing >97.5% 16S rRNA gene nucleotide sequence similarity can be described as separate '*Candidatus*' species if they are clearly distinguished by evident molecular diversity and ecological niche. These latter criteria were employed for delineating '*Ca.* P. mali' (16SrX-A), '*Ca.* P. prunorum' (16SrX-B), and '*Ca.* P. pyri' (16SrX-C), sharing 99% 16S rDNA sequence identity, but clearly distinct based on distinct insect vectors and plant hosts' features (Seemüller and Schneider 2004).

Multiple Gene Analyses for Differentiation Within '*Ca.* P. mali', '*Ca.* P. pyri', and '*Ca.* P. prunorum' Strains Due to the ecological complexity of '*Ca.* P. mali', '*Ca.* P. pyri', and '*Ca.* P. prunorum' biological cycles, including plant hosts and insect vectors, several studies focused on the identification of molecular markers allowing the distinction of strains related to specific biological features and the tracing their route of propagation. Analysis of the nitroreductase- and rhodanese-like genes (PR-1, PR-2, and PR-3 region) proved the existence of at least three '*Ca.* P. mali' genotypes (AT-1, AT-2, and AP15) differently distributed in orchards in southwestern Germany and north-eastern Italy (Cainelli et al. 2004; Jarausch et al. 2004). Further works based on molecular characterization of *rplV-rpsC* genes identified at least six '*Ca.* P. mali' genotypes according to geographical and, in some cases, also with epidemical distribution in northern Italy, Hungary, Czech Republic, Poland, Bulgaria, and Serbia (Martini et al. 2008; Paltrinieri et al. 2010; Casati et al. 2011, Cieślińska et al. 2015; Fránová et al. 2013, 2018). To gain an insight into the genetic diversity among '*Ca.* P. mali' populations in north-western Italy and Bulgaria, a multiple gene analysis was performed on four distinct chromosome segments: two ribosomal (16S/23S rDNA and *rplV-rpsC* genes) and two extra-ribosomal (nitroreductase- and rhodanese-like genes and *secY* gene) regions. Collective RFLP patterns, obtained by multiple gene sequence analyses, revealed the presence of 9 and 12 distinct '*Ca.* P. mali' lineages in Bulgaria and Italy, respectively (Casati et al. 2011; Fránová et al. 2018). Moreover, based on sequence identity and phylogenetic analyses, eight *aceF*, five *pnp*, seven *secY*, and ten *inmp* genotypes were distinguished among '*Ca.* P. mali' strains from European countries (Danet et al. 2011; Križanac et al. 2017; Dermastia et al. 2018). Furthermore, based on the sequence analysis of *hflB* gene, encoding a membrane-associated ATP- and Zn^{2+}-dependent protease, it was possible to distinguish '*Ca.* P. mali' strains with different degrees of virulence (Seemüller et al. 2011). Based on sequence identity and phylogenetic analyses, seven *aceF*, eight *pnp*, five *secY*, and 17 *inmp* genotypes were distinguished among '*Ca.* P. pyri' strains from European countries (Danet et al. 2007, 2011; Jernej et al. 2014; Bohunická et al. 2018), and in '*Ca.* P. prunorum', it was possible to distinguish 11 *aceF* (Fig. 9.4), two *pnp*, three *secY*, and 14 *inmp*

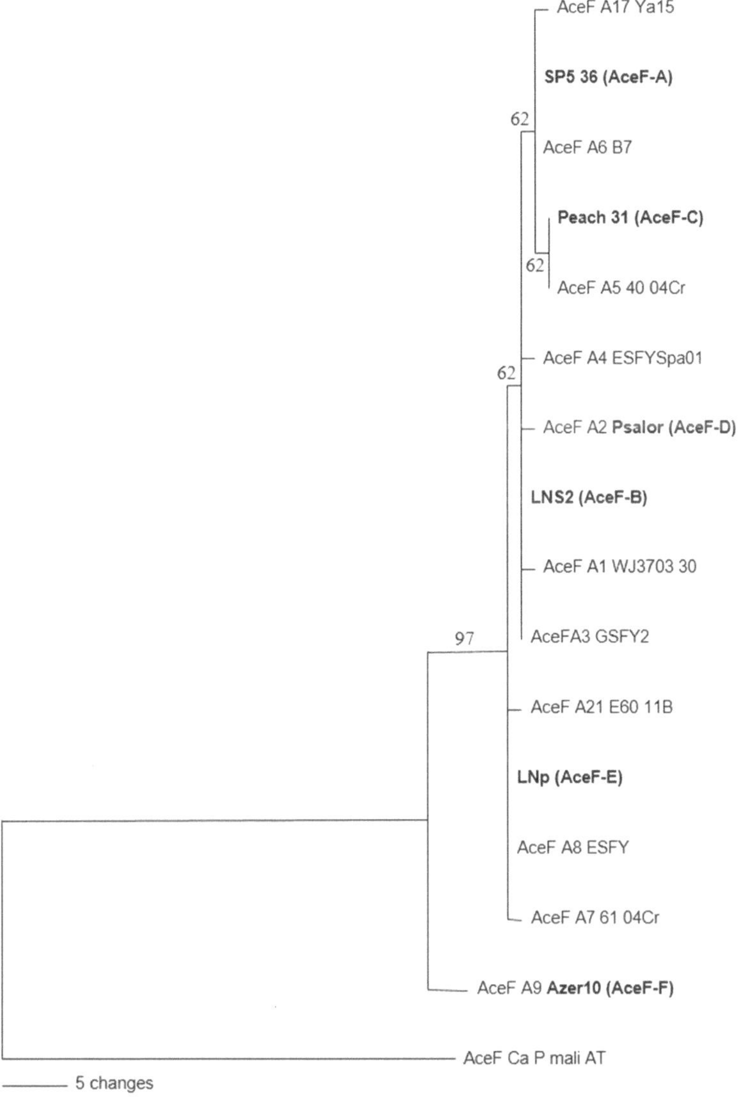

Fig. 9.4 Phylogenetic tree reconstructed by parsimony analyses of *AceF* gene sequences from '*Ca.* P. prunorum' strains described by Danet et al. 2011 and Martini et al. 2010. '*Ca.* P. mali' strain AT was used as the outgroup to root the tree. Branch lengths are proportional to the number of inferred character state transformations and bootstrap values are shown on the main branches

genotypes (Danet et al. 2007, 2011; Martini et al. 2010). This MLST analysis contributed to identify molecular markers specifically harboured by hypovirulent '*Ca.* P. prunorum' and suggested the existence of interspecies recombination between '*Ca.* P. pyri' and '*Ca.* P. prunorum' strains, which can happen in common hosts (*i.e.* peach and *Cacopsylla pyri*) (Danet et al. 2007, 2011). A study on one symptomatic apricot tree and asymptomatic infected wild blackthorn (*Prunus spinosa*) plants in Belorussia allowed to detect '*Ca.* P. prunorum' strains that were diversified based on their *imp* and *hflB* genes (Valasevich and Schneider 2016).

Group 16SrXI Phytoplasmas in the 16SrXI group enclose 16SrXI-A, 16SrXI-B, 16SrXI-C, 16SrXI-D, 16SrXI-E, and 16SrXI-F subgroups and were almost only detected in Asian and in some case in African countries mainly in gramineous plants. The historical representative of the group is the '*Ca.* P. oryzae' that is infecting economically relevant crops such as rice and sugarcane in Asia (Jung et al. 2003). The subgroups detected on these species are 16SrXI-A, 16SrXI-B, 16SrXI-D, and 16SrXI-F: the 16SrXI-B, 16SrXI-D, and 16SrXI-F were mainly detected in some sugarcane phytoplasma diseases and in palm with lethal yellowing-like disease in India (Zhang et al. 2016, Yadav et al. 2017). However, two subgroups, the 16SrXI-C and 16SrXI-E, were detected in insect and dicotyledonous plants, respectively, in Europe. The latter one was also described as '*Ca.* P. cirsii' (Šafářová et al. 2016).

***LeuS*, *secA*, *secY*, *poC*, *gyrA*, *gyrB*, and *dnaB* Genes** Phytoplasmas that infect gramineous plants, and associated with diseases including napier grass stunt, sugarcane white leaf, sugarcane grassy shoot, coconut and arecanut palm yellowing, in southern India and Sri Lanka have been classified into the 16SrXI ribosomal group. However, the 16S rRNA gene gives relatively poor resolution therefore, universal phytoplasma primers that amplify approximately 1 kb of the leucyl transfer RNA synthetase (*leuS*) in nested PCR and validated on a broad range of phytoplasma ribosomal groups (Leufor1: 5′-GAT ATG TTT CCT TAT CCT TCT-3′/Leurev1: 5′-TAC CAA GAR CTT CCW GC-3′ followed by Leufor2: 5′-CAT CCT TTT GGT TGG GAT TCT-3′/Leurev2: 5′-CTS CCC AAT ATC TTT GRC G-3′) were used, along with partial sequences of the *secA* gene to verify and better define the identity of phytoplasmas in group 16SrXI. Based on this data, the sugarcane white leaf and grassy shoot phytoplasmas appear to be the same phytoplasma. The phytoplasmas associated with coconut and arecanut in southern India and Sri Lanka, which are in the same 16SrXI group, appear in different groups based on *secA* gene analysis (Abeysinghe et al. 2016). Moreover, the *secA* sequences of '*Ca.* P cirsii' strains showed 83% identity with the *secA* gene of sugarcane grassy shoot phytoplasmas (strains Maharashtra and Assam, GenBank accession numbers KC347001 and KC347002) and 82% identity with the napier grass stunt phytoplasma (GenBank accession number EU168750), confirming its differences with other members of the same ribosomal group (Šafářová et al. 2016). Pairwise sequence comparison, phylogenetic and *in silico* RFLP analysis of partial 16S rRNA and *secA* gene sequences of eight strains of sugarcane grassy shoot phytoplasmas confirmed the association of '*Ca.* P. oryzae'-related strains (16SrXI-B) with symptomatic sugarcane varieties in India. Multilocus genes such as *secA*, *secY*, *poC*, *gyrA*, *gyrB*, and *dnaB* were also utilized for the characterization of the same phytoplasmas (Kumar et al. 2017; Rao et al. 2018).

Group 16SrXII Phytoplasmas classified in 16S rRNA group 16SrXII ("stolbur") infect a wide range of wild and cultivated plants worldwide and are transmitted by polyphagous plant hoppers of the family Cixiidae. Based on actual and virtual RFLP analysis of the 16S rDNA fragment amplified using primer pair R16F2n/R16R2, ten 16SrXII subgroups (16SrXII-A to 16SrXII-K) have been described (Bertaccini and Lee 2018). Five '*Candidatus* Phytoplasma' species have formally been described within group 16SrXII: '*Ca.* P. australiense' (16SrXII-B), infecting grapevine and other plant hosts in Australia and New Zealand (Davis et al. 1997); '*Ca.* P. japonicum' (16SrXII-D), infecting *Hydrangea* sp. in Japan and *Sophora japonica* in China (Sawayanagi et al. 1999; Duduk et al. 2010); '*Ca.* P. fragariae' (16SrXII-E), infecting strawberry in Europe (Valiunas et al. 2006); '*Ca.* P. convolvuli' (16SrXII-H), infecting bindweed in Europe (Martini et al. 2012); and '*Ca.* P. solani', associated with grapevine "bois noir" (16SrXII-A, 16SrXII-F, 16SrXII-G, 16SrXII-I, 16SrXII-J, 16SrXII-K) and "stolbur" (16SrXII-A) diseases in Europe, Asia, and America (Quaglino et al. 2013).

Multiple Gene Analyses for Differentiation Within '*Ca.* P. solani', '*Ca.* P. australiense, and '*Ca.* P. fragariae' In the Euro-Mediterranean regions, the '*Ca.* P. solani' is transmitted by *Hyalesthes obsoletus* Signoret (Homoptera: Cixiidae), a polyphagous insect vector living preferentially on nettle (*Urtica dioica* L.), bindweed (*Convolvulus arvensis* L.), mugwort (*Artemisia vulgaris* L.), and chaste tree (*Vitex agnus-castus* L.) inside and/or around the agrosystems (Langer and Maixner 2004; Sharon et al. 2005). Generally, the plant crop host (*e.g.* grapevine, potato, lavender, tomato) represents a dead-end host for '*Ca.* P. solani', which is only incidentally transmitted by *H. obsoletus* from other host plants to the crops during its feeding probing (Weintraub and Beanland 2006). Moreover, *Reptalus panzeri* and other insects have been reported as vectors of '*Ca.* P. solani' in Serbia and Italy (Cvrković et al. 2014; Mori et al. 2018). These evidences indicate that this phytoplasma exists in various ecosystems, where selection conceivably alters the strain populations (Quaglino et al. 2013, 2017). This hypothesis implies that ecological relationships can be reflected in intraspecies strain diversity. Thus, in the last years, numerous studies focused on distinguishing genetic structure of '*Ca.* P. solani' strains with the aim to identify strain-specific molecular markers associable to distinct biological features. Such studies were carried out by nucleotide sequence analyses of the genes *tuf* (Schneider et al. 1997), *secY* (Fialová et al. 2009), *vmp1*, and *stamp*, encoding membrane proteins presumably involved in the interaction among the '*Ca.* P. solani' strains and its plant and insect hosts (Cimerman et al. 2009; Fabre et al. 2011; Atanasova et al. 2015; Mitrović et al. 2016; Radonjić et al. 2016; Sémétey et al. 2018). Information obtained from the *tuf* gene characterization allowed the identification of two main *tuf* types associated with herbaceous plant

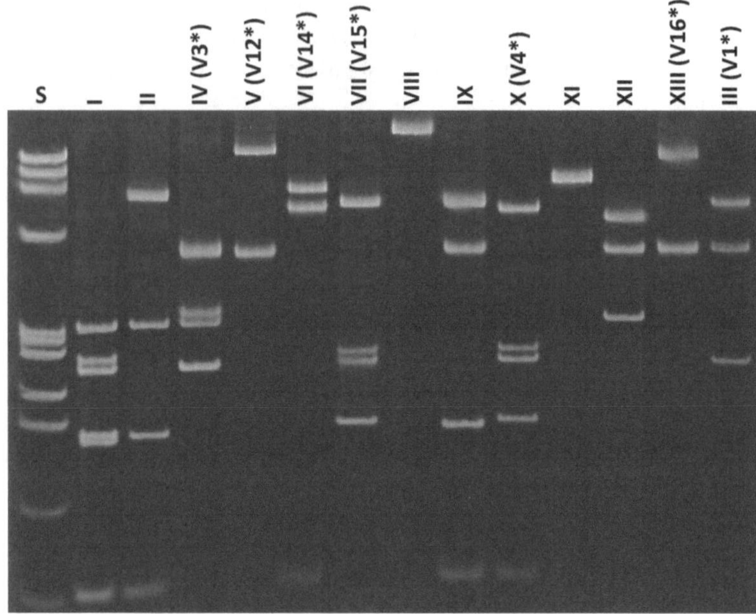

Fig. 9.5 Thirteen *vmp1* RFLP types differentiated by analyses with *Rsa*I of nested PCR products TYPH10F/TYPH10R amplified from more than 70 '*Ca.* P. solani' strains obtained from infected grapevine, crops, and weeds in Friuli Venezia Giulia (Italy) and from reference strains (Martini et al. 2011). Lane S: Φ174 *Hae*III digested. *Tentative affiliation to published vmp1 RFLP types

hosts related to distinct epidemiological systems in Europe (Langer and Maixner 2004; Belli et al. 2010; Maixner 2011): (i) tuf-type a, associated with *U. dioica* and prevalent in western Europe and northern Italy (Quaglino et al. 2013), and (ii) tuf-type b, mainly associated with *C. arvensis* and many other herbaceous hosts, and prevalent in the central-southern Italy (Pacifico et al. 2009; Murolo and Romanazzi 2016). Recently, a different tuf-type b, named tuf-type b2, was associated with a diverse strain detected in epidemic from nettle in Austria (Aryan et al. 2014). Furthermore, the analyses of nucleotide sequences of the genes *secY*, *stamp*, and *vmp1* evidenced a larger variability among "bois noir" strains within the two main tuf types (Foissac et al. 2013; Kostadinovska et al. 2014; Murolo et al. 2010; Murolo and Romanazzi 2015). Currently, 23 Vmp types can be distinguished by *vmp1* gene RFLP profiles (Fig. 9.5).

Moreover, based on sequence identity on *vmp1* (available for 161 '*Ca.* P. solani' strains) and *stamp* (available for 195 strains) gene sequences retrieved from NCBI GenBank, it was possible to determine the presence of 80 *vmp1* (Vmp1 to Vmp80) and 46 *stamp* (St1 to St46) gene sequence variants within the '*Ca.* P. solani' strain populations. The overall ratio between the non-synonymous and the synonymous mutations (dN/dS) was >1.0 for *vmp1* (dN/dS = 4.567; P = 0.000) and *stamp* (dN/dS = 2.436; P = 0.008) genes, indicating a high number of non-silent (dN) mutations.

Based on phylogenetic analysis of concatenated nucleotide sequences of the genes *vmp1* and *stamp* (available for 76 '*Ca.* P. solani' strains), 49 vmp1/stamp sequence variants were grouped in five clusters. The cluster vmp1/stamp-4 included strains (tuf-type a) associated with a nettle-related biological cycle, whilsts the other four clusters (vmp1/stamp-1, vmp1/stamp-2, vmp1/stamp-3, vmp1/stamp-5) included strains (tuf-type b) associated with a bindweed-related biological cycle (Quaglino et al. 2016; Pierro et al. 2018).

Molecular epidemiology approaches, using vmp1- and stamp-based molecular markers, allowed increasing the knowledge of the '*Ca.* P. solani' population structure and dynamics and the '*Ca.* P. solani' transmission routes in the Mediterranean area (Murolo and Romanazzi 2015). Moreover, recent studies reported the direct epidemiological role of *Vitex agnus-castus* in the *H. obsoletus*-mediated transmission of '*Ca.* P. solani' to grapevine (Kosovac et al. 2016) and the vectoring activity of *Reptalus panzeri* able to transmit '*Ca.* P. solani' also from corn, affected by corn reddening disease, to grapevine (Cvrković et al. 2014). The '*Ca.* P. solani' associated with the "bois noir" disease in two Azerbaijanian vineyards were characterized by MLSA on *tuf*, *secY*, *stamp*, and *mleP1* (encoding the 2-hydroxycarboxylate transporter) genes. Three *tuf* and *secY* genotypes were differentiated, but using the *mleP1* gene, five genetic variants were found. Finally, *stamp* gene allowed differentiating four new genotypes in grapevine among the ten new *stamp* genotypes detected in various plant species. Moreover, in *Hyalesthes obsoletus* and *Reptalus noahi*, new '*Ca.* P. solani' genotypes phylogenetically distant from the described genetic clusters were identified (Balakishiyeva et al. 2018).

'*Ca.* P. australiense' is associated with a diverse range of plant species, in which it is associated with a range of symptoms, and is geographically widespread throughout Australia and New Zealand. Sequence analysis of the *tuf* gene and rp operon showed that '*Ca.* P. australiense' strains can be differentiated into four subgroups, strawberry lethal yellows, strawberry green petal, Australian grapevine yellows, pumpkin yellow leaf curl, and cotton bush witches' broom phytoplasmas were designated members of the subgroup tuf-Australia I, rp-A. Strawberry lethal yellows and cotton bush witches' broom phytoplasmas were assigned to the subgroup tuf-New Zealand II, rp-B (Streten and Gibb 2005). Moreover, phylogenetic analysis of *tuf* gene nucleotide sequences from Australian and New Zealand of some '*Ca.* P. australiense' strains, revealed three distinct clades (tuf 1, tuf 2, and tuf 3): one found solely in Australia, one found solely in New Zealand, and a third with representatives from both countries. Based on synonymous substitution rate using a calibration date of 110 million years, it was hypothesized that the three clades diverged from a common ancestor during the Miocene, after the geological separation of New Zealand and Australia (Andersen et al. 2006). Recently, potato diseases with characteristic symptoms of phytoplasma infections were found in potato fields in China. RFLP and nucleotide sequence analysis of 16S rDNA amplicons (R16F2n/R2 fragments) showed that five phytoplasma strains identified in diseased potatoes belong to the species '*Ca.* P. fragariae' (16SrXII-E) and 11 strains were designated as belonging to a new 16SrXII subgroup, 16SrXII-I. The genetic diversity of these strains was corroborated by sequence analysis of their ribosomal protein *tuf*, and *sec*Y genes (Cheng et al. 2015).

Group 16SrXIII The group 16SrXIII was for longtime believed to be restricted to Mexico in periwinkle, but recently several subgroups were defined mainly in South America, and now it encloses seven ribosomal subgroups (16SrXIII-A to 16SrXIII-I). In this group are enclosed '*Ca.* P. hispanicum' from Mexico (16SrXIII-A) and '*Ca.* P. meliae' from Argentina (16SrXIII-G) infecting herbaceous hosts (periwinkle, strawberry) and woody plants, respectively (Davis et al. 2016; Fernández et al. 2016). The Mexican periwinkle phytoplasma group has been found mainly in countries within Latin America, from Mexico to Argentina (Lee et al. 1998; Fernández et al. 2015). Subgroups 16SrXIII-B and 16SrXIII-F have been detected affecting strawberry plants in Florida, and Argentina, respectively (Jomantiene et al. 1998; Fernández et al. 2015); subgroup 16SrXIII-A was detected in periwinkle plants in Mexico (Lee et al. 1998); subgroup 16SrXIII-C was found in chinaberry plants in Bolivia (Harrison et al. 2003); and subgroup 16SrXIII-D was detected in samples collected from potato plants in Mexico (Santos-Cervantes et al. 2010). Recently the diversity within this group has been expanded with the identification of subgroups 16SrXIII-G and 16SrXIII-H (Pérez-López et al. 2016).

***RpoB* and *groEL* Genes** Strawberry plants showing symptoms associated with strawberry green petal disease and from two periwinkle plants showing virescence, sampled in different areas of Mexico, showed the presence of distinct 16S rRNA gene sequences along with a single chaperonin-60 (*cpn60* gene) sequence and a single *rpoB* gene sequence, suggesting that these strains display a 16S rRNA gene sequence heterogeneity, but are in reality infected by the same phytoplasmas (Pérez-López and Dumonceaux 2016).

Group 16SrXIV This ribosomal group encloses subgroups 16SrXIV-A, 16SrXIV-B, and 16SrXIV-C and the '*Ca.* P. cynodontis' (Bertaccini and Lee 2018) infecting mainly the Bermuda grass in Europe and in Asia. It has been reported recently to infect also corn in Turkey where it appears as seed transmitted (Çağlar et al. 2019).

***GroEL* Gene** A system for amplification of fragments containing the '*Ca.* P. cynodontis' *groEL* gene enabled to study its variability in related strains, belonging to different 16SrXIV subgroups. Despite the fact that the *groEL* gene exhibited a greater sequence variation than 16S rRNA gene, the phylogenetic tree based on *groEL* gene sequence analysis was highly congruent with the 16S rDNA-based tree. The *groEL* gene analyses also supported the differentiation of the subgroup 16SrXIV-C, moreover the phylogenetic analyses confirmed the presence of distinct lineages among the studied strains belonging to diverse 16SrXIV subgroups (Mitrovic et al. 2015).

***SecA* and *LeuS* Genes** Assays based on the *secA* gene have been successfully used to detect phytoplasmas associated with diseases in Poaceae species differentiating between the phytoplasmas infecting the Gramineae in Africa and Asia (Bekele et al. 2011). Moreover the phytoplasmas associated with lethal yellowing-like disease of palms in Sri Lanka were shown to have the *secA* gene in the same cluster of

16SrXIV phytoplasmas, whilst the characterization on *leuS* gene allowed to clearly separate the Bermuda grass white leaf phytoplasma (16SrXIV) from those infecting sugarcane in the Indian areas enclosing Sri Lanka classified in diverse 16SrXI subgroups (Abeysinghe et al. 2016).

Group 16SrXXII The 16SrXXII group only encloses phytoplasmas detected in palm with lethal yellowing disease in some regions of Africa; despites the restricted number of plant host species, up to three subgroups were described (16SrXXII-A, 16SrXXII-B, and 16SrXXII-C) with the latter two present only in western Africa (Ghana and Ivory Coast) and the first one resulting present in both the eastern and western Africa (Nigeria and Mozambique) (Harrison et al. 2014, Kra et al. 2017).

***LeuS*and *secA* Genes** The use of these genes is resulting in reliable resolution of geographic differences also in strains from Ghana (CSPWD) and Ivory Coast (CILY) (Dickinson et al. 2019; Arocha-Rosete et al. 2016, 2017). Analysis based on the *secA* and *rp* genes distinguished the CILY phytoplasma strains from others of the same subgroup (16SrXXII-B); in particular, the use of the nested secA PCR in combination with *Mbo*II RFLP clearly distinguished these strains and is a useful alternative to support the epidemiologic screening. The sequencing of the *leuS* gene allowed the differentiation of phytoplasma strains in the Western and Central regions of Ghana from strain in the Volta region where the epidemic of lethal yellowing started (Cape Saint Paul) (Dickinson et al. 2019).

9.4 Conclusions

The studies on housekeeping genes, such as *rp* and *secY*, allowed not only to readily delineate subgroups that are consistent with 16S ribosomal subgroups but also to identify, within some subgroups, additional distinct strains that could not be resolved by the analysis of 16S rRNA gene sequences. In several cases, these housekeeping genes have more phylogenetically informative characters than the 16S rRNA gene, which significantly increase the resolving power for distinguishing genetic closely related phytoplasma strains within a given 16S ribosomal group or '*Candidatus* Phytoplasma' species. However, whilst the work for the first 22 identified ribosomal groups was exploited according to their epidemiologic relevance in some areas of the world, for the remaining ribosomal groups this information, even if potentially very relevant, is still lacking. It is very important to continue to exploit phytoplasma genes for their relevance in molecular epidemiology, paying strong attention to their characteristics of strain genetic conservation in order to avoid producing unnecessary taxa that have just academic relevance. The characterization on alternative genes must be implemented and applied keeping in mind the biological and possibly phytopathological features of the studied phytoplasma strains (Bertaccini 2015) to be useful for the research and management of the most devastating phytoplasma-associated diseases.

References

Abeysinghe S, Abeysinghe PD, Kanatiwela-de Silva C, Udagama P, Warawichanee K, Aljafar N, Praphat K, Dickinson M (2016) Refinement of the taxonomic structure of 16SrXI and 16SrXIV phytoplasmas of gramineous plants using multilocus sequence typing. *Plant Disease* **100**, 2001–2010.

Al-Subhi A, Hogenhout SA, Al-Yahyai RA, Al-Sadi AM (2017) Classification of a new phytoplasmas subgroup 16SrII-W associated with *Crotalaria* witches' broom diseases in Oman based on multigene sequence analysis. *BMC Microbiology* **17**, 221.

Alvarez E, Mejía JF, Llano GA, Loke JB, Calari A, Duduk B, Bertaccini A (2009) Detection and molecular characterization of a phytoplasma associated with frogskin disease in Cassava. *Plant Disease* **93**, 1139–1145.

Alvarez E, Mejia JF, Contaldo N, Paltrinieri S, Duduk B, Bertaccini A (2014) '*Candidatus* Phytoplasma asteris' strains associated with oil palm lethal wilt in Colombia. *Plant Disease* **98**, 311–318.

Andersen MT, Newcomb RD, Liefting LW, Beever RE (2006) Phylogenetic analysis of '*Candidatus* Phytoplasma australiense' reveals distinct populations in New Zealand. *Phytopathology* **96**, 838–845.

Angelini E, Clair D, Borgo M, Bertaccini A, Boudon-Padieu E (2001) "Flavescence dorée" in France and Italy – occurrence of closely related phytoplasma isolates and their near relationships to Palatinate grapevine yellows and an alder yellows phytoplasma. *Vitis* **40**, 79–86.

Angelini E, Constable F, Duduk B, Fiore N, Quaglino F, Bertaccini A (2018) Grapevine phytoplasmas. In: Phytoplasmas: Plant Pathogenic Bacteria-I. Characterization and Epidemiology of Phytoplasma-Associated Diseases. Eds GP Rao, A Bertaccini, N Fiore, L Liefting. Chapter 5. Springer, Singapore, 123–151 pp.

Arnaud G, Malembic-Maher S, Salar P, Maixner M, Marcone C, Boudon-Padieu E, Foissac X (2007) Multilocus sequence typing confirms the close genetic inter-relatedness between three distinct "flavescence dorée" phytoplasma strain clusters and group 16SrV phytoplasmas infecting grapevine and alder in Europe. *Applied and Environmental Microbiology* **73**, 4001–4010.

Arocha-Rosete Y, Diallo HA, Konan Konan JL, Assiri PK, Séka K, Daniel KK, Toualy MN, Koffi EK, Daramcoum MP, Beugré NI, Ouattara VM, Allou K, Fursy-Rodelec ND, Doudjo ON, Yankey N, Dery S, Maharaj A, Saleh M, Summerbell R, Contaldo N, Paltrinieri S, Bertaccini A, Scott J (2016) Detection and identification of the coconut lethal yellowing phytoplasma in weeds growing in coconut farms in Côte d'Ivoire. *Canadian Journal of Plant Pathology* **38**, 164–173.

Arocha-Rosete Y, Atta Diallo H, Konan Konan JL, Yankey N, Saleh M, Pilet F, Contaldo N, Paltrinieri S, Bertaccini A, Scott J (2017) Detection and differentiation of the coconut lethal yellowing phytoplasma in coconut-growing villages of Grand-Lahou, Côte d'Ivoire. *Annals of Applied Biology* **170**, 333–347.

Atanasova B, Spasov D, Jakovljević M, Jović J, Krstić O, Mitrović M, Cvrković T (2014) First report of alder yellows phytoplasma associated with common alder (*Alnus glutinosa*) in the Republic of Macedonia. *Plant Disease* **98**, 1268.

Atanasova B, Jakovljević M, Spasov D, Jović J, Mitrović M, Toševski I, Cvrković T (2015) The molecular epidemiology of "bois noir" grapevine yellows caused by '*Candidatus* Phytoplasma solani' in the Republic of Macedonia. *European Journal of Plant Pathology* **142**, 759–770.

Aryan A, Brader G, Mörtel J, Pastar M, Riedle-Bauer M (2014) An abundant '*Candidatus* Phytoplasma solani' tuf-b strain is associated with grapevine, stinging nettle and *Hyalesthes obsoletus*. *European Journal of Plant Pathology* **140**, 213–227.

Bagadia PG, Polashock J, Bottner-Parker KD, Zhao Y, Davis RE, Lee I-M (2013) Characterization and molecular differentiation of 16SrI-E and 16SrIX-E phytoplasmas associated with blueberry stunt disease in New Jersey. *Molecular and Cellular Probes* **27**, 90–97.

Bai X, Zhang J, Ewing A, Miller SA, Radek AJ, Shevchenko DV, Tsukerman K, Walunas T, Lapidus A, Campbell JW, Hogenhout SA (2006) Living with genome instability: the adaptation of phytoplasmas to diverse environments of their insect and plant hosts. *Journal of Bacteriology* **188**, 3682–3696.

Balakishiyeva G, Bayramova J, Mammadov A, Salar P, Danet J-L, Ember I, Verdin E, Foissac X, Huseynova I (2018) Important genetic diversity of '*Candidatus* Phytoplasma solani'-related strains associated with "bois noir" grapevine yellows and planthoppers in Azerbaijan. *European Journal of Plant Pathology* **151**, 937–946.

Bekele B, Abeysinghe A, Hoat T, Hodgetts J, Dickinson M (2011) Development of specific secA-based diagnostics for the 16SrXI and 16SrXIV phytoplasmas of the Gramineae. *Bulletin of Insectology* **64**(Supplement), S15–S16.

Belli G, Bianco PA, Conti M (2010) Grapevine yellows in Italy: past, present and future. *Journal of Plant Pathology* **92**, 303–326.

Bertaccini A (2015) Phytoplasma research between past and future: what directions? *Phytopathogenic Mollicutes* **5**(1-Supplement), S1–S4.

Bertaccini A, Lee I-M (2018). Phytoplasmas: an update. In: Phytoplasmas: Plant Pathogenic Bacteria-I. Characterization and Epidemiology of Phytoplasma-Associated Diseases. Ed GP Rao, A Bertaccini, N Fiore, L Liefting. Chapter 1. Springer, Singapore, 1–29 pp.

Bila J, Mondjana A, Samils B, Högberg N (2015) High diversity, expanding populations and purifying selection in phytoplasmas causing coconut lethal yellowing in Mozambique. *Plant Pathology* **64**, 597–604.

Bohunická M, Valentová L, Suchá J, Nečas T, Eichmeier A, Kiss T, Cmejla R (2018) Identification of 17 '*Candidatus* Phytoplasma pyri' genotypes based on the diversity of the imp gene sequence. *Plant Pathology* **67**, 971–977.

Botti S, Bertaccini A (2003) Variability and functional role of chromosomal sequences in phytoplasmas of 16SrI-B subgroup (aster yellows and related strains). *Journal of Applied Microbiology* **94**, 103–110.

Botti S, Bertaccini A (2006) FD-related phytoplasmas and their association with epidemic and non epidemic situations in Tuscany (Italy). 15th ICVG, Stellenbosch, South Africa, 3–7 April, 163–164.

Botti S, Bertaccini A (2007) Grapevine yellows in Northern Italy: molecular identification of "flavescence dorée" phytoplasma strains and of "bois noir" phytoplasmas. *Journal of Applied Microbiology* **103**, 2325–2330.

Çağlar BK, Satar S, Bertaccini A, Elbeaino T (2019) Detection and seed transmission of Bermudagrass phytoplasma in maize in Turkey. *Journal of Phytopathology*, **167**, 248–255.

Cai H, Wei W, Davis RE, Chen H, Zhao Y (2008) Genetic diversity among phytoplasmas infecting *Opuntia* species: virtual RFLP analysis identifies new subgroups in the peanut witches' broom phytoplasma group. *International Journal Systematic and Evolutionary Microbiology* **58**, 1448–1457.

Cai H, Wang L, Mu W, Wan Q, Wei W, Davis RE, Chen H, Zhao Y (2016) Multilocus genotyping of a '*Candidatus* Phytoplasma aurantifolia'-related strain associated with cauliflower phyllody disease in China. *Annals of Applied Biology* **169**, 64–74.

Cainelli C, Bisognin C, Vindimian ME, Grando MS (2004) Genetic variability of phytoplasmas detected in the apple growing area of Trentino (North Italy). *Acta Horticulturae* **657**, 425–430.

Casati P, Quaglino F, Stern AR, Tedeschi R, Alma A, Bianco PA (2011) Multiple gene analyses reveal extensive genetic diversity among '*Candidatus* Phytoplasma mali' populations. *Annals of Applied Biology* **158**, 257–266.

Casati P, Jermini M, Quaglino F, Corbani G, Schaerer S, Passera A, Bianco PA, Rigamonti IE (2017) New insights on "flavescence dorée" phytoplasma ecology in the vineyard agro-ecosystem in southern Switzerland. *Annals of Applied Biology* **171**, 37–51.

Cheng M, Dong J, Lee I-M, Bottner-Parker KD, Zhao Y, Davis RE, Laski PJ, Zhang Z, McBeath JH (2015) Group 16SrXII phytoplasma strains, including subgroup 16SrXII-E ('*Candidatus* Phytoplasma fragariae') and a new subgroup, 16SrXII-I, are associated with diseased potatoes (*Solanum tuberosum*) in the Yunnan and Inner Mongolia regions of China. *European Journal of Plant Pathology* **142**, 305–318.

Cieślińska M, Hennig E, Kruczyńska D, Bertaccini A (2015) Genetic diversity of '*Candidatus* Phytoplasma mali' strains in Poland. *Phytopathologia Mediterranea* **54**, 477–487.

Cimerman A, Pacifico D, Salar P, Marzachì C, Foissac X (2009) Striking diversity of *vmp1*, a variable gene encoding a putative membrane protein of the "stolbur" phytoplasma. *Applied Environmental Microbiology* **75**, 2951–2957.

Clark GW, Tillier ER (2010) Loss and gain of GroEL in the *Mollicutes*. *Biochemistry and Cell Biology* **88**, 185–194.

Cvrković T, Jović J, Mitrović M, Krstić O, Toševski I (2014) Experimental and molecular evidence of *Reptalus panzeri* as a natural vector of "bois noir". *Plant Pathology* **63**, 42–53.

Contaldo N, Canel A, Makarova O, Paltrinieri S, Bertaccini A, Nicolaisen M (2011). Use of a fragment of the *tuf* gene for phytoplasma 16Sr group/subgroup differentiation. *Bulletin of Insectology* **64**(Supplement), S45–S46.

Contaldo N, Mejia JF, Paltrinieri S, Calari A, Bertaccini A (2012) Identification and *groEL* gene characterization of green petal phytoplasma infecting strawberry in Italy. *Phytopathogenic Mollicutes* **2**, 59–62.

Contaldo N, Paltrinieri S, Makarova O, Bertaccini A, Nicolaisen M (2015) Q-bank phytoplasma: a DNA barcoding tool for phytoplasma identification. In: Plant Pathology. Humana Press, New York, NewYork, United States of America, 123–135.

Contaldo N., Satta E, Zambon Y, Paltrinieri S, Bertaccini A (2016) Development and evaluation of different complex media for phytoplasma isolation and growth. *Journal of Microbiological Methods* **127**, 105–110.

Daire X, Clair D, Larrue J, Boudon-Padieu E (1997) Detection and differentiation of grapevine yellows phytoplasmas belonging to elm yellows group and to the "stolbur" subgroup by PCR amplification of non-ribosomal DNA. *European Journal of Plant Pathology* **103**, 507–514.

Danet J-L, Bonnet P, Jarausch W, Carraro L, Skoric D, Labonne G, Foissac X (2007) Imp and secY, two new markers for MLST (multilocus sequence typing) in the 16SrX phytoplasma taxonomic group. *Bulletin of Insectology* **60**, 339–340.

Danet J-L, Balakishiyeva G, Cimerman A, Sauvion N, Marie-Jeanne V, Labonne G, Lavina A, Batlle A, Krizanac I, Skoric D, Ermacora P, Ulubas Serce C, Caglayan K, Jarausch W, Foissac X (2011) Multilocus sequence analysis reveals the genetic diversity of European fruit tree phytoplasmas and supports the existence of inter-species recombination. *Microbiology* **157**, 438–450.

Davino S, Calari A, Davino M, Bertaccini A, Bellardi MG (2007) Virescence of tenweeks stock associated to phytoplasma. infection in Sicily. *Bulletin of Insectology* **60**, 279–280.

Davis RE, Dally EL, Gundersen DE, Lee I-M, Habili N (1997) '*Candidatus* Phytoplasma australiense', a new phytoplasma taxon associated with Australian grapevine yellows. *International Journal of Systematic Bacteriology* **47**, 262–269.

Davis RE, Zhao Y, Dally EL, Jomantiene R, Lee I-M, Wei W, Kitajima EW (2012) '*Candidatus* Phytoplasma sudamericanum', a novel taxon, and strain PassWB-Br4, a new subgroup 16SrIII-V phytoplasma, from diseased passion fruit (*Passiflora edulis* f. *flavicarpa* Deg.). *International Journal Systematic and Evolutionary Microbiology* **62**, 984–989.

Davis RE, Zhao Y, Dally EL, Lee I-M, Jomantiene R, Douglas SM (2013) '*Candidatus* Phytoplasma pruni', a novel taxon associated with X-disease of stone fruits, *Prunus* spp.: multilocus characterization based on 16S rRNA, *secY*, and ribosomal protein genes. *International Journal Systematic and Evolutionary Microbiology* **63**, 766–776.

Davis RE, Dally EL, Zhao Y, Lee I-M, Wei W, Wolf TK, Beanland L, LeDoux DG, Johnson DA, Fiola JA, Walter-Peterson H, Dami I, Chien M (2015) Unraveling the etiology of north American grapevine yellows (NAGY): novel NAGY phytoplasma sequevars related to '*Candidatus* Phytoplasma pruni'. *Plant Disease* **99**, 1087–1097.

Davis RE, Harrison NA, Zhao Y, Wei W, Dally EL (2016) '*Candidatus* Phytoplasma hispanicum', a novel taxon associated with Mexican periwinkle virescence disease of *Catharanthus roseus*. *International Journal Systematic and Evolutionary Microbiology* **66**, 3463–3467.

Dickinson M, Hodgetts J (2013) PCR analysis of phytoplasmas based on the *secA* gene. In: Phytoplasma. Humana Press, Totowa, New Jersey, United States of America, 205–215 pp.

Dickinson M, Brown H, Yankey EN, Andoh-Mensah S, Bremang F (2019) Genetic differentiation of the 16SrXXII-B phytoplasmas in Ghana based on the gene leucyl tRNA synthetase gene. *Phytopathogenic Mollicutes* **9**, 195–196.

Dermastia M, Dolanc D, Mlinar P, Mehle N (2018) Molecular diversity of '*Candidatus* Phytoplasma mali' and '*Ca.* P. prunorum' in orchards in Slovenia. *European Journal of Plant Pathology* **152**, 791–800.

Desai AR, Musil KM, Carr AP, Hill JE (2009) Characterization and quantification of feline fecal microbiota using cpn60 sequence-based methods and investigation of animal-to-animal variation in microbial population structure. *Veterinary Microbiology* **137**, 120–128.

Duduk B, Calari A, Paltrinieri S, Duduk N, Bertaccini A (2009) Multigene analysis for differentiation of aster yellows phytoplasmas infecting carrots in Serbia. *Annals of Applied Biology* **154**, 219–229.

Duduk B, Tian JB, Contaldo N, Fan XP, Paltrinieri S, Chen QF, Zhao QF, Bertaccini A (2010) Occurrence of phytoplasmas related to "stolbur" and to '*Candidatus* Phytoplasma japonicum' in woody host plants in China. *Journal of Phytopathology* **158**, 100–104.

Duduk B, Bertaccini A (2011) Phytoplasma classification: taxonomy based on 16S ribosomal gene, is it enough? *Phytopathogenic Mollicutes* **1**, 1–13.

Dumonceaux T, Green M, Hammond C, Pérez-López E, Olivier C (2014) Molecular diagnostic tools for detection and differentiation of phytoplasmas based on chaperonin-60 reveal differences in host plant infection patterns. *Plos One* **9**, e116039.

El-Sisi Y, Omar AF, Sidaros SA, Elsharkawy MM, Foissac X (2018) Multilocus sequence analysis supports a low genetic diversity among '*Candidatus* Phytoplasma australasia' related strains infecting vegetable crops and periwinkle in Egypt. *European Journal of Plant Pathology* **150**, 779–784.

Fabre A, Danet J-L, Foissac X (2011) The "stolbur" phytoplasma antigenic membrane protein gene *stamp* is submitted to diversifying positive selection. *Gene* **472**, 37–41.

Fernández FD, Meneguzzi NG, Guzmán FA, Kirschbaum DS, Conci VC, Nome CF, Conci LR (2015) Detection and identification of a novel 16SrXIII subgroup phytoplasma associated with strawberry red leaf disease in Argentina. *International Journal of Systematic and Evolutionary Microbiology* **65**, 2741–2747.

Fernández FD, Galdeano E, Kornowski MV, Arneodo JD, Conci LR (2016) Description of '*Candidatus* Phytoplasma meliae', a phytoplasma associated with Chinaberry (*Melia azedarach* L.) yellowing in South America. *International Journal of Systematic and Evolutionary Microbiology* **66**, 5244–5251.

Fialová R, Válová P, Balakishiyeva G, Danet J-L, Šafárová D, Foissac X, Navrátil M (2009) Genetic variability of "stolbur" phytoplasma in annual crop and wild plant species in south Moravia. *Journal of Plant Pathology* **91**, 411–416.

Fiore N, Bertaccini A, Bianco PA, Cieślińska M, Ferretti L, Hoat TX, Quaglino F (2018) Fruit crop phytoplasmas. In: Phytoplasmas: Plant Pathogenic Bacteria-I. Characterization and Epidemiology of Phytoplasma-Associated Diseases. Chapter 6. Eds GP Rao, A Bertaccini, N Fiore, L Liefting. Springer, Singapore, 153–190 pp.

Foissac X, Carle P, Fabre A, Salar P, Danet JL, Stolbureuromed Consortium (2013) '*Candidatus* Phytoplasma solani' genome project and genetic diversity in the Euro-Mediterranean basin. 3rd European "Bois Noir" Workshop, Barcelona, Spain, 11–13.

Fránová J, Ludvíková H, Paprštein F, Bertaccini A (2013) Genetic diversity of Czech '*Candidatus* Phytoplasma mali' strains based on multilocus gene analyses. *European Journal of Plant Pathology* **136**, 675–688.

Fránová J, de Sousa E, Mimoso C, Cardoso F, Contaldo N, Paltrinieri S, Bertaccini A (2016) Multigene characterization of a new '*Candidatus* Phytoplasma rubi'-related strain associated with blackberry witches' broom in Portugal. *International Journal of Systematic and Evolutionary Microbiology* **66**, 1438–1446.

Fránová J, Koloniuk I, Lenz O, Sakalieva D (2018) Molecular diversity of '*Candidatus* Phytoplasma mali' strains associated with apple proliferation disease in Bulgarian germplasm collection. *Folia microbiologica* **64**, 373–382.

Galdeano E, Guzmán FA, Fernández F, Conci RG (2013) Genetic diversity of 16SrIII group phytoplasmas in Argentina. Predominance of subgroups 16SrIII-J and B and two new subgroups 16SrIII-W and X. *European Journal of Plant Pathology* **137**, 753–764.

Goh SH, Potter S, Wood JO, Hemmingsen SM, Reynolds RP, Chow AW (1996) HSP60 gene sequences as universal targets for microbial species identification: studies with coagulase-negative staphylococci. *Journal of Clinical Microbiology* **34**, 818–823.

Granata G, Paltrinieri S, Botti S, Bertaccini A (2006) Aetiology of *Opuntia ficus-indica* malformations and stunting disease. *Annals of Applied Biology* **149**, 317–325.

Gundersen DE, Lee I-M, Schaff DA, Harrison NA, Chang CJ, Davis RE, Kinsbury DT (1996) Genomic diversity among phytoplasma strains in 16S rRNA group I (aster yellows and related phytoplasmas) and III (X-disease and related phytoplasmas). *International Journal of Systematic Bacteriology* **46**, 64–75.

Harrison N, Boa E, Carpio M (2003) Characterization of phytoplasmas detected in Chinaberry trees with symptoms of leaf yellowing and decline in Bolivia. *Plant Pathology* **52**, 147–157.

Harrison NA, Davis RE, Oropeza C, Helmick EE, Narváez M, Eden-Green S, Dollet M, Dickinson M (2014) '*Candidatus* Phytoplasma palmicola', associated with a lethal yellowing-type disease of coconut (*Cocos nucifera* L.) in Mozambique. *International Journal of Systematic and Evolutionary Microbiology* **64**, 1890–1899.

Hodgetts J, Boonham N, Mumford R, Harrison N, Dickinson M (2008) Phytoplasma phylogenetics based on analysis of *secA* and 23S rRNA gene sequences for improved resolution of candidate species of '*Candidatus* Phytoplasma'. *International Journal of Systematic and Evolutionary Microbiology* **58**, 1826–1837.

Hodgetts J, Dickinson M (2010) Phytoplasma phylogeny and detection based on genes other than 16S rRNA. In: Phytoplasmas-Genomes, Plant Hosts and Vectors. Eds Weintraub PG, Jones P. CAB International. Wallingford, United Kingdom, 93–113 pp.

Iriti M, Quaglino F, Maffi D, Casatti P, Bianco PA, Faoro F (2008) *Solanum malacoxylon*, a new natural host of "stolbur" phytoplasma. *Journal of Phytopathology* **156**, 8–14.

IRPCM (2004) '*Candidatus* Phytoplasma', a taxon for the wall-less, non-helical prokaryotes that colonise plant phloem and insects. *International Journal of Systematic and Evolutionary Microbiology* **54**, 1243–1255.

Jarausch W, Schwind N, Jarausch B, Krczal G (2004) Analysis of the distribution of apple proliferation phytoplasma subtypes in a local fruit growing region in Southwest Germany. *Acta Horticulturae* **657**, 421–424.

Jernej P, Nataša M, Petra N, Marina D (2014) Molecular diversity of '*Candidatus* Phytoplasma pyri' isolates in Slovenia. *European Journal of Plant Pathology* **139**, 801–809.

Jomantiene R, Davis RE, Maas J, Dally EL (1998) Classification of new phytoplasmas associated with diseases of strawberry in Florida, based on analysis of 16S rRNA and ribosomal protein gene operon sequences. *International Journal of Systematic Bacteriology* **48**, 269–277.

Jović J, Cvrković T, Mitrović M, Petrović A, Krstić O, Krnjajić S, Toševski I (2011) Multigene sequence data and genetic diversity among '*Candidatus* Phytoplasma ulmi' strains infecting *Ulmus* spp. in Serbia. *Plant Pathology* **60**, 356–368.

Jung H-Y, Sawayanagi T, Wongkaew P, Kakizawa S, Nishigawa H, Wei W, Oshima K, Miyata S, Ugaki M, Hibi T, Namba S (2003) '*Candidatus* Phytoplasma oryzae', a novel phytoplasma taxon associated with rice yellow dwarf disease. *International Journal of Systematic and Evolutionary Microbiology* **53**, 1925–1929.

Kakizawa S, Oshima K, Kuboyama T, Nishigawa H, Jung HY, Sawayanagi T, Tsuchizaki T, Miyata S, Ugaki M, Namba S (2001) Cloning and expression analysis of *Phytoplasma* protein translocation genes. *Molecular Plant Microbe Interactions* **14**, 1043–1050.

Kakizawa S, Oshima K, Namba S (2006) Diversity and functional importance of phytoplasma membrane proteins. *Trends in Microbiology* **14**, 254–256.

Katanić Z, Krstin L, Ježić M, Zebec M, Ćurković-Perica M (2016) Molecular characterization of elm yellows phytoplasmas in Croatia and their impact on *Ulmus* spp. *Plant Pathology* **65**, 1430–1440.

Kim KS, Ko KS, Chang MW, Hahn TW, Hong SK, Kook YH (2003) Use of rpoB sequences for phylogenetic study of *Mycoplasma* species. *FEMS Microbiology Letters* **226**, 299–305.

Kosovac A, Radonjić S, Hrnčić S, Krstić O, Toševski I, Jović J (2016) Molecular tracing of the transmission routes of "bois noir" in Mediterranean vineyards of Montenegro and experimental evidence for the epidemiological role of *Vitex agnus-castus* (Lamiaceae) and associated *Hyalesthes obsoletus* (Cixiidae). *Plant Pathology* **65**, 285–298.

Kostadinovska E, Quaglino F, Mitrev S, Casati P, Bulgari D, Bianco PA (2014) Multiple gene analyses identify distinct "bois noir" phytoplasma genotypes in the Republic of Macedonia. *Phytopathologia Mediterranea* **53**, 491–501.

Kra KD, Toualy MN, Kouamé AEP, Séka K, Kwadjo KE, Diallo HA, Bertaccini A, Arocha Rosete A (2017) New phytoplasma subgroup identified from Arecaceae palm species in Grand-Lahou, Côte d'Ivoire. *Canadian Journal of Plant Pathology* **39**, 297–306.

Križanac I, Plavec J, Budinšćak Ž, Ivić D, Škorić D, Šeruga-Musić M (2017) Apple proliferation disease in Croatian orchards: a molecular characterization of '*Candidatus* Phytoplasma mali'. *Journal of Plant Pathology* **99**, 95–101.

Kumar S, Jadon VS, Rao GP (2017) Use of *secA* gene for characterization of phytoplasmas associated with sugarcane grassy shoot disease in India. *Sugar Tech* **19**, 632–637.

Lai F, Song CS, Ren ZG, Lin CL, Xu QC, Li Y, Piao CG, Yu SS, Guo MW, Tian GZ (2014) Molecular characterization of a new member of the 16SrV group of phytoplasma associated with *Bischofia polycarpa* (Levl.) Airy Shaw witches' broom disease in China by a multiple gene-based analysis. *Australasian Plant Pathology* **43**, 557–569.

Langer M, Maixner M (2004) Molecular characterization of grape-vine yellows associated phytoplasmas of the "stolbur"-group based on RFLP-analysis of non-ribosomal DNA. *Vitis* **43**, 191–199.

Lee I-M, Gundersen-Rindal DE, Davis RE, Bartoszyk IM (1998) Revised classification scheme of phytoplasmas based on RFLP analyses of 16S rRNA and ribosomal protein gene sequences. *International Journal of Systematic and Evolutionary Microbiology* **48**, 1153–1169.

Lee I-M, Martini M, Bottner KD, Dane RA, Black MC, Troxclair N (2003) Ecological Implications from a molecular analysis of phytoplasmas involved in an aster yellows epidemic in various crops in Texas. *Phytopathology* **93**, 1368–1377.

Lee I-M, Gundersen-Rindal D, Davis RE, Bottner KD, Marcone C, Seemüller E (2004a) '*Candidatus* Phytoplasma asteris', a novel taxon associated with aster yellows and related diseases. *International Journal of Systematic and Evolutionary Microbiology* **54**, 1037–1048.

Lee I-M, Martini M, Marcone C, Zhu SF (2004b) Classification of phytoplasma strains in the elm yellows group (16SrV) and proposal of '*Candidatus* Phytoplasma ulmi' for the phytoplasma associated with elm yellows. *International Journal of Systematic and Evolutionary Microbiology* **54**, 337–347.

Lee I-M, Zhao Y, Bottner KD (2006) *SecY* gene sequence analysis for finer differentiation of diverse strains in the aster yellows phytoplasma group. *Molecular and Cellular Probes* **20**, 87–91.

Lee I-M, Bottner-Parker KD, Zhao Y, Davis RE, Harrison NA (2010) Phylogenetic analysis and delineation of phytoplasmas based on *secY* gene sequences. *International Journal of Systematic and Evolutionary Microbiology* **60**, 2887–2897.

Lee I-M, Bottner-Parker KD, Zhao Y, Bertaccini A, Davis RE (2012) Differentiation and classification of phytoplasmas in the pigeon pea witches' broom group (16SrIX): an update based on multiple gene sequence analysis. *International Journal of Systematic and Evolutionary Microbiology* **62**, 2279–2285.

Li Y, Piao CG, Tian GZ, Liu ZX, Guo MW, Lin CL, Wang XZ (2014) Multilocus sequences confirm the close genetic relationship of four phytoplasmas of peanut witches' broom group 16SrII-A. *Journal of Basic Microbiology* **54**, 818–827.

Lim PO, Sears BB (1991) DNA sequence of the ribosomal protein genes *rpl2* and *rps19* from a plant-pathogenic mycoplasma-like organism. *FEMS Microbiology Letters* **84**, 71–74.

Lim PO, Sears BB (1992) Evolutionary relationship of plant pathogenic mycoplasma-like organism and *Acholeoplasma laidlawii* deduced from two ribosomal protein gene sequences. *Journal of Bacteriology* **174**, 2606–2611.

Links MG, Dumonceaux TJ, Hemmingsen SM, Hill JE (2012) The chaperonin-60 universal target is a barcode for bacteria that rnables *de novo* assembly of metagenomic sequence data. *Plos One* **7**, e49755.

Makarova OV, Contaldo N, Paltrinieri S, Kawube G, Bertaccini A, Nicolaisen M (2012) DNA barcoding for universal identification of '*Candidatus* Phytoplasmas' using a fragment of the elongation factor Tu gene. *Plos One* **7**, e52092.

Makarova O, Contaldo N, Paltrinieri S, Bertaccini A, Nyskjold H, Nicolaisen M (2013) DNA barcoding for phytoplasma identification. In: Phytoplasma. Humana Press, Totowa, New Jersey, United States of America, 301–317 pp.

Maixner M (2011) Recent advances in "bois noir" research. *Petria* **21**, 95–108.

Malembic-Maher S, Salar P, Filippin L, Carle P, Angelini E, Foissac X (2011) Genetic diversity of European phytoplasmas of the 16SrV taxonomic group and proposal of '*Candidatus* Phytoplasma rubi'. *International Journal of Systematic and Evolutionary Microbiology* **61**, 2129–2134.

Marcone C, Lee I-M, Davis RE, Ragozzino A, Seemüller E (2000) Classification of aster yellows-group phytoplasmas based on combined analyses of rRNA and *tuf* gene sequences. *International Journal of Systematic and Evolutionary Microbiology* **50**, 1703–1713.

Marcone C, Franco-Lara L, Tosevski I (2018) Major phytoplasma diseases of forest and urban trees. In: Phytoplasmas: Plant Pathogenic Bacteria-I. Characterization and Epidemiology of Phytoplasma-Associated Diseases. Chapter 10. Eds GP Rao, A Bertaccini, N Fiore, L Liefting. Springer, Singapore, 298–312 pp.

Martini M, Murari E, Mori N, Bertaccini A (1999) Identification and epidemic distribution of two "flavescence dorée"-related phytoplasmas in Veneto (Italy). *Plant Disease* **83**, 925–930.

Martini M, Botti S, Marcone C, Marzachì C, Casati P, Bianco PA, Benedetti R, Bertaccini A (2002) Genetic variability among "flavescence dorée" phytoplasmas from different origins in Italy and France. *Molecular and Cellular Probes* **16**, 197–208.

Martini M, Lee I-M, Bottner KD, Zhao Y, Botti S, Bertaccini A, Harrison NA, Carraro L, Marcone C, Khan J, Osler R (2007) Ribosomal protein gene-based phylogeny for finer differentiation and classification of phytoplasmas. *International Journal of Systematic and Evolutionary Microbiology* **57**, 2037–2051.

Martini M, Ermacora P, Falginella L, Loi N, Carraro L (2008) Molecular differentiation of '*Candidatus* Phytoplasma mali' and its spreading in Friuli Venezia Giulia region (North-East Italy). *Acta Horticulturae* **781**, 395–402.

Martini M, Ferrini F, Danet J-L, Ermacora P, Sertkaya G, Delić D, Loi N, Foissac X, Carraro L (2010) PCR/RFLP-based method for molecular characterization of '*Candidatus* Phytoplasma prunorum' strains using the *aceF* gene. *Julius-Kühn-Archiv* **427**, 386–391.

Martini M, Ermacora P, Tosone N, Loschi A, Pavan F, Loi N (2011) Genetic diversity of "stolbur" phytoplasma strains from different host plants in Friuli Venezia Giulia. *Petria* **21**, 148–149.

Martini M, Marcone C, Mitrović J, Maixner M, Delić D, Myrta A, Ermacora P, Bertaccini A, Duduk B (2012) '*Candidatus* Phytoplasma convolvuli', a new phytoplasma taxon associated with bindweed yellows in four European countries. *International Journal of Systematic and Evolutionary Microbiology* **62**, 2910–2915.

Mitrovic J, Kakizawa S, Duduk B, Oshima K, Namba S, Bertaccini A (2011) The *cpn60* gene as an additional marker for finer differentiation of '*Candidatus* Phytoplasma asteris'-related strains. *Annals of Applied Biology* **159**, 41–48.

Mitrovic J, Smiljković M, Seemüller E, Reinhardt R, Hüttel B, Büttner C, Bertaccini A, Kube M, Duduk B (2015) Differentiation of '*Candidatus* Phytoplasma cynodontis' based on 16S rRNA and *groEL* genes and identification of a new subgroup, 16SrXIV-C. *Plant Disease* **99**, 1578–1583.

Mitrović M, Jakovljević M, Jović J, Krstić O, Kosovac A, Trivellone V, Tosevski I, Cvrković T (2016) '*Candidatus* Phytoplasma solani' genotypes associated with potato "stolbur" in Serbia and the role of *Hyalesthes obsoletus* and *Reptalus panzeri* (Hemiptera, Cixiidae) as natural vectors. *European Journal of Plant Pathology* **144**, 619–630.

Miyazaki A, Shigaki T, Koinuma H, Iwabuchi N, Rauka G, Kembu A, Saul J, Watanabe K, Nijo T, Maejima K, Yamaji Y, Namba S (2018) '*Candidatus* Phytoplasma novoguineense', a novel taxon associated with Bogia coconut syndrome and banana wilt disease on the island of New Guinea. *International Journal of Systematic and Evolutionary Microbiology* **68**, 170–175.

Mori N, Quaglino F, Sanna F, Filisetti S, Faccincani M, Bianco PA (2018) Potential role of *Euscelis incisus* Kirschbaum and *Dicranotropis hamata* Boheman in the transmission of '*Candidatus* Phytoplasma solani' to grapevine. 19th Meeting of ICVG, Santiago, Chile April 9–12, 92–93.

Murolo S, Romanazzi G (2015) In-vineyard population structure of '*Candidatus* Phytoplasma solani' using multilocus sequence typing analysis. *Infection, Genetics and Evolution* **31**, 221–230.

Murolo S, Romanazzi G (2016) Multilocus sequence analysis as a powerful tool to monitor molecular epidemiology of '*Candidatus* Phytoplasma solani' at vineyard scale. *Mitteilungen Klosterneuburg* **66**, 40–73.

Murolo S, Marcone C, Prota V, Garau R, Foissac X, Romanazzi G (2010) Genetic variability of the "stolbur" phytoplasma *vmp1* gene in grapevines, bindweeds and vegetables. *Journal of Applied Microbiology* **109**, 2049–2059.

Myrie W, Oropeza C, Sáenz L, Harrison NA, Roca MM, Córdova I, Ku S, Douglas L (2011) Reliable improved molecular detection of coconut lethal yellowing phytoplasma and reduction of associated disease through field management strategies. *Bulletin of Insectology* **64**(Supplement), S203–S204.

Oshima K, Kakizawa S, Nishigawa H, Jung HY, Wei W, Suzuki S, Arashida R, Nakata D, Miyata S, Ugaki M, Namba S (2004) Reductive evolution suggested from the complete genome sequence of a plant-pathogenic phytoplasma. *Nature Genetics* **36**, 27–29.

Pacifico D, Foissac X, Veratti F, Marzachì C (2007) Genetic diversity of Italian and French "bois noir" phytoplasma isolates. *Bulletin of Insectology* **60**, 345–346.

Pacifico D, Alma A, Bagnoli B, Foissac X, Pasquini G, Tessitori M, Marzachì C (2009) Characterization of "bois noir" isolates by restriction fragment length polymorphism of a "stolbur"-specific putative membrane protein gene. *Phytopathology* **99**, 711–715.

Paltrinieri S, Duduk B, Dal Molin F, Mori N, Comerlati G, Bertaccini A (2010) Molecular characterization of '*Candidatus* Phytoplasma mali' strains in outbreaks of apple proliferation in north eastern Italy, Hungary, and Serbia. *Julius-Kühn-Archiv* **427**, 178–182.

Paltrinieri S, Contaldo N, Duduk B, Bertaccini A (2012) Strain differentiation in "flavescence dorée" phytoplasmas on *secY* and *tuf* genes. 17th Meeting of ICVG, Davis California, USA, October 7–14, 236–237.

Paredes-Tomás C, Satta E, Paltrinieri S, Oropeza Salín C, Myrie W, Bertaccini A Maritza Luis-Pantoja, (2019) '*Candidatus* Phytoplasma' species detection in coconuts in Cuba. *Phytopathogenic Mollicutes* **9**, 191–192.

Pérez-López E, Dumonceaux TJ (2016) Detection and identification of the heterogeneous novel subgroup 16SrXIII-(A/I) I phytoplasma associated with strawberry green petal disease and Mexican periwinkle virescence. *International Journal of Systematic and Evolutionary Microbiology* **66**, 4406–4415.

Pérez-López E, Luna-Rodríguez M, Olivier CY, Dumonceaux TJ (2016) The underestimated diversity of phytoplasmas in Latin America. *International Journal of Systematic and Evolutionary Microbiology* **66**, 492–513.

Pérez-López E, Wei W, Wang J, Davis RE, Luna-Rodríguez M, Zhao Y (2017) Novel phytoplasma strains of X-disease group unveil genetic markers that distinguish North American and South

American geographic lineages within subgroups 16SrIII-J and 16SrIII-U. *Annals of Applied Biology* **171**, 405–416.

Pierro R, Passera A, Panattoni A, Casati P, Luvisi A, Rizzo D, Bianco PA, Quaglino F, Materazzi A (2018) Molecular typing of "bois noir" phytoplasma strains in the Chianti Classico area (Tuscany, central Italy) and their association with symptom severity in *Vitis vinifera* "Sangiovese". *Phytopathology* **108**, 362–373.

Quaglino F, Zhao Y, Bianco PA, Wei W, Casati P, Durante G, Davis RE (2009) New 16Sr subgroups and distinct single nucleotide polymorphism lineages among grapevine "bois noir" phytoplasma populations. *Annals of Applied Biology* **154**, 279–289.

Quaglino F, Casati P, Bianco PA (2010) Distinct rpsC single nucleotide polymorphism lineages of "flavescence dorée" subgroup 16SrV-D phytoplasma co-infect *Vitis vinifera* L. *Folia Microbiologica* **55**, 251–257.

Quaglino F, Zhao Y, Casati P, Bulgari D, Bianco PA, Wei W, Davis RE (2013) '*Candidatus* Phytoplasma solani', a novel taxon associated with "stolbur"- and "bois noir"-related diseases of plants. *International Journal of Systematic and Evolutionary Microbiology* **63**, 2879–2894.

Quaglino F, Kube M, Jawhari M, Abou-Jawdah Y, Siewert C, Choueiri E, Sobh H, Casati P, Tedeschi R, Molino Lova M, Alma A, Bianco PA (2015) '*Candidatus* Phytoplasma phoenicium' associated with almond witches' broom disease: from draft genome to genetic diversity among strain populations. *BMC Microbiology* **15**, 148.

Quaglino F, Maghradze D, Casati P, Chkhaidze N, Lobjanidze M, Ravasio A, Passera, Venturini AG, Failla O, Bianco PA (2016) Identification and characterization of new '*Candidatus* Phytoplasma solani' strains associated with "bois noir" disease in *Vitis vinifera* L. cultivars showing a range of symptoms severity in Georgia, the Caucasus region. *Plant Disease* **100**, 904–915.

Quaglino F, Murolo S, Zhao Y, Casati P, Durante G, Wei W, Bianco PA, Romanazzi G, Davis RE (2017) Identification of new -J and -K 16SrXII subgroups and distinct single nucleotide polymorphism genetic lineages among '*Candidatus* Phytoplasma solani' strains associated with "bois noir" in Central Italy. *Australasian Plant Pathology* **46**, 31–34.

Radonjić S, Hrnčić S, Kosovac A, Krstić O, Mitrović M, Jović J, Toševski I (2016) First report of '*Candidatus* Phytoplasma solani' associated with potato "stolbur" disease in Montenegro. *Plant Disease* **100**, 1775.

Rao GP, Alvarez E, Yadav A. 2018. Phytoplasma diseases of industrial crops. In: Phytoplasmas: Plant Pathogenic Bacteria-I. Characterization and Epidemiology of Phytoplasma-Associated Diseases. Chapter 4. Eds GP Rao, A Bertaccini, N Fiore, L Liefting. Springer, Singapore, 91–121 pp.

Ren ZG, Lin CL, Li Y, Song CS, Wang XZ, Piao CG, Tian GZ (2014) Comparative molecular analyses of phytoplasmas infecting *Sophora japonica* cv. golden and *Robinia pseudoacacia*. *Journal of Phytopathology* **162**, 98–106.

Riolo P, Landi L, Nardi S, Isidoro N (2007) Relationships among *Hyalesthes obsoletus*, its herbaceous host plants and "bois noir" phytoplasma strains in vineyard ecosystems in the Marche region (central-eastern Italy). *Bulletin of Insectology* **60**, 353–354.

Šafářová D, Zemánek T, Válová P, Navrátil M (2016) '*Candidatus* Phytoplasma cirsii', a novel taxon from creeping thistle [*Cirsium arvense* (L.) Scop]. *International Journal of Systematic and Evolutionary Microbiology* **66**, 1745–1753.

Salehi M, Salehi E, Siampour M, Quaglino F, Bianco PA (2018) Apricot yellows associated with '*Candidatus* Phytoplasma phoenicium' in Iran. *Phytopathologia Mediterranea* **57**, 269–283.

Santos-Cervantes ME, Chávez-Medina JA, Acosta-Pardini J, Flores-Zamora GL, Méndez-Lozano J, Leyva-López NE (2010) Genetic diversity and geographical distribution of phytoplasmas associated with potato purple top disease in Mexico. *Plant Disease* **94**, 388–395.

Satta E, Contaldo N, Mejia JF, Paltrinieri S, Bertaccini A, Bellardi MG (2013) Characterization on six genes of '*Candidatus* Phytoplasma asteris'-related phytoplasmas infecting cyclamen. *Phytopathogenic Mollicutes* **3**, 72–76.

Sawayanagi T, Horikoshi N, Kanheira T, Shinhoara M, Bertaccini A, Cousin M-T, Hiruki C, Namba S (1999) '*Candidatus* Phytoplasma japonicum', a new phytoplasma taxon associated with Japanese Hydrangea phyllody. *International Journal of Systematic Bacteriology* **49**, 1275–1285.

Schneider B, Gibb KS, Seemüller E (1997) Sequence and RFLP analysis of the elongation factor Tu gene used in differentiation and classification of phytoplasmas. *Microbiology* **143**, 3381–3389.

Schneider B, Seemüller E (2009) Strain differentiation of '*Candidatus* Phytoplasma mali' by SSCP- and sequence analyses of the *hflB* gene. *Journal of Plant Pathology* **91**, 103–112.

Seemüller E, Schneider B (2004) Taxonomic description of '*Candidatus* Phytoplasma mali' sp. nov., '*Candidatus* Phytoplasma pyri' sp. nov. and '*Candidatus* Phytoplasma prunorum' sp. nov., the causal agents of apple proliferation, pear decline and European stone fruit yellows, respectively. *International Journal of Systematic and Evolutionary Microbiology* **54**, 1217–1226.

Seemüller E, Kampmann M, Kiss E, Schneider B (2011) *HflB* gene-based phytopathogenic classification of '*Candidatus* Phytoplasma mali' strains and evidence that strain composition determines virulence in multiply infected apple trees. *Molecular Plant-Microbe Interactions* **24**, 1258–1266.

Sémétey O, Gaudin J, Danet J-L, Salar P, Theil S, Fontaine M, Krausz M, Chaisse E, Eveillard S, Verdin E, Foissac X (2018) Lavender decline in France is associated with chronic infection by lavender-specific strains of '*Candidatus* Phytoplasma solani'. *Applied Environmental Microbiology* **84**, e01507–18.

Shao J, Jomantiene R, Dally EL, Zhao Y, Lee I-M, Nuss DL, Davis RE (2006) Phylogeny and characterization of phytoplasmal NusA and use of the *nusA* gene in detection of group 16SrI strains. *Journal of Plant Pathology* **88**, 193–201.

Sharon R, Soroker V, Wesley SD, Zahavi T, Harari A, Weintraub PG (2005) *Vitex agnus-castus* is a preferred host plant for *Hyalesthes obsoletus*. *Journal of Chemical Ecology* **31**, 1051–1063.

Streten C, Gibb KS (2005) Genetic variation in '*Candidatus* Phytoplasma australiense'. *Plant Pathology* **54**, 8–14.

Tolu G, Botti S, Garau R, Prota VA, Sechi A, Prota U, Bertaccini A (2006) Identification of 16SrII-E phytoplasmas in *Calendula arvensis* L., *Solanum nigrum* L. and *Chenopodium* spp.. *Plant Disease* **90**, 325–330.

Valasevich N, Schneider B (2016) Detection, identification and molecular diversity of '*Candidatus* Phytoplasma prunorum' in Belarus. *Journal of Plant Pathology* **98**, 625–629.

Valiunas D, Staniulis J, Davis RE (2006) '*Candidatus* Phytoplasma fragariae', a novel phytoplasma taxon discovered in yellows diseased strawberry, *Fragaria* x *ananassa*. *International Journal of Systematic and Evolutionary Microbiology* **56**, 277–281.

Valiunas D, Jomantiene R, Davis RE (2013) Evaluation of the DNA-dependent RNA polymerase β-subunit gene (*rpoB*) for phytoplasma classification and phylogeny. *International Journal of Systematic and Evolutionary Microbiology* **63**, 3904–3914.

Verdin E, Salar P, Danet J-L, Choueiri E, Jreijiri F, El Zammar S, Gèlie B, Bové J, Garnier M (2003) '*Candidatus* Phytoplasma phoeniceum', a new phytoplasma associated with an emerging lethal disease of almond trees in Lebanon and Iran. *International Journal of Systematic and Evolutionary Microbiology* **53**, 833–838.

Vermette CJ, Russell AH, Desai AR, Hill JE (2010) Resolution of phenotypically distinct strains of *Enterococcus* spp. in a complex microbial community using cpn60 universal target sequencing. *Microbial Ecology* **59**, 14–24.

Wang J, Liu Q, Wei W, Davis RE, Tan Y, Lee I-M, Zhu D, Wei H, Zhao Y (2018) Multilocus genotyping identifies a highly homogeneous phytoplasma lineage associated with sweet cherry virescence disease in China and its carriage by an erythroneurine leafhopper. *Crop Protection* **106**, 13–22.

Wei W, Davis RE, Lee I-M, Zhao Y (2007) Computer-simulated RFLP analysis of 16S rRNA genes: identification of ten new phytoplasma groups. *International Journal of Systematic and Evolutionary Microbiology* **57**, 1855–1867.

White DT, Blackall LL, Scott PT, Walsh KB (1998) Phylogenetic positions of phytoplasmas associated with dieback, yellow crinkle and mosaic diseases of papaya, and their proposed inclusion in '*Candidatus* Phytoplasma australiense' and a new taxon, '*Candidatus* Phytoplasma australasia'. *International Journal of Systematic Bacteriology* **48**, 941–951.

Win NKK, Lee S-Y, Bertaccini A, Namba S, Jung H-Y (2013) '*Candidatus* Phytoplasma balanitae' associated with witches' broom disease of *Balanites triflora*. *International Journal of Systematic and Evolutionary Microbiology* **63**, 636–640.

Weintraub PG, Beanland L (2006) Insect vectors of phytoplasmas. *Annual Review of Entomology* **51**, 91–111.

Yadav A, Thorat V, Deokule S, Shouche Y, Prasad DT (2017) New subgroup 16SrXI-F phytoplasma strain associated with sugarcane grassy shoot (SCGS) disease in India. *International Journal of Systematic and Evolutionary Microbiology* **67**, 374–378.

Zambon Y, Canel A, Bertaccini A, Contaldo N (2018) Molecular diversity of phytoplasmas associated with grapevine yellows disease in north-eastern Italy. *Phytopathology* **108**, 206–214.

Zhang RY, Li WF, Huang YK, Wang XY, Shan HL, Luo Z-M, Yin J (2016) Group 16SrXI phytoplasma strains, including subgroup 16SrXI-B and a new subgroup, 16SrXI-D, are associated with sugarcane white leaf. *International Journal of Systematic and Evolutionary Microbiology* **66**, 487–491.

Zreik L, Carle P, Bové J-M, Garnier M (1995) Characterization of the mycoplasmalike organism associated with witches' broom disease of lime and proposition of a '*Candidatus*' taxon for the organism, '*Candidatus* Phytoplasma aurantifolia'. *International Journal of Systematic Bacteriology* **45**, 449–453.

Chapter 10
Host Metabolic Interaction and Perspectives in Phytoplasma Research

Govind Pratap Rao, Ramaswamy Manimekalai, Manish Kumar, Hemavati Ranebennur, Shigeyuki Kakizawa, and Assunta Bertaccini

Abstract Phytoplasmas are plant pathogenic bacteria that have large economic impacts on crops and landscape plants. Knowledge of their biology is limited also because they are still not easily cultured in media. It is still a mystery how phytoplasmas use the sugar-rich phloem sap and how they interact with the hosts. It is agriculturally important to identify the factors involved in their pathogenicity and to discover effective measures to control phytoplasma-associated diseases. The knowledge about host-pathogen interaction during the infection process can help to elucidate the processes leading to symptom expression. Transcriptomics studies paved the way for analysing the gene expression pattern in phytoplasma-infected plants and revealed the up-regulation of genes responsible for hormonal balance, transcription factors, and signalling. Recent studies have identified potential virulence factors that induce some of the typical phytoplasma disease symptoms and have started the annotation of their genomes having unique reductive evolution features. The novel manipulation tool represented by the potential of the synthetic biology can be helpful for its potential application in studying efficient management strategies to reduce the agricultural impact of the diseases associated with the phytoplasma presence.

Keywords Host-pathogen interactions · Carbohydrate metabolism · Protein metabolism · Molecular responses · Transcriptomics · Synthetic approach

G. P. Rao (✉) · M. Kumar · H. Ranebennur
Plant Virology Unit, Division of Plant Pathology, ICAR-Indian Agricultural Research Institute, New Delhi, India

R. Manimekalai
Sugarcane Breeding Institute, Coimbatore, India

S. Kakizawa
National Institute of Advanced Industrial Science and Technology, Tsukuba, Ibaraki, Japan

A. Bertaccini
Department of Agricultural and Food Sciences, *Alma Mater Studiorum* – University of Bologna, Bologna, Italy

A. Bertaccini et al. (eds.), *Phytoplasmas: Plant Pathogenic Bacteria - III*,
https://doi.org/10.1007/978-981-13-9632-8_10

10.1 Introduction

Phytoplasmas are responsible for several hundred plant diseases worldwide, including many diseases of economically important plants. Little is known about the underlying molecular mechanisms for the symptoms observed in the host plants. However, results emerging from whole-genome sequencing projects provide an initial insight into the physiology and biological requirements of these bacteria. Phytoplasmas colonize the phloem tissues of plant hosts and are transmitted by insect vectors and propagation materials enclosing seeds (Bertaccini et al. 2019). Although the unique features of phytoplasmas have long made them a subject of interest and of numerous studies, the difficulty of their cultivation in artificial media (Contaldo et al. 2012, 2016, 2019) has hindered their biological and functional characterization to a great extent. In the past decade, whole-genome sequences have been completed for several phytoplasma strains (Oshima et al. 2004; Bai et al. 2006; Kube et al. 2008; Tran-Nguyen et al. 2008; Andersen et al. 2013; Orlovskis et al. 2017; Wang et al. 2018a) enabling some better understanding of the molecular mechanism underlying virulence and host interaction (Bai et al. 2009). Recently the synthetic biology offers constructive approaches for biological researches enabling to reconstruct and understand biological phenomena or to express a variety of organic compounds by adding whole metabolic pathway genes into some model organisms. In this chapter the progress in phytoplasma research with a major emphasis on phytoplasma-host metabolism interaction and possible application of synthetic biology tools for future research are summarized.

10.2 General Phytoplasma Metabolic Knowledge

Although the phytoplasma genome contains genes for basic cellular functions such as DNA replication, transcription, translation, and protein translocation (Kakizawa et al. 2001; Jung et al. 2003), it lacks many genes normally present in the prokaryote metabolic pathways. In general the mycoplasmas depend on their host to supply them with the products of missing pathways (Razin et al. 1998); similarly the phytoplasmas lack genes for numerous biosynthetic pathways and seem to have lost more of such genes than the mycoplasmas (Oshima et al. 2004; Bai et al. 2006), including those of the pentose phosphate pathway, which is parallel to the glycolytic pathway. However, the phytoplasmas harbour multiple copies of transporter-related genes that are not found in the mycoplasmas (Oshima et al. 2004). As with other mollicutes, phytoplasmas lack several genes that autonomous bacteria such as *Escherichia coli* need for metabolism; for example, they do not possess genes for the *de novo* synthesis of aminoacids, fatty acids, or nucleotides. Mycoplasmas and spiroplasmas have a salvage pathway for nucleotide synthesis and an ATP synthase, and spiroplasmas can synthesize a few aminoacids, but none of these features are found in the sequenced phytoplasma genomes. Thus, phytoplasmas stand out lacking many genes otherwise considered to be essential for the cell metabolism, and must rely strongly on the uptake of nutrients by membrane transport processes. This is also reflected by the fact that many important membrane transporters are still retained in phytoplasma genomes, compared with those present in other Gram-positive bacteria. An intriguing absence is, for example, the lack of the tubulin-like protein FtsZ, which is thought to be essential for cell division (Møller-Jensen and Löwe 2005); this raises the question about how these mollicutes divide without this cytoskeleton protein. The minimal genome content makes the phytoplasmas the simplest known natural self-replicating life form, on the border between living cellular organisms and viruses. Having hosts in two kingdoms imposes constraints on the rapidity of their evolution: the genome of the onion yellows phytoplasma strain (OY) contains some incomplete pathways (Oshima et al. 2004), whereas the smaller genome of the aster yellows witches' broom phytoplasma strain (AYWB) appears to be more streamlined (Bai et al. 2006), lacking sequences that are truncated in OY.

10.3 Plant Metabolic Aspects in Presence of Phytoplasmas

Choi et al. (2004) studied the metabolomic profile of *Catharanthus roseus* infected by 10 phytoplasmas using NMR spectroscopy. The analysis of phytoplasma-infected leaves showed increased level of metabolites related to the phenylpropanoids/terpenoids/alkaloids biosynthetic pathways. Abundant levels of glucose, polyphenols, succinic acid, and sucrose were therefore detected. This analysis suggested that the activation of the biosynthetic pathway of phenylpropanoids might be part of the defence system that the plant elicits to alleviate the phytoplasma infection. Kube et al. (2012) determined the protein content of *Nicotiana occidentalis* infected by '*Candidatus* Phytoplasma mali' using a shotgun proteomics approach. The study reported 102 out of 497 predicted expressed phytoplasma proteins as present in the shoot tissues and 940 proteins of *N. occidentalis* expressed in response to the phytoplasma infection. A high portion of proteins with unknown function was identified besides the proteins involved in transportation. Several proteins with unknown function contain a signal peptide suggesting their potential pathogen-host interaction. Ji et al. (2009) studied the response of mulberry (*Morus alba* L.) to the mulberry dwarf (MD) phytoplasma and conducted a comparative proteomic analysis study using 2-DE on infected and healthy leaves. Among the 500 protein spots that were reproducibly detected, 20 were down-regulated and 17 were up-regulated. Mass spectrometry (MS) identified 16 differentially expressed proteins. The photosynthetic proteins rubisco large subunit, rubisco activase, and sedoheptulose-1,7-bisphosphatase showed enhanced degradation in the infected leaves. In another study the response of the Mexican lime tree to '*Ca.* P. aurantifolia' infection was studied by using the same technique. The results revealed among 800 leaf proteins that were reproducibly detected in eight replicates of healthy and infected plants, 55 showed a significant response to the disease. MS analyses resulted in the identification of 39 regulated proteins, which included proteins involved in oxidative stress defence, photosynthesis, metabolism, and the stress response (ascorbate peroxidase 2, Cu/Zn superoxide dismutase, miraculin-like proteins, and annexin p35) (Taheri et al. 2011).

10.4 Carbohydrate Metabolism and Glycolysis

Embden-Meyerhof-Parnas (= glycolysis) pathway (Fig. 10.1) was suggested to be the major energy-yielding pathway in phytoplasmas (Oshima et al. 2004) despite the apparent lack of the hexokinase (glucose phosphorylating) and a sugar-specific phosphotransferase system (PTS) mediating a phosphorylated hexose to enter glycolysis. One promising candidate is the glucosyltransferase GtfA, which was first predicted in '*Ca.* P. australiense'. The assignment of the deduced protein sequence and the ortholog of the onion yellows phytoplasma strain OY-M was confirmed by InterPro (IPR022527 sucrose phosphorylase, GftA). The GtfA, which is probably better described as disaccharide glucosyltransferase or sucrose phosphorylase, allows the formation of α-D-glucose-1-phosphate from phosphate and sucrose (Voet and Abeles 1970), which is often a predominant sugar in the phloem and is the entry compound of glycolysis. GtfA may compensate the absence of a hexokinase and PTS system. However, GtfA or similar phosphorylases are not a general trait of phytoplasmas, since they are absent in the genomes of closely related phytoplasmas such as aster yellows witches' broom strain AY-WB and in '*Ca.* P. mali'. Thus, the observed differences in glycolysis among phytoplasmas may arise from genome plasticity as suggested by the close proximity of prophage-related elements and GtfA in '*Ca.* P. australiense'. Theoretically, the uptake of phosphorylated hexoses would overcome this problem. Candidates for such sugar phosphates are trehalose-6-phosphate, sucrose-6-phosphate, and β-D-fructose-6-phosphate. The transporter complements of the phytoplasmas may contain uptake systems for importing these phosphorylated di- and monosaccharides from the environment. For example, trehalose-6-phosphate has to be monomerized prior to entry into the glycolysis.

The breakdown would occur within the phytoplasmas mediated by hydrolases/phosphatases. At least one phosphatase each of subfamily IIIb and of subfamily IIa is encoded within the four phytoplasma genomes annotated. These phosphatases are poorly characterized so far. A single copy of the subfamily IIIa of the haloacid dehalogenase (HAD) superfamily of hydrolases, representing hypothetical proteins, is encoded in each phytoplasma genome. The most characterized members of this subfamily and of the HAD superfamily are phosphatases. This protein family consists of sequences from fungi, plants, cyanobacteria, and Gram-positive bacteria.

Functional interpretation of the second hydrolase of the HAD superfamily hydrolase subfamily IIb may be more straightforward. Members of the HAD superfamily hydrolase subfamily IIb are encoded by at least one gene ('*Ca.* P. mali') in each genome. They encompass trehalose-6-phosphatase, plant and cyanobacterial sucrose phosphatase, and a closely related group of bacterial and archaeal orthologs, eukaryotic phosphomannomutase. If these proteins function as

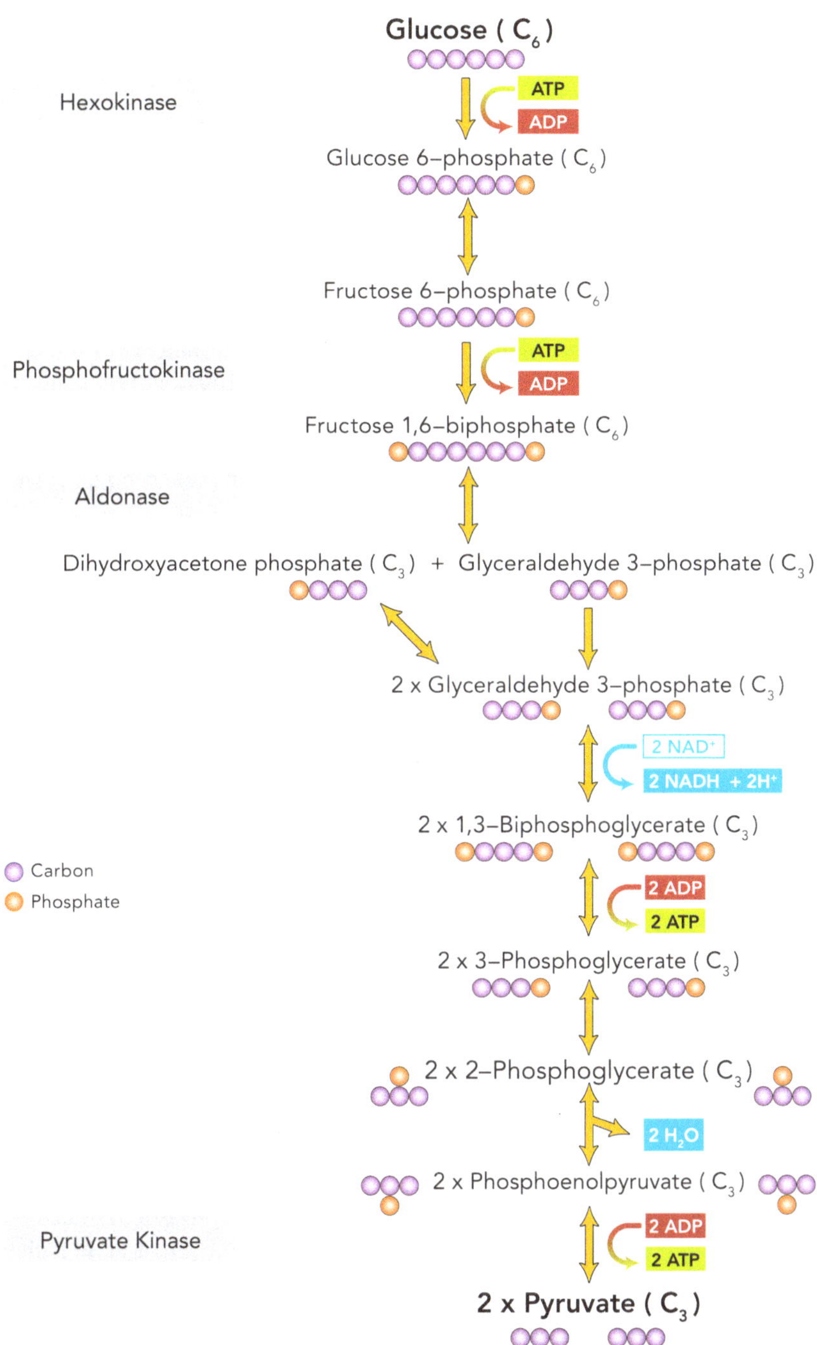

Fig. 10.1 Glycolysis pathway

trehalose-6-phosphatase, phytoplasmas could use α-trehalose-6-phosphate, and phosphate to produce glucose-6-phosphate and β-D-glucose-1-phosphate, which are the entry molecules of glycolysis. If sucrose-6-phosphate is used as a substrate, the sucrose-6-phosphatase may generate glucose-6-phosphate and fructose (Brückner et al. 1993). The utilization of trehalose-6-phosphate and/or sucrose-6-phosphate appears to be likely due to the phosphoglucose isomerase encoded in the phytoplasma genomes. This step would be unnecessary if the fructose-6-phosphate is available. However, it should be considered that only trace amounts of trehalose-6-phosphate are present in general in higher plants. The impact of a phytoplasma infection on plant metabolism cannot be estimated, but trehalose-6-phosphate is a signalling molecule in plants with strong regulatory effects on metabolism, growth, and development (Eastmond and Graham 2003; Zhang et al. 2009; Paul et al. 2010). The sucrose-6-phosphate concentration in the phloem is unclear. It is an interesting scenario that sucrose and trehalose compounds could be utilized depending on their availability in the phloem of the infected plants and in the haemolymph of insect vectors.

The general upper part of the glycolysis (energy demanding) is encoded within all four phytoplasma genomes (Oshima et al. 2004; Bai et al. 2006; Kube et al. 2008; Tran-Nguyen et al. 2008) starting with α-D-glucose-6-phosphate converted to β-D-fructose-6-phosphate by phosphoglucose isomerase and ATP-dependent formation of two molecules β-D-fructose-1,6-bisphosphate by phosphofructokinase (PfkA). Subsequently, fructose-bisphosphate aldolase (Fba) catalyses the formation of D-glyceraldehyde-3-phosphate and dihydroxyacetone phosphate; this is suggested to enter glycerophospholipid metabolism, while D-glyceraldehyde-3-phosphate is channelled into the energy-yielding part of the glycolysis. The interconversion of these C3-intermediates is performed by a triosephosphate isomerase. Dihydroxyacetone phosphate can also be generated by a conserved kinase related to dihydroxyacetone kinase. However, it remains unclear if this kinase can also act in the opposite direction as a transferase.

Except for '*Ca.* P. mali' (Kube et al. 2008), the protein components of the lower part of glycolysis (energy yielding) from D-glyceraldehyde-3-phosphate to pyruvate are encoded in the analysed phytoplasma genomes, owing to the presence of five enzymes: glyceraldehyde phosphate dehydrogenase, phosphoglycerate kinase, mutase, enolase, and pyruvate kinase. Notably, gluconeogenic phosphoenolpyruvate synthase (PpsA), the enzyme that catalyses the conversion of pyruvate and ATP to phosphoenolpyruvate (PEP), adenosine monophosphate (AMP), and phosphate (which is thought to function in the gluconeogenesis) are lacking in all four phytoplasmas. Other strategies to obtain host-derived ATP such as ATP/ADP translocase (Oshima et al. 2004; Kube et al. 2008) and also the arginine dihydrolase pathway (Pollack et al. 1989, 1997) have not been identified in phytoplasmas. Carbohydrate metabolism is one of the most important physiological traits of the phytoplasmas. A pathway for malate conversion to acetate is potentially encoded in all four genomes. Uptake of malate is enabled by the symporter MleP (Kube et al. 2008; Oshima et al. 2004). Malate can be oxidatively decarboxylated to pyruvate by the malic enzyme

ScfA. Pyruvate would then also be oxidatively decarboxylated to acetyl-CoA by the pyruvate dehydrogenase multienzyme complex. In the case of '*Ca*. P. mali', pyruvate might be generated by an additional way. Here, an aldolase (Eda) is predicted, serving two possible functions: a 4-hydroxy-2-oxoglutarate aldolase (EC: 4.1.3.16) (KHG-aldolase) would catalyse the interconversion of 4-hydroxy-2-oxoglutarate into pyruvate and glyoxylate, and a phospho-2-dehydro-3-deoxygluconate aldolase would catalyse the interconversion of 6-phospho-2-dehydro-3-deoxy-D-gluconate into pyruvate and glyceraldehyde-3-phosphate. In both cases, pyruvate would be formed independently from glycolysis.

The host shift of phytoplasmas is between organisms that have the sugar concentration at a similar high level, consisting of up to 0.9 M trehalose in the haemolymph of phloem-sucking homoptera (Weisberg et al. 1989) and up to 1 M sucrose in the sieve tubes of plants. These different forms of sugars serve as potential source of food supply. Intriguingly, however, the genomes of two phytoplasmas do not contain genes coding phosphotransferase system (PTS) (Oshima et al. 2004), which most bacteria use as an energy-efficient way of simultaneously importing and phosphorylating sugars such as sucrose, glucose, and fructose. Phytoplasmas have a maltose ABC transporter system and, therefore, might import maltose, trehalose, sucrose, and palatinose and utilize them as energy source (Carle et al. 1995). In light of the lack of sucrose-degrading enzymes and the lack of the PTS system for hexoses, it appears that phytoplasmas depend on the uptake of phosphorylated hexoses for their carbon source, which can then enter the glycolytic pathway. The high sugar concentration in both plant host and insect vector subjects phytoplasmas to a high osmolarity. It can be postulated that the pathogen cells adapt their water potential to isotonic conditions, making use of membrane channels and transporters to cope with osmotic water loss.

10.5 Carbohydrate Intake and Energy Generation

Phytoplasma genome data revealed that they do not possess any genes encoding enzymes involved in aerobic metabolism of carbohydrates, indicating that glycolysis is the main route for sugar metabolism and energy generation. In other bacteria, the F1F0-ATP synthase complex catalyses ATP synthesis from ADP and phosphate driven by a cross-membrane proton gradient. The F1F0-ATP synthase complex can also hydrolyse ATP and generate a proton electrochemical gradient in the opposite direction for locomotion, nutrient uptake, and other functions. Prior to the completion of the first phytoplasma genome sequencing, the F1F0-ATP synthase complex had been found in the genomes of all sequenced bacteria, including mollicutes; therefore, it was believed indispensable to cellular life. The lack of the genes encoding components of ATP synthase complex implied that some alternative ATP synthesis and energy-coupling mechanisms must exist in phytoplasmas (Oshima et al. 2004). In addition the gene *nox* encoding NADH oxidase (NADH dehydrogenase), which transfers electrons along a chain of acceptors and releases energy for ATP formation, is also absent in the phytoplasma genomes. Furthermore, the genes encoding components of PTS, the phosphoenolpyruvate-dependent sugar phosphotransferase system (*ptsG*, *ptsH*, and *ptsI*), are absent in the genomes of phytoplasmas. PTS is a distinct sugar uptake system essential to free-living bacteria. It utilizes energy stored in phosphoenolpyruvate (PEP), powering the cross-membrane transport of glucose, mannose, fructose, cellobiose, and other sugar molecules.

The lack of PTS in phytoplasmas indicates the presence of other transmembrane system(s) responsible for transport of sugars into their cells. Given that genes encoding ABC-type sugar transporter systems, for example, *malEGFK* and *ugpBEAC*, are present in phytoplasmas (Kube et al. 2012). Some sugars could be imported through ABC-type sugar transporters, and in addition, sugars could possibly be transported into phytoplasma cells through cation symporter(s) or by other membrane channels in the form of complexes with other classes of compounds. The selection of appropriate carbon source(s) is particularly crucial, since phytoplasmas also lack genes encoding various enzymes required for the pentose phosphate pathway (deoxyribose-phosphate aldolase and ribose-5-phosphate isomerase), for pentose and glucuronate interconversions (UTP-glucose-1-phosphate uridylyltransferase), for galactose metabolism (UDP-glucose-4-epimerase), for fructose and mannose metabolism (phosphomannomutase), and for pyruvate metabolism (phosphotransacetylase); the lack of these genes further indicates that phytoplasmas have a relatively limited resource for energy generation. The genes that encode enzymes necessary for converting malate to pyruvate (phosphate acetyltransferase), and then pyruvate to acetyl CoA (pyruvate dehydrogenase multienzyme complex), are present in phytoplasmas. However, a gene encoding phosphotransacetylase, the enzyme that catalyses the conversion of acetyl CoA to acetyl phosphate, is absent; therefore, phytoplasmas may not be capable of utilizing the energy carried in acetyl CoA to produce ATP.

The phytoplasma genomes contain no genes encoding enzymes for the citrate cycle (tricarboxylic acid cycle: TCA) or oxidative phosphorylation. Based on metabolic pathway genes identified in the phytoplasma genomes, it has been suggested that the glycolysis may be their main energy-yielding process (Oshima et al. 2004). In phytoplasma-infected Mexican lime, the energy production by the TCA cycle was significantly enhanced (Monavarfeshani et al. 2013). In other bacterial cells, the entry point of glycolysis is glucose-6-phosphate imported through the phosphoenolpyruvate-dependent sugar phosphotransferase system (PTS). As phytoplasmas lack the PTS for sugar phosphate uptake and lack the hexokinase for glucose phosphorylation, the presence of a *gtfA* gene in OY-M and '*Ca.* P. australiense' raises the possibility that some phytoplasmas may obtain sugar phosphate for glycolysis through the catalytic activity of the *gtfA* gene products, which allow the formation of α-D-glucose-1-phosphate from sucrose and phosphate. The most unique feature of the phytoplasma genome may be the absence of the gene encoding F1F0-type ATP synthase. In general, bacteria use F1F0-type ATP synthases to synthesize and hydrolyse ATP using ATP-proton motive force interconversion. Since the genes encoding ATP synthase have been found in most of the fully sequenced bacteria, '*Ca.* P. asteris' OY-M strain was the first example of a naturally occurring organism with no ATP synthase genes (Oshima et al. 2004). Despite this there is a considerable membrane potential in phytoplasmas, as has been demonstrated with potentiometric dye (Christensen et al. 2004). Some phytoplasma strains have five genes encoding P-type ATPases, which are similar to animal Na+/K+ and H+/K+ pumps (Bai et al. 2006), implying that these ATPases may generate electrochemical gradients across the membrane in these cases (Christensen et al. 2005).

10.6 Protein and Nucleotide Metabolism

Phytoplasma chromosomes encode the glutamine-dependent NAD synthetase (NadE). In this ATP-dependent process, nicotinate adenine dinucleotide (deamido-NAD), L-glutamine, and H_2O are used to form L-glutamate, AMP, diphosphate, H+, and NAD. Other proteins involved in the generation of NAD were not identified in the phytoplasma genomes. In contrast, *Acholeplasma laidlawii*, a phytoplasma ancestor, contains *nad*D and *pnc*B genes, allowing the formation of NAD from nicotinate. It appears likely that phytoplasmas import deamido-NAD or its precursor from their environment. All four genomes encode a superoxide dismutase (SOD), which converts O^{2-} to the less toxic H_2O_2. The formation of Mn SOD is reported for the acholeplasmas (O'Brien et al. 1981). The catalase dismutase responsible for the conversion of H_2O_2 to nontoxic products was not identified in phytoplasmas. It remains unclear, whether the release of H_2O_2 will weaken or damage the plant host due to the presence of phytoplasmas inside the sieve elements. Lipoic acid derivates act as cofactors in enzymatic systems, such as the pyruvate dehydrogenase (Reed and Cronan 1993) in the phytoplasmas. The phytoplasma chromosomes encode the ATP-dependent lipoyl-protein ligase (LplA), which preferentially utilizes imported lipoate to form lipoyl adenylate or an octanylated protein from octanylate (Morris et al. 1994). Since phytoplasmas possess no genetic repertoire for *de novo* synthesis of purine or pyrimidine bases (Oshima et al. 2004), they depend, like many other mollicutes, on environmentally derived nucleotide precursors (Bizarro and Schuck 2007). However, nucleobase or nucleoside transporters were not identified in phytoplasmas (Oshima et al. 2004), but it was suggested for mycoplasmas that the limited repertoire of transporters could be tuned to a wider variety of substrates (Pollack et al. 2002). Membrane-associated nucleases are suggested for the nucleotide precursor uptake and were identified in 20 mycoplasma species (Minion et al. 1993). However, it was not possible to identify similar candidates in phytoplasmas. AMP and dAMP as well as GMP and dGMP represent the entry points in the purine metabolism in all sequenced phytoplasma genomes. The (deoxy)nucleoside diphosphate kinase (Ndk) is absent from the sequenced phytoplasmas, and this was also observed in many mycoplasmas (Pollack et al. 2002). Cytidine/uridine kinase (Udk), involved in the pyrimidine pathway, participates in pyrimidine metabolism and phosphorylates both uridine and cytidine, using ATP, GTP, or dGTP as phosphate donor. The gene encoding uridine kinase (*udk* gene) is present in the genomes of OY-M and AY-WB but is missing in '*Ca.* P. australiense' and '*Ca.* P. mali'. As in the case of other mollicutes (Bizarro and Schuck 2007), phytoplasmas lack a gene set essential for *de novo* synthesis of pyrimidine bases. As outlined by Kube et al. (2012), phytoplasmas may use imported uridine and cytidine as the entry point for the synthesis of UTP/dTTP and CTP/dCTP, respectively, building blocks for DNA and RNA. Since *udk* gene is required for converting uridine to UMP and cytidine to CMP, its lack in some phytoplasmas indicates that they may import UMP and CMP from the host cells. While CMP may also be available as a by-product from the phospholipid metabolism (Kube et al. 2012), its limited source seems unlikely to be sufficient to support the needs for DNA and RNA biosynthesis. Phytoplasmas lack most of the genes encoding enzymes required for aminoacid biosynthesis and metabolism, while a few genes are present in a strain-specific manner. Asparagine synthase which catalyses the ATP-dependent conversion of aspartic acid to asparagine, using either glutamine or ammonia as the nitrogen source, is present in OY-M phytoplasma which has a full-length *asnB* gene (PAM594). In AY-WB, the *asnB* gene has decayed substantially, and only a partial C-terminal AsnB domain remains recognizable. The s-adenosylmethionine synthetase (MetK) and C-5 cytosine-specific DNA methylase (C5 Mtase) are considered key components of cysteine and methionine metabolism and impact multiple metabolic pathways. Among the sequenced phytoplasma genomes, the *metK* gene is present only in the genome of '*Ca.* P. mali', and the gene encoding C5 Mtase is present only in the genome of '*Ca.* P. australiense'. Acetylornithine deacetylase present in OY-M and AY-WB is a zinc-dependent hydrolase that catalyses the deacylation of N2-acetyl-L-ornithine, yielding ornithine and acetate (Javid-Majd and Blanchard 2000). However it is still doubtful that *arg*E gene (acetylornithine deactylase) actually participates in arginine biosynthesis in OY-M and AY-WB, since none of the three genes (*argF/I*, *argG*, and *argH*), encoding the enzymes necessary for the downstream conversion of ornithine to arginine (Caldara et al. 2008), is present in any of the phytoplasma genomes.

Since phytoplasmas have no cell wall and reside inside of host cells, their membrane proteins and secreted proteins function in the cytoplasm of the host plant or insect cell and are predicted to have important roles in host-parasite interactions and/or virulence. Thus, the identification of both a secretion system and secreted proteins in the phytoplasma genomes is important. Phytoplasmas possess two secretion systems: the YidC system for the integration of membrane proteins and the Sec system for the integration and secretion of proteins into the host cell cytoplasm. The Sec protein translocation system is essential for viability in many bacteria (Tjalsma et al. 2000). SecA, SecY, and SecE are essential for protein translocation and cell viability in *E. coli* (Economou 1999). The genes encoding SecA, SecY, and SecE proteins have also been identified (Kakizawa et al. 2001, 2004), and their expression has been confirmed in phytoplasma-infected plants (Kakizawa et al. 2001; Wei et al. 2004; Bai et al. 2006; Lee et al. 2006; Kube et al. 2008; Tran-Nguyen et al. 2008). Antigenic membrane protein (Amp), a major surface membrane protein of phytoplasmas (Barbara et al. 2002), has been reported to be a substrate of the Sec system. Amp has a Sec signal sequence (Kakizawa et al. 2004), suggesting that the phytoplasma Sec system utilizes recognition and cleavage of a signal sequence, as other bacterial Sec systems, whereas YidC mediates integration of newly synthesized membrane proteins (Dalbey and Kuhn 2000), and it is sufficient to promote insertion of membrane proteins *in vitro*, suggesting that its function is independent from the Sec system (Serek et al. 2004). YidC is encoded in the phytoplasma genomes suggesting that they should have a YidC integration system, which is an essential protein in *E. coli* (Samuelson et al. 2000).

Zinc and manganese are essential nutrients for bacteria since they play important catalytic and structural roles in a variety of enzymes and metallo-regulatory proteins (Moore and Helmann 2005). Manganese is crucial in detoxification of superoxide free radicals (Abreu and Cabelli 2010). Zinc is found in at least 100 enzymes that cover all enzyme classes (Trumbo et al. 2001). Phytoplasmas are the only cell wall-less bacteria known to have genes encoding ABC-type Mn/Zn transport systems that, perhaps significantly, reside for the majority in the SVM (sequence variable mosaic)/phage-derived genomic islands (Kube et al. 2012).

The sequenced phytoplasma genomes lack many genes encoding enzymes responsible for metabolism of aminoacids, especially cysteine, methionine, histidine, lysine, arginine, proline, and tyrosine. The presence of phytoplasma-unique ABC-type aminoacid and dipeptide/oligopeptide transport systems may reflect a necessity for the aminoacid importation from the host cells (Manimekalai et al. 2014, 2015). The presence of a malate/citrate symporter gene in all studied phytoplasma genomes indicates that malate is imported and possibly utilized as an alternative carbon source for energy generation. Since malate is abundant in the phytoplasma plant host cells (Ziegler 1975), and since the genes encoding most of the enzymes required for converting malate to acetate and generation of ATP are present in the phytoplasma genomes (Kube et al. 2012), it is quite possible that phytoplasmas use this alternative carbon source.

Besides having limited biosynthetic capacity, phytoplasmas also have limited catabolic capacity. Conceivably, instead of degrading toxic chemicals and other cellular wastes metabolically, they may rely on membrane transport systems to discharge the harmful substances; the genomes of sequenced phytoplasma strains encode an ABC-type multidrug/protein/lipid transport system that can be used for this discharge.

10.7 Lipid, Phosphate, and Vitamin Synthesis

All sequenced phytoplasma genomes encode a common biosynthetic pathway for essential phospholipids, that is, for the formation of cytidine di phosphate (CDP) diacylglycerol from an acyl phosphate and dihydroxyacetone phosphate. The only detected putative protein involved in the riboflavin syntheses is an ATP-dependent riboflavin kinase in the strain OYM. Folic acid (or vitamin B9) is a precursor of the coenzyme tetrahydrofolate (THF). All phytoplasma genomes encode dihydrofolate reductase (FolA), which uses 7,8-dihydrofolate monoglutamate and NADPH + H+ for the generation of THF. Only the strain OYM encodes a gene set which may allow generating THF from other precursors (Bai et al. 2006). In the clover phyllody phytoplasma, genes assigned as *folP* and *folK* encode frameshifts and represent pseudogenes (Davis et al. 2003). It remains unclear how phytoplasmas obtain 7,8-dihydrofolate monoglutamate and how they process 5,10-methylenetetrahydrofolate. The incomplete folate synthesis may indicate the loss of this genetic modules and a folate dependence on the host. This may influence the host because folate is also involved in photorespiration, amino acid metabolism, and chloroplastic protein biosynthesis in plants (Hanson and Gregory 2002). Phytoplasma genomes encode the ATP-dependent thiamine phosphate phosphotransferase (ThiJ), which forms the coenzyme thiamine pyrophosphate from thiamine phosphate. ThiJ, which is encoded in the chromosomes except for '*Ca.* P. mali', is required for thiazole synthesis in the thiamine biosynthesis pathway (Webb et al. 1998). The kinase PdxK of the vitamin B6 metabolism is only encoded in the genome of '*Ca.* P. asteris' OYM. This ATP-dependent kinase catalyses the formation of pyridoxine-5-phosphate from pyridoxal, pyridoxine, and pyridoxamine. NifU-like proteins are predicted to occur in all four phytoplasmas. Proteins such as NifS and NifU are required for the formation of metalloclusters of nitrogenase in *Azotobacter vinelandii* and the maturation of other FeS proteins (Hwang et al. 1996).

A set of six proteins involved in the metabolism of phosphorous compounds is encoded in the phytoplasma genomes. The inorganic pyrophosphatases (Ppa) catalyse the hydrolysis of inorganic pyrophosphate (PPi) to two orthophosphates (Pi), regulating the PPi pool which is replenished by various metabolic processes (Cooperman et al. 1992). Additional predicted proteins related to phosphate metabolism are phosphohydrolases, metallophosphoesterases, and lysophospholipases. Several genes with a weak catalytic or unclear pathway assignment are shared between some of the phytoplasmas. An example is the predicted endoglucanase FrvX (M42 peptidase family). Considering that endo-1,4-beta-glucanase proteins in *Arabidopsis thaliana* are associated with the plant growth (in particular elongation), xylem development, and cell wall thickening (Shani et al. 2006), the phytoplasma FrvX protein could potentially contribute to the virulence.

The sequenced phytoplasma genomes possess a gene annotated as encoding CDP-diglyceride synthetase or phosphatidate cytidylyltransferase (CdsA) (PAM169, AYWB_550, PA0515, and ATP_00236) (Kube et al. 2012). The CdsA is crucial to the lipid biosynthesis: by catalysing the activation of phosphatidic acid, the CdsA reaction yields CDP diglyceride, the phosphatidyl moiety donor for biosynthesis of various phospholipids (Sparrow and Raetz 1985). It is interesting to note that a conserved domain database (CDD) search using the amino acid sequences of the four presumed phytoplasma CdsA as queries revealed that the phytoplasma CdsA proteins contain only a partial CdsA domain at the C-terminus. Thus far, whether or not phytoplasmas have a functional *cdsA* gene remains an open question. If phytoplasmas do not have it, they need to import from the host cells activated phosphatidic acid, *i.e.* CDP diglyceride, for phospholipid biosynthesis. Phytoplasmas also lack a gene encoding glycerol kinase (GlpK) that provides glycerol-3-phosphate for phospholipid synthesis. Without GlpK, phytoplasma cells will need either to draw glycerol-3-phosphate from glycolysis or to import it from host cells. Also missing in phytoplasma genomes is the gene encoding the prolipoprotein diacylglyceryl transferase (LGT). LGTs are integral membrane proteins that catalyse transfer of the n-acyl diglyceride group onto proteins that then become anchored in the phytoplasma membrane. In diverse bacteria, LGT is required for membrane genesis, for membrane transport activity, and for normal growth (Pailler et al. 2012; Chimalapati et al. 2012). The lack of LGT in phytoplasma genomes will conceivably impact membrane genesis and homeostasis of phytoplasma cells. Genes encoding putative membrane-associated phospholipid phosphatases are present in the genomes of OYM (PAM005) and '*Ca.* P. mali' (ATP_00085) (Oshima et al. 2004; Kube et al. 2008). The amino acid sequences deduced from these phytoplasma genes each possess a PAP2-like domain. PAP2 is a superfamily of histidine phosphatases that includes type 2 phosphatidic acid phosphatase, phosphatidylglycerophosphatase B, glucose-6-phosphatase, and bacterial acid phosphatases (PFAM01569). As reviewed by Kube et al. (2012), a common set of genes encoding enzymes for biosynthesis of essential phospholipids are present in the sequenced phytoplasma genomes. These enzymes include PlsX, GpsA, PlsY, PlsC, and CdsA. The presence of such a gene set indicates that phytoplasmas may be able to synthesize CDP-diacylglycerol from acyl phosphate and dihydroxyacetone phosphate. Besides membrane biogenesis, phospholipid biosynthetic pathways play important roles in the virulence of diverse pathogens including fungi (Chen et al. 2010) and bacteria (Conde-Alvarez et al. 2006; Bukata et al. 2008).

10.8 Host Plant Response to Phytoplasma Infection

Biochemical analysis of phytoplasma-infected tobacco plants indicated that soluble carbohydrates and starch accumulate in source leaves, while sink organs showed a marked decrease in sugar levels (Lepka et al. 1999; Maust et al. 2003). Initial knowledge was achieved about phytoplasma/plant interaction in polyphenol production, sugar and aminoacid transportation, and comprehensive differences in gene expression mainly reported in the experimental host plant periwinkle (Musetti et al. 2000; Jagoueix-Eveillard et al. 2001). Bertamini et al. (2003) studied the effect of phytoplasma infection on growth and photosynthesis in leaves of field-grown apple and revealed that the contents of chlorophyll, leaf biomass, and soluble proteins were markedly decreased. Similar results were also observed for CO_2 fixation, ribulose-1,5-bisphosphate carboxylase, and nitrate reductase activity. In contrast, the contents of sugars, starch, amino acids, and total saccharides were significantly increased in the phytoplasma-infected leaves. In isolated chloroplasts, phytoplasma presence was associated with a marked inhibition of the photosystem II activity. Enhanced levels of polyamines and phenolics were reported in juice from infected sugarcane (Fontaniella et al. 2007). Yellow leaf syndrome affecting sugarcane plants showed lower sucrose content but had high content of reducing sugars. Infected plants had higher polyamine (PA) contents, especially putrescine, cadaverine, spermidine, and spermine either free or conjugated with high molecular mass glycoproteins. The disease was associated with an increase in the total PA fraction. Enhanced arginase and ornithine decarboxylase activities were detected followed by significant levels of chlorogenic, syringic, and ferulic acids in the juice of yellow leaf syndrome-infected sugarcanes (Fontaniella et al. 2003).

Musetti and Favali (2004) carried out cytochemical and biochemical analysis to assess the localization of H_2O_2, peroxidase activity, and levels of glutathione and malondialdehyde in healthy, apple proliferation-infected and recovered apple trees. H_2O_2 was detected in the membrane of the phloem in leaves of recovered apple trees, but not in healthy or infected leaves. In all the cultivars, the peroxidase activity in the tissues of diseased trees was higher than in the recovered and healthy trees; however, glutathione content of the leaves decreased in the reverse order. Ultrastructural changes observed in plants with symptoms of yellow leaf and healthy sugarcane plants revealed that the abaxial epidermis of diseased leaves shows a large amount of adhered superficial bodies mainly located between parallel thorny ribs. Some stomata were occluded by plugging material, whereas the perimeter of the stomata pore was more or less winding in the other. In contrast, healthy leaves showed normal well-defined stomata and regular granules or ribbons. Mesophyll cells between veins showed no significant changes in diseased leaves, although a thick layer of an amorphous material, similar to wax, displaced subepidermal tissues towards deeper zones. Bundle sheath cells surrounding the bottom of the phloem of diseased leaves were separated from conducting tissues by a large layer of wax followed by partial destruction of cell walls.

Maust et al. (2003) studied the changes in the carbohydrate metabolism in coconut palms infected with the lethal yellowing phytoplasma (LY). The CO_2 exchange rate in the intermediate leaves was found to decrease with the disease severity. Stomata conductance decreased gradually through the later stages of the disease. Chlorophyll fluorescence parameters, such as the maximum quantum use efficiency (Fv/FmorfPo) of photosystem II (PSII), the efficiency of moving electrons beyond QA (*f*Eo), and the proportion of QB non-reducing PSII reaction centres, remained unaltered by the severity of LY disease. However, the number of active PSII reaction centres per cross section (RC CSm-1), the pool size of electron acceptors (QA) on the reducing side of PSII, and the structural functional index of events leading to electron transport within photosynthesis (SFIPo) decreased progressively with the progression of the severity of the infection. The sugar and starch concentrations increased slowly in freshly expanded leaves with the development of the disease, and significantly decreased in later stages of sugarcane leaf yellows. This increase was more rapid in intermediate leaves with the advance of the disease before decreasing in the later stages. Root respiration decreased in the trees with the development of LY symptoms followed by a reduction of total sugar concentrations in the primary roots. Starch concentrations decreased significantly in primary roots initially, but remained unaltered in later disease stages. In a similar study, reduced production of phloem sap and phloem sugar transportation was observed in cut trunks of coconut palm trees having LY infection (Eden-Green and Waters 1982).

Reduced transport of ^{14}C-assimilated sugars to the roots was observed in phloem necrosis-diseased elm (Braun and Sinclair 1978), resulting in reduced root starch content. Similar results were also observed in phytoplasma-infected plants where the phloem necrosis is sometime a symptom shown by infected pear (Batjer and Schneider 1960), apple (Kartte and Seemüller 1991), and elm (Braun and Sinclair 1978), and it seems to be the result of excessive production of reactive oxygen species (ROS) in the phloem.

The molecular mechanisms involved in the interaction between pathogenic phytoplasmas and their plant hosts are largely unknown, and the genes associated with pathogenicity have begun to be identified (Garnier et al. 2001). Carginale et al. (2004) showed that two groups of genes were affected by phytoplasma infection in apricot leaves; there was up-regulation of heat-shock protein HSP70 and metallothionein and down-regulation of an aminoacid transporter homologous to one of *A. thaliana*. Hren et al. (2009) demonstrated that sucrose synthase (*SuSy*) and *adh1* gene expression were higher in the leaves of infected grapevine, this gene controls the reduced demanding for degradation of sucrose to hexoses. Recently Kawar et al. (2009) studied the sugarcane plant response to phytoplasma infection by cDNA-SSH and identified 274 genes up-regulated under SCGS infection. Three transcripts involved in the carbohydrate metabolism were induced, one encoding a sucrose phosphate synthase III and another glucose-6-phosphate isomerase and phosphoglucomutase. The phytoplasmas colonize the phloem sieve cells and act as an additional sink for the photosynthesized carbohydrates by blocking sugar transport from the source, *i.e.* infected leaves, to the sink tissues; therefore, sugars are found depleted in the roots (Lepka et al. 1999; Maust et al. 2003; Choi et al. 2004). It is well known that phytoplasmas lack enzymes for sucrose utilization (Oshima et al. 2004), and they might use fructose or glucose as a source of energy. This confirms the hypothesis that the increase in demand for simple sugars is probably compensated by the up-regulation of plant sucrose synthase gene in the infected tissues (Hren et al. 2009). Infection of tobacco or periwinkle plants with various phytoplasmas resulted in a marked increase in soluble carbohydrates in the leaves and lower levels of carbohydrates in the sink organs (Lepka et al. 1999).

A putative receptor protein kinase was identified in a sugarcane grassy shoot infected strain subtractive library. In addition to specific stress signalling, other factors related to cellular signalling such as protein Rab-6A, brassinosteroid insensitive1-associated receptor kinase1, proline-rich protein, cellulose, phytoene synthase (Y1), catalase CAT-2, horcolin (mannose-specific lectin), and jacalin-like lectin domain containing protein were reported as up-regulated by the phytoplasma presence. In another study 55 candidate genes involved in the interaction of Mexican lime trees with '*Ca.* P. aurantifolia' were identified. Among 36 lime tree transcripts, 70% were down-regulated during the infection and were homologous to known resistance genes; they tend to be repressed in response to the infection. These include the genes for modifier of snc1 and autophagy protein 5 and genes involved in metabolism, transcription, transport, and cytoskeleton, which included the genes for formin, importin b3, transducin, L-asparaginase, glycerophosphoryl diester phosphodiesterase, and RNA polymerase b. In contrast, genes that encode a proline-rich protein, ubiquitin-protein ligase, phosphatidyl glycerol-specific phospholipase C-like, and serine/threonine-protein kinase were up-regulated during the infection (Zamharir et al. 2011).

It was speculated that the growth aberrations induced by phloem-restricted bacteria are the result of alteration in the plant hormonal balance (Chang et al. 1994). The gene coding a putative sterol C-methyltransferase, an enzyme involved in phytosterol biosynthesis, can play such a role, and it was found down-regulated in "stolbur" phytoplasma-infected plants. Phytoplasma infection has been shown to affect phloem function, impairing the translocation of carbohydrates with subsequent accumulation of soluble sugars in leaves (Lepka et al. 1999; Maust et al. 2003). However, accumulation of glucose was observed also in periwinkle infected by *S. citri,* when no characteristic flower malformations were observed (André et al. 2005). In tomato flower abnormalities induced by the "stolbur" phytoplasma presence, changes in the expression of flower development genes were detected. The pathways from stress perception due to phytoplasma multiplication in the phloem sieve tubes and its specific response towards deregulation of the flower meristem gene expression are starting to be deciphered (Pracros et al. 2006).

10.9 Transcriptomics in Plant-Phytoplasma Interaction

Transcriptomics is the study of the genome-wide expression, and RNA isolation from phytoplasma-infected plants can be a challenging task, given the high levels of polyphenol contents and accumulation of sucrose and starch in the different plant tissues (Pacifico et al. 2019). Wang et al. (2018b) studied the response of the jujube tree to '*Ca.* P. ziziphi' infection (jujube witches' broom disease). In the first infection stage (0–2 weeks after grafting), some differentially expressed genes (DEGs) related to abscissic acid (ABA) and cytokinin (CTK) were down-regulated, while some others, related to jasmonic acid (JA) and salicylic acid (SA), were up-regulated. In the second infection stage (37–39 weeks after grafting), when the witches' broom symptoms were visible, genes involved in biosynthesis and signal transduction of ABA, brassinosteroid (BR), CTK, ethylene (ET), and auxins such as indole acetic acid (IAA), gibberellic acid (GA), JA, and SA and flavonoid biosynthesis were significantly deregulated. Meanwhile, DEGs involved in photosynthesis, chlorophyll and peroxisome biosynthesis, and carbohydrate metabolism were down-regulated. These results suggested that phytoplasma infection had completely destroyed the jujube trees' defence system and had modified the chlorophyll synthesis and photosynthetic activity in the infected leaves at the late stage, resulting in yellow leaves and other symptoms.

Analysis of a peanut witches' broom (PnWB) phytoplasma genome highlights the importance of improving the sampling when investigating phytoplasma genome modifications (Chung et al. 2013). Wheat blue dwarf (WBD) disease comparative genome analyses showed that this phytoplasma had strongly reduced metabolic capabilities. However, 46 transporters were identified involved with dipeptides/oligopeptides, spermidine/putrescine, cobalt, and Mn/Zn transport. Among the 37 secreted proteins encoded, three were similar to the reported phytoplasma virulence factors TENGU, SAP11, and SAP54. In addition, WBD phytoplasma possessed several proteins that were predicted to play a role in its adaptation to diverse environments (Chen et al. 2014).

A multi-omics approach was taken by Ye et al. (2017) during graft infection of jujube by the witches' broom phytoplasmas through the analysis of the plant transcriptome, proteome, and phytohormone levels. The phytoplasma infection resulted in reduced auxin and increased jasmonate contents, indicating that auxin and JA have important roles in regulating the jujube responses to the '*Ca.* P. jujube' infection. At later stages of phytoplasma infection, genes and proteins involved in carbon metabolism and fixation in photosynthetic organisms were down-regulated, indicating that the photosynthesis was affected. The cDNA samples from healthy and diseased plants of *Ziziphus jujuba* infected by witches' broom phytoplasma were sequenced and the gene expression and metabolism regulation were compared (Fan et al. 2017). In the two libraries obtained from the leaves, 4,266 DEGs were identified. Among them, 2,070 genes were up-regulated, and 2,196 genes were down-regulated. The comparison between the libraries from flowers showed 3,800 DEGs, with 1,965 up-regulated and 1,835 down-regulated genes. Among all these some specific genes were identified to be functionally related to the aminoacid metabolism and the carotenoid pathway, possibly related to the lack of nutrition and the main symptoms of yellowing, brooming, and small leaves.

Mardi et al. (2015) conducted high-throughput transcriptome sequencing for the Mexican lime trees infected with phytoplasmas and reported that the up-regulated DEGs were categorized into pathways with possible implication in plant-pathogen interaction, including cell wall biogenesis and degradation, sucrose metabolism, secondary metabolism, hormone biosynthesis and signalling, and amino acid and lipid metabolism, while down-regulated DEGs were predominantly enriched in ubiquitin proteolysis and oxidative phosphorylation pathways.

Nejat et al. (2015) evaluated the whole transcriptome profiles of naturally infected leaves of *Cocos nucifera* ecotype Malayan Red Dwarf in response to the yellow decline disease associated with phytoplasmas from group 16SrXIV, using RNA-Seq technique. The transcriptomics-based analysis identified genes involved in the coconut innate immunity. The number of down-regulated genes in response to phytoplasma infection exceeded the number of genes up-regulated. Comparative analysis revealed that genes associated with defence signalling against biotic stimuli were significantly overexpressed in phytoplasma-infected leaves versus healthy coconut leaves. Genes involving cell rescue and defence, cellular transport, oxidative stress, hormone stimulus and metabolism, photosynthesis reduction, transcription, and biosynthesis of secondary metabolites were differentially expressed. A core set of genes associated with defence of coconut in response to phytoplasma presence was identified, although several novel defence response candidate genes with unknown function have also been detected.

Paulownia witches' broom is one of the most destructive diseases threatening the paulownia tree production. A transcriptomics-assisted proteomic technique was used to analyse the protein changes in phytoplasma-infected *Paulownia tomentosa* and healthy seedlings. A total of 2,051 proteins were obtained, 879 of which were found to be differentially abundant. Among these 43 were related to paulownia witches' broom disease, and many of them were annotated to be involved in photosynthesis, expression of dwarf symptom, energy production, and cell signal pathways (Cao et al. 2017).

10.10 Synthetic Approach for Phytoplasma Research

Recently the so called synthetic biology approach has been commonly applied, and it could offer a constructive tool to understand molecular mechanisms for biological researches. This new approach enables to reconstruct and understand a lot of biological phenomena or to express a variety of organic compounds by adding whole metabolic pathway genes into some model organisms like *E. coli*. Therefore, this approach would be useful to a wide range of research fields. Numerous huge projects associated with the synthetic biology have run, and some of these projects reported important achievements, *e.g.* the production of mycoplasma cells with synthetic genomes (Gibson et al. 2008, 2010; Lartigue et al. 2009) or a minimal genome (Hutchison et al. 2016) (Fig. 10.2), the production of synthetic yeast by the Yeast 2.0 project (Annaluru et al. 2014; Pretorius and Boeke 2018), the GP-Write project aiming to make human cells with synthetic genome (Boeke et al. 2016), to make *Bacillus subtilis* genome vector (BGV) (Itaya et al. 2008) and for the reconstruction of the whole gene set of several metabolic pathways on the BGV (Nishizaki et al. 2007), cloning the whole genome of a photosynthetic *Cyanobacterium* in the BGV to make a *Cyano-Bacillus* (Itaya et al. 2005). Among them, the minimal mycoplasma cells (Hutchison et al. 2016) must be a groundbreaking result in bacteriology, and it will be expected to accelerate the phytoplasma research in the future. For example, comparing genomes of the mycoplasma minimal cell and several phytoplasmas would be informative to investigate important genes for cultivation on media or to clarify differences of these prokaryotes' lifestyles. In addition, the minimal genome is thought to be a minimal gene set for the existence of life; thus, to understand all gene functions of the minimal cell would be important to understand also bacterial life (Mariscal et al. 2018).

The whole genome of *Cyanobacteria* was cloned into the *Bacillus genome* (Itaya et al. 2005), and this approach might be useful for phytoplasma research, *i.e.* whole cloning of phytoplasma genomes in other organisms. This approach could enable a lot of variety of further analyses, including complete genome sequencing, subcloning of specific genomic regions, stable genome preservation, expression of phytoplasma genes in other organisms, reconstruction of metabolic pathways of phytoplasmas, and engineering of the potential mobile units (PMU, synonim of sequences variable mosaic, SMV) genes. In addition, the complete synthesis of

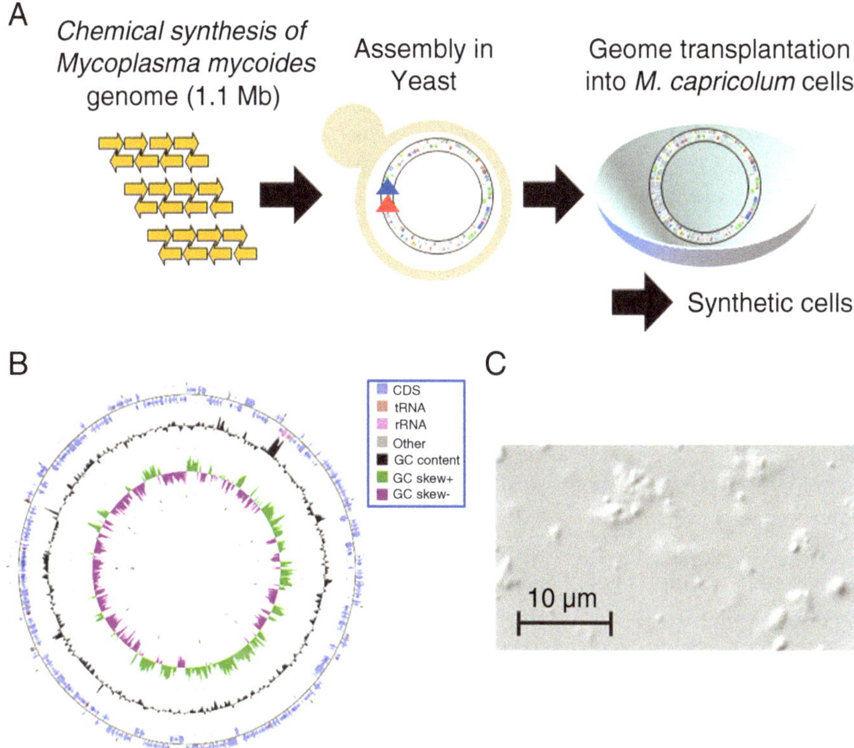

Fig. 10.2 The minimum mycoplasma cells. A. Procedures of whole-genome synthesis, genome assembly in yeast, genome transplantation for making the synthetic cells. B. Whole-genome map of the minimum genome bacteria (synthetic bacterium JCVI-Syn3.0, GenBank accession number: CP014940). The genome size is 531 kbp containing 473 genes. This organism has the smallest genome among all known culturable microorganisms. This map was created by the GC viewer server (Grant and Stothard 2008). C. Microscopic photograph of the minimal mycoplasma cells. This cell could be easily cultured both in liquid media and on agar plates

a phytoplasma genome would be possible. So far, the synthesis of mega-base genomes is still very expensive, but it is expected that with some new developed methods, the cost to synthesize genomes might be greatly decreased. Then, complete synthesis of whole genomes of microorganisms might be easily possible, and this approach would bring great change in all the research fields including phytoplasma research.

10.11 Conclusions and Perspectives

Phytoplasmas have adapted intimately to their host cells, which makes them experts in sieve element manipulation. Substrate specificity shows that the adaptation of phytoplasmas and spiroplasmas to the sieve elements must be different. Whereas phytoplasmas seem to mobilize nucleotides and phosphorylated hexoses from the companion cells, possibly altering the pore-plasmodesma units, spiroplasmas are thought to initiate shifts in the distribution of hexoses between sieve element and companion cell. The release of hexoses from the companion cells might follow the strong sink of hexose consumption of spiroplasmas. Regulation of plasmodesma conductance is thought to involve the phosphorylation and dephosphorylation of plasmodesma proteins. It appears that the effects of phytoplasmas on long-distance transport are secondary and dependent on the immediate impact of phytoplasmas on companion cells. Thus, understanding the interactions between phytoplasmas and their host environment carries the potential to understand the biology not only of the pathogens, but also of the most specialized cells of the phloem and their symplasmic connections. Phytoplasma-related diseases are expected to increase because the global climate warming is advantageous to the cold-sensitive insect vectors of the phytoplasmas. Therefore, pest control and detection of phytoplasmas are important. Further analysis of phytoplasmas at the molecular level will increase the understanding of these economically important and biologically fascinating microorganisms.

Future research should experimentally validate the functions of the key proteins identified in phytoplasmas. The labelling of substrates and heterologous expression in plant hosts are possible strategies, but besides this basic research, at least two further fields should be developed for phytoplasmas. First, the '*Ca.* Phytoplasma' species closer as affiliation with the *Acholeplasma* should be searched since studies on such ancestors could provide insights into the evolution of the genomes and relationships of these prokaryotes with their plant hosts. Second, the complete gene expression from a pathogenic phytoplasma strain and a host plant for which the whole genome is available should be exploited. To date, most insights were obtained from these pathogens, but the data need to be integrated and compared with the metabolic situation in healthy plants having the same genome, to learn more about the overall plant response that will help in carry out an appropriate management of the phytoplasma associated diseases.

References

Abreu IA, Cabelli DE (2010) Superoxide dismutases – a review of the metal-associated mechanistic variations. *Biochimica et Biophysica Acta – Proteins and Proteomics* **1804**, 263–274.

Andersen MT, Liefting LW, Havukkala I, Beever RE (2013) Comparison of the complete genome sequence of two closely related isolates of '*Candidatus* Phytoplasma australiense' reveals genome plasticity. *BMC Genomics* **14**, 529.

André A, Maucourt M, Moing A, Rolin D, Renaudin J (2005) Sugar import and phytopathogenicity of *Spiroplasma citri*: glucose and fructose play distinct roles. *Molecular Plant-Microbe Interactions* **18**, 33–42.

Annaluru N, Muller H, Mitchell LA, Ramalingam S, Stracquadanio G, Richardson SM, Dymond JS, Kuang Z, Scheifele LZ, Cooper EM,Cai Y, Zeller K, Agmon N, Han JS, Hadjithomas M, Tullman J, Caravelli K, Cirelli K, Guo Z, London V, Yeluru A, Murugan S, Kandavelou K, Agier N, Fischer G, Yang K, Martin JA, Bilgel M, Bohutski P, Boulier KM, Capaldo BJ, Chang

J, Charoen K, Choi WJ, Deng P, DiCarlo JE, Doong J, Dunn J, Feinberg JI, Fernandez C, Floria CE, Gladowski D, Hadidi P, Ishizuka I, Jabbari J, Lau CY, Lee PA, Li S, Lin D, Linder ME, Ling J, Liu J, Liu J, London M, Ma H, Mao J, McDade JE, McMillan A, Moore AM, Oh WC, Ouyang Y, Patel R, Paul M, Paulsen LC, Qiu J, Rhee A, Rubashkin MG, Soh IY, Sotuyo NE, Srinivas V, Suarez A, Wong A, Wong R, Xie WR, Xu Y, Yu AT, Koszul R, Bader JS, Boeke JD, Chandrasegaran S (2014) Total synthesis of a functional designer eukaryotic chromosome. *Science* **344**, 55–58.

Bai XD, Zhang JH, Ewing A, Miller SA, Radek AJ, Shevchenko DV, Tsukerman K, Walunas T, Lapidus A, Campbell JW, Hogenhout S (2006) Living with genome instability: the adaptation of phytoplasmas to diverse environments of their insect and plant hosts. *Journal of Bacteriology* **188**, 3682–3696.

Bai XD, Correa VR, Toruno TY, Ammar ED, Kamoun S, Hogenhout SA (2009) AY-WB phytoplasma secretes a protein that targets plant cell nuclei. *Molecular Plant-Microbe Interactions* **22**, 18–30.

Barbara DJ, Morton A, Clark MF, Davies DL (2002) Immunodominant membrane proteins from two phytoplasmas in the aster yellows clade (chlorante aster yellows and clover phyllody) are highly divergent in the major hydrophilic region. *Microbiology* **148**, 157–167.

Batjer LP, Schneider H (1960) Relation of pear decline to rootstocks and sieve-tube necrosis. *Proceedings of American Society for Horticultural Science* **76**, 85–97.

Bertaccini A, Weintraub P, Rao GP, Mori N (2019) Phytoplasmas: Plant Pathogenic Bacteria-II. Transmission and Management of Phytoplasma Associated Diseases. Springer, Singapore, 258 pp.

Bertamini M, Grando MS, Nedunchezhian N (2003) Effects of phytoplasma infection on pigments, chlorophyll-protein complex and photosynthetic activities in field grown apple leaves. *Biologia Plantarum* **47**, 237–242.

Bizarro CV, Schuck DC (2007) Purine and pyrimidine nucleotide metabolism in *Mollicutes*. *Genetics and Molecular Biology* **30**, 190–201.

Boeke JD, Church G, Hessel A, Kelley NJ, Arkin A, Cai Y, Carlson R, Chakravarti A, Cornish VW, Holt L,Isaacs FJ, Kuiken T, Lajoie M, Lessor T, Lunshof J, Maurano MT, Mitchell LA, Rine J, Rosser S, Sanjana NE, Silver PA, Valle D, Wang H, Way JC, Yang L. 2016. The genome project-write. *Science* **353**, 126–127.

Braun EJ, Sinclair WA (1978) Translocation in phloem necrosis-diseased American elm seedlings. *Phytopathology* **68**, 1733–1737.

Brückner R, Wagner E, Götz F (1993) Characterization of a sucrase gene from *Staphylococcus xylosus*. *Journal of Bacteriology* **175**, 851–857.

Bukata L, Altabe S, de Mendoza D, Ugalde RA, Comerci DJ (2008) Phosphatidyl ethanolamine synthesis is required for optimal virulence of *Brucella abortus*. *Journal of Bacteriology* **190**, 8197–8203.

Caldara M, Dupont G, Leroy F, Goldbeter A, De Vuyst L, Cunin R (2008) Arginine Biosynthesis in *Escherichia coli*: experimental perturbation and mathematical modeling. *The Journal of Biological Chemistry* **283**, 6347–6358.

Cao X, Fan G, Dong Y, Zhao Z, Deng M, Wang Z, Liu W (2017) Proteome profiling of paulownia seedlings infected with phytoplasma. *Frontiers in Plant Science* **8**, 342.

Carginale V, Maria G, Capasso C, Ionata E, La Cara F, Pastore M, Bertaccini A, Capasso A (2004) Identification of genes expressed in response to phytoplasma infection in leaves of *Prunus armeniaca* L. by messenger RNA differential display. *Gene* **12**, 29–34.

Carle P, Laigret F, Tully JG, Bové J-M (1995) Heterogeneity of genome sizes within the genus *Spiroplasma*. *International Journal of Systematic Bacteriology* **45**, 178–181.

Chang C-J, Donaldson R, Wilkinson RE (1994) Growth comparison between *Spiroplasma citri*-infected and aster yellows mycoplasma like organism-infected periwinkles. *IOM Letters* **3**, 276–277.

Chen C, Sun Q, Narayanan B, Nuss DL, Herzberg O (2010) Structure of oxalacetate acetylhydrolase, a virulence factor of the chestnut blight fungus. *Journal of Biological Chemistry* **285**, 26685–26696.

Chen W, Li Y, Wang Q, Wang N, Wu Y (2014) Comparative genome analysis of wheat blue dwarf phytoplasma, an obligate pathogen that causes wheat blue dwarf disease in China. *Plos One* **9**, e96436.

Chimalapati S, Cohen JM, Camberlein E, MacDonald N, Durmort C, Vernet T, Brown JS. (2012) Effects of deletion of the *Streptococcus pneumoniae* lipoprotein diacylglyceryl transferase gene *lgt* on ABC transporter function and on growth *in vivo*. *Plos One* **7**, e41393.

Choi YH, Tapias EC, Kim HK, Lefeber AW, Erkelens C, Verhoeven JTJ, Verpoorte R (2004) Metabolic discrimination of *Catharanthus roseus* leaves infected by phytoplasma using 1H-NMR spectroscopy and multivariate data analysis. *Plant Physiology* **135**, 2398–2410.

Christensen NM, Nicolaisen M, Hansen M, Schulz A (2004) Distribution of phytoplasmas in infected plants as revealed by real-time PCR and bioimaging. *Molecular Plant-Microbe Interactions* **17**, 1175–1184.

Christensen NM, Axelsen KB, Nicolaisen M, Schulz A (2005) Phytoplasmas and their interactions with hosts. *Trends in Plant Science* **10**, 526–535.

Chung W, Chen L, Lo W, Lin C, Kuo C (2013) Comparative analysis of the peanut witches' broom phytoplasma genome reveals horizontal transfer of potential mobile units and effectors. *Plos One* **8**, e62770.

Conde-Alvarez R, Grilló MJ, Salcedo SP, De Miguel MJ, Fugier E, Gorvel JP, Iriarte M (2006) Synthesis of phosphatidylcholine, a typical eukaryotic phospholipid, is necessary for full virulence of the intracellular bacterial parasite *Brucella abortus*. *Cellular Microbiology* **8**, 1322–1335.

Contaldo N, Bertaccini A, Paltrinieri S, Windsor HM, Windsor DG (2012) Axenic culture of plant pathogenic phytoplasmas. *Phytopathologia Mediterranea* **51**, 607–617.

Contaldo N, Satta E, Zambon Y, Paltrinieri S, Bertaccini A (2016) Development and evaluation of different complex media for phytoplasma isolation and growth. *Journal of Microbiological Methods* **127**, 105–110.

Contaldo N, D'Amico G, Paltrinieri S, Diallo HA, Bertaccini A, Arocha Rosete Y (2019) Molecular and biological characterization of phytoplasmas from coconut palms affected by the lethal yellowing disease in Africa. *Microbiological Research* **223–225**, 51–57.

Cooperman BS, Baykov AA, Lahti R (1992) Evolutionary conservation of the active site of soluble inorganic pyrophosphatase. *Trends in Biochemical Sciences* **17**, 262–266.

Dalbey RE, Kuhn A (2000) Evolutionarily related insertion pathways of bacterial, mitochondrial, and thylakoid membrane proteins. *Annual Review of Cell Biology* **16**, 51–87.

Eastmond PJ, Graham IA (2003) Trehalose metabolism: a regulatory role for trehalose-6-phosphate? *Current Opinion in Plant Biology* **6**, 231–235.

Economou A (1999) Following the leader: bacterial protein export through the Sec pathway. *Trends in Microbiology* **7**, 315–320.

Eden-Green SJ, Waters H (1982) Collection and properties of phloem sap from healthy and lethal yellowing-diseased coconut palms in Jamaica. *Phytopathology* **72**, 667–672.

Fan XP, Liu W, Qiao YS, Shang YJ, Wang GP, Tian X, Han YH, Bertaccini A (2017) Comparative transcriptome analysis of *Ziziphus jujuba* healthy and infected by jujube witches' broom phytoplasmas. *Scientia Horticulturae* **226**, 50–58.

Fontaniella B, Vicente C, Estrella Legaz M, de Armas R, Rodríguez CW, Martínez M, Piñón D, Acevedo R, Solas MT (2003) Yellow leaf syndrome modifies the composition of sugarcane juices in polysaccharides, phenols, and polyamines. *Plant Physiology and Biochemistry* **41**, 1027–1036.

Fontaniella B, Vicente C, de Armas R, Legaz ME (2007) Effect of leaf scald (*Xanthomonas albilineans*) on polyamine and phenolic acid metabolism of two sugarcane cultivars. *European Journal of Plant Pathology* **119**, 401–409.

Garnier M, Foissac X, Gaurivaud P, Laigret F, Renaudin J, Saillard C, Bové J-M (2001) Mycoplasmas, plants, insects vectors: a matrimonial triangle. *Life Sciences* **32**, 923–928.

Gibson DG, Benders GA, Andrews-Pfannkoch C, Denisova EA, Baden-Tillson H, Zaveri J, Stockwell TB, Brownley A, Thomas DW, Algire MA, Merryman C, Young L, Noskov VN,

Glass JI, Venter JC, Hutchison CA, Smith HO (2008) Complete chemical synthesis, assembly, and cloning of a *Mycoplasma genitalium* genome. *Science* **319**, 1215–1220.

Gibson DG, Glass JI, Lartigue C, Noskov VN, Chuang RY, Algire MA, Benders GA, Montague MG, Ma L, Moodie MM, Merryman C, Vashee S, Krishnakumar R, Assad-Garcia N, Andrews-Pfannkoch C, Denisova EA, Young L, Qi ZQ, Segall-Shapiro TH, Calvey CH, Parmar PP, Hutchison CA, Smith HO, Venter JC (2010) Creation of a bacterial cell controlled by a chemically synthesized genome. *Science* **329**, 52–56.

Grant JR, Stothard P (2008) The CGView Server: a comparative genomics tool for circular genomes. *Nucleic Acids Research* **36**, 181–184.

Hanson AD, Gregory JF (2002) Synthesis and turnover of folates in plants. *Current Opinion in Plant Biology* **5**, 244–249.

Hren M, Nikolić P, Rotter A, Blejec A, Terrier N, Ravnikar M, Gruden K (2009) "Bois noir" phytoplasma induces significant reprogramming of the leaf transcriptome in the field grown grapevine. *BMC Genomics* **10**, 460.

Hutchison CA, Chuang RY, Noskov VN, Assad-Garcia N, Deerinck TJ, Ellisman MH, Gill J, Kannan K, Karas BJ, Ma L, Pelletier JF, Qi ZQ, Richter RA, Strychalski EA, Sun L, Suzuki Y, Tsvetanova B, Wise KS, Smith HO, Glass JI, Merryman C, Gibson DG, Venter JC (2016) Design and synthesis of a minimal bacterial genome. *Science* **351**, 6253.

Hwang DM, Dempsey A, Tan KT, Liew CC (1996) A modular domain of NiFu, a nitrogen fixation cluster protein, is highly conserved in evolution. *Journal of Molecular Evolution* **43**, 536–540.

Itaya M, Tsuge K, Koizumi M, Fujita K (2005) Combining two genomes in one cell: stable cloning of the *Synechocystis* PCC6803 genome in the *Bacillus subtilis* 168 genome. *Proceedings of the National Academy of Sciences of the United States of America* **102**, 15971–15976.

Itaya M, Fujita K, Kuroki A, Tsuge K (2008) Bottom-up genome assembly using the *Bacillus subtilis* genome vector. *Nature Methods* **5**, 41–43.

Jagoueix-Eveillard S, Tarendeau F, Guolter K, Danet J-L, Bové J-M, Garnier M (2001) *Catharanthus roseus* genes regulated differentially by mollicute infections. *Molecular Plant-Microbe Interactions* **14**, 225–233.

Javid-Majd F, Blanchard JS (2000) Mechanistic analysis of the argE-encoded N-acetylornithine deacetylase. *Biochemistry* **39**, 1285–1293.

Ji X, Gai Y, Zheng C, Mu Z (2009) Comparative proteomic analysis provides new insights into mulberry dwarf responses in mulberry (*Morus alba* L.). *Proteomics* **9**, 5328–5339.

Jung HY, Miyata SI, Oshima K, Kakizawa S, Nishigawa H, Wei W, Suzuki S, Ugaki M, Hibi T, Namba S (2003) First complete nucleotide sequence and heterologous gene organization of the two rRNA operons in the phytoplasma genome. *DNA and Cell Biology* **22**, 209–215.

Kakizawa S, Oshima K, Kuboyama T, Nishigawa H, Jung H-Y, Sawayanagi T, Namba S. (2001) Cloning and expression analysis of phytoplasma protein translocation genes. *Molecular Plant-Microbe Interactions* **14**, 1043–1050.

Kakizawa S, Oshima K, Nishigawa H, Jung H-Y, Wei W, Suzuki S, Namba S (2004) Secretion of immunodominant membrane protein from onion yellows phytoplasma through the Sec protein-translocation system in *Escherichia coli*. *Microbiology* **150**, 135–142.

Kartte S, Seemüller E (1991) Susceptibility of grafted *Malus* taxa and hybrids to apple proliferation disease. *Journal of Phytopathology* **131**, 137–148.

Kawar PG, Devarumath RM, Nerkar YS (2009) Use of RAPD markers for assessment of genetic diversity in sugarcane cultivars. *Indian Journal of Biotechnology* **8**, 67–71.

Kube M, Schneider B, Kuhl H, Dandekar T, Heitmann K, Migdoll AM, Reinhardt R, Seemüller E (2008) The linear chromosome of the plant-pathogenic mycoplasma '*Candidatus* Phytoplasma mali'. *BMC Genomics* **9**, 306.

Kube M, Mitrovic J, Duduk B, Rabus R, Seemüller E (2012) Current view on phytoplasma genomes and encoded metabolism. *Scientific World Journal* **2012**, 185942.

Lartigue C, Vashee S, Algire MA, Chuang RY, Benders GA, Ma L, Noskov VN, Denisova EA, Gibson DG, Assad-Garcia N, Alperovich N, Thomas DW, Merryman C, Hutchison 3rd CA,

Smith HO, Venter JC, Glass JI (2009) Creating bacterial strains from genomes that have been cloned and engineered in yeast. *Science* **325**, 1693–1696.

Lee I-M, Zhao Y, Bottner KD (2006) *SecY* gene sequence analysis for finer differentiation of diverse strains in the aster yellows phytoplasma group. *Molecular and Cellular Probes* **20**, 87–91.

Lepka P, Stitt M, Moll E, Seemüller E (1999) Effect of phytoplasmal infection on concentration and translocation of carbohydrates and amino acids in periwinkle and tobacco. *Physiological and Molecular Plant Pathology* **55**, 59–68.

Manimekalai R, Anil Kumar NC, Roshna OM, Satyamoorthy K (2014) Isolation and comparative analysis of *potc* gene of ABC-transporter system from coconut and sugarcane – 16SrXI group phytoplasmas. *Journal of Plant Pathology* **96**, 35–42.

Manimekalai R, Roshna OM, Ganga Raj KP, Viswanathan R, Rao GP (2015) ABC transporter from sugarcane grassy shoot phytoplasma: gene sequencing and sequence characterization. *Sugar Tech* **18**, 407–413.

Mardi M, Farsad LK, Gharechahi J, Salekdeh GH (2015) In-depth transcriptome sequencing of Mexican lime trees infected with '*Candidatus* Phytoplasma aurantifolia'. *Plos One* **10**, e0130425.

Mariscal AM, Kakizawa S, Hsu JY, Tanaka K, Gonzalez-Gonzalez L, Broto A, Querol E, Lluch-Senar M, Pinero-Lambea C, Sun L,Weyman PD, Wise KS, Merryman C, Tse G, Moore AJ, Hutchison 3rd CA, Smith HO, Tomita M, Venter JC, Glass JI, Pinol J, Suzuki Y (2018) Tuning gene activity by inducible and targeted regulation of gene expression in minimal bacterial cells. *ACS Synthetic Biology* **7**, 1538–1552.

Maust BE, Espadas F, Talavera C, Aguilar M, Santamaría JM, Oropeza C (2003) Changes in carbohydrate metabolism in coconut palms infected with the lethal yellowing phytoplasma. *Phytopathology* **93**, 976–981.

Minion FC, Jarvill-Taylor KJ, Billings DE, Tigges E (1993) Membrane-associated nuclease activities in mycoplasmas. *Journal of Bacteriology* **175**, 7842–7847.

Møller-Jensen J, Löwe J (2005) Increasing complexity of the bacterial cytoskeleton. *Current Opinion in Cell Biology* **17**, 75–81.

Monavarfeshani A, Mirzaei M, Sarhadi E, Amirkhani A, Nekouei MK, Haynes PA, Mardi M, Hosseini Salekdeh G (2013). Shotgun proteomic analysis of the Mexican Lime tree infected with '*Candidatus* Phytoplasma aurantifolia'. *Journal of Proteome Research* **12**, 785–795.

Moore CM, Helmann JD (2005) Metal ion homeostasis in *Bacillus subtilis*. *Current opinion in Microbiology* **8**, 188–195.

Musetti R, Favali MA (2004) Microscopy techniques applied to the study of phytoplasma diseases: traditional and innovative methods. *Current Issues on Multidisciplinary Microscopy Research and Education* **2**, 72–80.

Musetti R, Favali MA, Pressacco L (2000) Histopathology and polyphenol content in plants infected by phytoplasmas. *Cytobios* **102**, 133–148.

Nejat N, David C, Vadamalai G, Ziemann M, James Neda N (2015) Transcriptomics-based analysis using RNA-Seq of the coconut (*Cocos nucifera*) leaf in response to yellow decline phytoplasma infection. *Molecular Genetics and Genomics* **290**, 1899–1910.

Nishizaki T, Tsuge K, Itaya M, Doi N, Yanagawa H (2007) Metabolic engineering of carotenoid biosynthesis in *Escherichia coli* by ordered gene assembly in *Bacillus subtilis*. *Applied and Environmental Microbiology* **73**, 1355–1361.

O'Brien SJ, Simonson JM, Grabowski MW (1981) Analysis of multiple isoenzyme expression among twenty-two species of *Mycoplasma* and *Acholeplasma*. *Journal of Bacteriology* **146**, 222–232.

Orlovskis Z, Canale MC, Haryono M, Lopes JRS, Kuo CH, Hogenhout SA (2017) A few sequence polymorphisms among isolates of maize bushy stunt phytoplasma associate with organ proliferation symptoms of infected maize plants. *Annals of Botany* **119**, 869–884.

Oshima K, Kakizawa S, Nishigawa H, Jung H-Y, Wei W, Suzuki S, Namba S (2004) Reductive evolution suggested from the complete genome sequence of a plant-pathogenic phytoplasma. *Nature Genetics* **36**, 27.

Pacifico D, Abbà S, Palmano S (2019) Transcriptomic analyses of phytoplasmas. *Methods in Molecular Biology* **1875**, 239–251.

Pailler J, Aucher W, Pires M, Buddelmeijer N (2012) Phosphatidyl glycerol: prolipoprotein diacylglyceryl transferase (Lgt) of *Escherichia coli* has seven transmembrane segments, and its essential residues are embedded in the membrane. *Journal of Bacteriology* **194**, 2142–2151.

Paul MJ, Jhurreea D, Zhang Y, Primavesi LF, Delatte T, Schluepmann H, Wingler A (2010) Upregulation of biosynthetic processes associated with growth by trehalose 6-phosphate. *Plant Signal Behaviours* **5**, 386–392.

Pollack JD, Mcelwain MC, Desantis D, Manolukas JT, Tully JG, Chang C-J, Whitcomb RF, Hackett KJ, Williams MW (1989) Metabolism of members of the Spiroplasmataceae. *International Journal of Systematic Bacteriology* **39**, 406–412.

Pollack JD, Williams MV, McElhaney RN (1997) The comparative metabolism of the mollicutes (mycoplasmas): the utility for taxonomic classification and the relationship of putative gene annotation and phylogeny to enzymatic function in the smallest free-living cells. *Critical Reviews in Microbiology* **23**, 269–354.

Pollack JD, Myers MA, Dandekar T, Herrmann R (2002) Suspected utility of enzymes with multiple activities in the small genome *Mycoplasma* species: the replacement of the missing "household" nucleoside diphosphate kinase gene and activity by glycolytic enzymes. *Omics* **6**, 247–258.

Pracros P, Renaudin J, Eveillard S, Mouras A, Hernould M (2006) Tomato flower abnormalities induced by stolbur phytoplasma infection are associated with changes of expression of floral development genes. *Molecular Plant-Microbe Interactions* **19**, 62–68.

Pretorius IS, Boeke JD (2018) Yeast 2.0-connecting the dots in the construction of the world's first functional synthetic eukaryotic genome. *FEMS Yeast Research* **18**, foy032.

Razin S, Yogev D, Naot Y (1998) Molecular biology and pathogenicity of mycoplasmas. *Microbiology and Molecular Biology Reviews* **62**, 1094–1156.

Reed KE, Cronan JE (1993) Lipoic acid metabolism in *Escherichia coli*: sequencing and functional characterization of the *lipA* and *lipB* genes. *Journal of Bacteriology* **175**, 1325–1336.

Samuelson JC, Chen MY, Jiang FL, Moller I, Wiedmann M, Kuhn A, Phillips G, Dalbey R (2000) YidC mediates membrane protein insertion in bacteria. *Nature* **406**, 637–641.

Serek J, Bauer-Manz G, Struhalla G, van den Berg L, Kiefer D, Dalbey R, Kuhn A (2004) *Escherichia coli* YidC is a membrane insertase for Sec-independent proteins. *EMBO Journal* **23**, 294–301.

Shani E, Yanai O, Ori N (2006) The role of hormones in shoot apical meristem function. *Current Opinion in Plant Biology* **9**, 484–489.

Sparrow CP, Raetz CR (1985) Purification and properties of the membrane-bound CDP-diglyceride synthetase from *Escherichia coli*. *Journal of Biological Chemistry* **5**, 12084–12091.

Taheri F, Nematzadeh G, Zamharir MG, Khayam Nekouei M, Naghavi M, Mardi M, Salekdeh GH (2011) Proteomic analysis of the Mexican lime tree response to '*Candidatus* Phytoplasma aurantifolia' infection. *Molecular Biology Systematics* **7**, 3028–3035.

Tjalsma H, Bolhuis A, Jongbloed JD, Bron S, van Dijl JM (2000) Signal peptide-dependent protein transport in *Bacillus subtilis*: a genome-based survey of the secretome. *Microbiological Molecular Biology Reviews* **64**, 515–547.

Tran-Nguyen LT, Kube M, Schneider B, Reinhardt R, Gibb KS (2008) Comparative genome analysis of '*Candidatus* Phytoplasma australiense' (subgroup tuf-Australia I; rp-A) and '*Ca.* Phytoplasma asteris' strains OY-M and AY-WB. *Journal of Bacteriology* **190**, 3979–3991.

Trumbo P, Yates AA, Schlicker S, Poos M (2001) Dietary reference intakes: vitamin A, vitamin K, arsenic, boron, chromium, copper, iodine, iron, manganese, molybdenum, nickel, silicon, vanadium, and zinc. *Journal of the Academy of Nutrition and Dietetics* **101**, 294.

Voet JG, Abeles RH (1970) Mechanism of action of sucrose phosphorylase. Isolation and properties of a β-linked covalent glucose-enzyme complex. *Journal of Biological Chemistry* **245**, 1020–1031.

Wang J, Song L, Jiao Q, Yang S, Gao R, Lu X, Zhou G (2018a) Comparative genome analysis of jujube witches' broom phytoplasma, an obligate pathogen that causes jujube witches' broom disease. *BMC Genomics* **19**, 689.

Wang H, Ye X, Li J, Tan B, Chen P, Cheng J, Wang W, Zheng X, Feng J (2018b) Transcriptome profiling analysis revealed co-regulation of multiple pathways in jujube during infection by '*Candidatus* Phytoplasma ziziphi'. *Gene* **665**, 82–95.

Webb E, Claas K, Downs D (1998) ThiBPQ encodes an ABC transporter required for transport of thiamine and thiamine pyrophosphate in *Salmonella typhimurium*. *Journal of Biological Chemistry* **273**, 8946–8950.

Wei W, Kakizawa S, Jung H-Y, Suzuki S, Tanaka M, Nishigawa H, Miyata S, Oshima K, Ugaki M, Hibi T, Namba S (2004) Antibody against the SecA membrane protein of one phytoplasma reacts with those of phylogenetically different phytoplasmas. *Phytopathology* **94**, 683–686.

Weisberg WG, Tully JG, Rose DLJ, Petzel P, Oyaizu H, Yang D, Mandelco L, Sechrest J, Lawrence TG, Van Etten J (1989) A phylogenetic analysis of the mycoplasmas: basis for their classification. *Journal of Bacteriology* **171**, 6455–6467.

Ye X, Wang H, Chen P, Fu B, Zhang M, Li J, Zheng X, Tan B, Feng J (2017) Combination of iTRAQ proteomics and RNA-seq transcriptomics reveals multiple levels of regulation in phytoplasma-infected *Ziziphus jujube* Mill. *Horticulture Research* **4**, 170-180.

Zamharir MG, Mardi M, Alavi SM, Hasanzadeh N, Nekouei MK, Zamanizadeh HR, Salekdeh GH (2011) Identification of genes differentially expressed during interaction of Mexican lime tree infected with '*Candidatus* Phytoplasma aurantifolia'. *BMC Microbiology* **11**, 1.

Zhang Y, Primavesi LF, Jhurreea D, Andralojc PJ, Mitchell RAC, Powers SJ, Schluepmann H, Delatte T, Wingler A, Matthew JP (2009) Inhibition of SNF1-related protein kinasel activity and regulation of metabolic pathways by trehalose-6-phosphate. *Plant Physiology* **149**, 1860–1871.

Ziegler H (1975) Nature of transported substances. In: Transport in Plants I. Springer, Berlin, Heidelberg, Germany, 59–100 pp.